Present day Preparations being made for manned expeditions to the Moon and Mars

1990 Ulysses launched to explore solar activity out of the ecliptic

1990 Magellan begins to map the surface of Venus
Hubble space telescope swung into orbit from space shuttle Discovery

1989 Voyager 2 reaches Neptune

1987 Discovery of Supernova 1987A in Large Magellanic Cloud

1986 Spacecraft Giotto's rendezvous with Halley's Comet. Launching of Soviet space station MIR

1986 Space shuttle Challenger explodes, killing its crew of seven
Voyager 2 reaches Uranus

1985 Return of Halley's Comet to the inner solar system

1982 La Palma Observatory comes into operation

1981 Voyager 2 reaches Saturn

1980 Voyager 1 reaches Saturn

1979 Voyagers 1 and 2 reach Jupiter

1977 The 6-metre Zelenchukskaya telescope becomes operational. Voyagers 1 and 2 begin their exploration of the outer solar system

1976 Viking 1 lander lands on the Martian surface

1975 Launch of two Viking spacecraft to Mars

1973 Pioneer 11 launched. Visited Jupiter and Saturn

1972 Pioneer 10 launched. Visited Jupiter and Saturn

1971 Mariner 9 becomes the first artificial Martian satellite

1781 William Herschel discovers the planet Uranus

1752 Melvill discovers the D-line in the sodium spectrum

1725 Bradley discovers stellar aberration (death of the Ptolemaic system)

1704 Publication of Newton's *Optics*

1687 Publication of Newton's *Principia*

1676 Founding of Greenwich Observatory by Charles II with Flamsteed as first Astronomer Royal

1609 Kepler publishes his first two laws

Development of modern instrumental astronomy

1608 Galileo builds his first telescope

1576 Tycho builds his Danish observatory

1543 Publication of Copernicus' theory of the solar system

Arabian Astronomy

1038 Al-Hasen studies refraction by the atmosphere

850 AD Al-Battānī studies and checks Greek discoveries

eek
onomy

140 AD Ptolemy's *Almagest* published

120 BC Hipparchus

380 BC Eudoxus' theory of the world

432 BC Meton discovers Metonic cycle

585 BC Thales forecasts an eclipse

720 BC Mesopotamians record eclipses

(present day) The space age

1957 First artificial satellite (Sputnik 1)

1609 Development of modern instrumental astronomy

3,000 years BC Development of astronomy

OXFORD
ILLUSTRATED
ENCYCLOPEDIA
of

THE UNIVERSE

OXFORD ILLUSTRATED ENCYCLOPEDIA *of* THE UNIVERSE

Volume Editor
Archie Roy

BCA
LONDON · NEW YORK · SYDNEY · TORONTO

This edition published 1992 by BCA by arrangement with
Oxford University Press

CN 6729

Text processed by Oxford University Press
Printed in Hong Kong

Foreword

Astronomy is one of the oldest sciences. The astronomical alignments of innumerable megalithic monuments in Western Europe and the planetary information found on baked-clay cuneiform tablets from ancient Nineveh, show that there have been people observing and pondering the heavens for at least five millennia and probably much longer, considering that the human brain has been much the same in intellectual capacity for the past 50,000 years. The 'computer' hasn't changed; only the 'programming' has. The daily and nightly pageantry of the cosmos, the yearly cycle of shifting sunrise and sunset points, the Moon's strange monthly change in shape, and the stately pavane of the planets against the field of fixed stars cannot have failed to fascinate our ancestors and spur them to attempt to understand their mysteries. Certainly they would have noticed early on the cosmic connection: that the sunrise and sunset points were related to the seasons, that the changing phases of the Moon dictated the timing of the spring and neap tides, and that all growth was linked to the Sun's yearly cycle, with birds migrating and creatures hibernating as the days became shorter and colder. If our species had developed on a planet where it never saw the sky because an ever-present and impenetrable cloud layer prevented even the Sun from being seen, our civilization and world outlook would have been entirely different. The millennia-long first phase of naked-eye astronomy brought navigational and timekeeping systems as well as religious, philosophical, and astronomical beliefs. The introduction of the telescope in AD 1609 revolutionized our appreciation of the size and majesty of the cosmos. In the same century Isaac Newton verified his Law of Gravitation and Laws of Motion in the test bed of the solar system: there began in a real sense our present scientific age. Dynamics and applied mathematics are only two of the disciplines indebted to our urge to understand the celestial phenomena seen through the telescope. Applying the camera and spectroscope to the telescope have turned this river of knowledge into a torrent. Einstein's theories, tested by astronomical observations early in this century, have recently been confirmed by observation of those strange pairs of stars called binary pulsars, and the search is on today to detect gravitational waves, a further prediction of his work. There are more astronomers and space scientists alive and at work today than in any previous generation, fascinated and excited by this flood of discovery. The advent of radio, X-ray, gamma-ray, ultraviolet, and infra-red telescopes, computers, charge-coupled devices, spacecraft, and other sophisticated equipment means that missions to the planets such as Voyager, Magellan, Galileo, and Cassini will enable the solar system's worlds to be studied more completely than ever before. Knowledge of the bleakness and lifelessness of those worlds will surely strengthen among human beings the realization that spaceship Earth must be preserved by more responsible housekeeping. In a strange way the wheel has turned full circle. In the remote past, the astrologer, by studying the seemingly parochial panorama of the changing heavens, made statements about the lives and futures of individual human beings. The modern astronomer, presenting his picture of the relationship of the human race to this ancient, enormous, evolving universe he inhabits, can now make statements about the origin, fate, and significance of the human race in the universe's gigantic fabric. Linked closely with the astronomer's search for understanding is his search for signs of extraterrestrial life and intelligence, perhaps in truth a wry cosmic joke. If we find life is present only on the tiny speck of matter we call the Earth, are we then a highly improbable, momentary accident in a lifeless universe? On the other hand if life as we know it is found to be abundant throughout the universe, shall we turn out to be the most primitive natives in it? The ancient Greeks chose man as a starting point for the knowledge of the universe, putting on the temple of Apollo at Delphi the inscription 'Know thyself'. Today we have reached the time in which we can say with truth, 'Know the universe and you will know yourself.' It is said that a new book on astronomy is published every day throughout the year, so rapidly is our knowledge of the universe increasing, and it is hoped that this present up-to-date volume, written by a number of experts working in astronomy, will paint for the reader a clear picture not only of what we now know but also depict the varied methods used by astronomers and astrophysicists in their continued exploration of the universe.

Finally, with great pleasure, I take this opportunity to thank all those who have helped me in creating this volume of the *Oxford Illustrated Encyclopedia*. In its early stages Christopher Riches guided me with his expert counsel; equally helpful have been Bridget Hadaway and Emma Morgan Williams of Oxford University Press. Mike Nugent has put me in his debt with his unstinting support as have those who have worked on the illustrations. And of course my thanks go to all the contributors who, in busy lives, responded to my request for material.

ARCHIE E. ROY

CONTRIBUTORS

Professor John C. Brown

Dr A. Carusi

Dr David Clarke

Dr R. J. Cohen

Professor R. D. Davies

Dr Paolo Farinella

Dr Robin D. Green

Dr David Hughes

John Isles

Professor Andrea Milani

Dr Anna Nobili

S. M. Nugent

Dr Ian Ridpath

Professor Archie Roy

Carole Stott

Professor William Tobin

Dr Giovanni Valsecchi

Emma Morgan Williams

General Preface

The *Oxford Illustrated Encyclopedia* is designed to be useful and to give pleasure to readers throughout the world. Particular care has been taken to ensure that it is not limited to one country or to one civilization, and that its many thousands of entries can be understood by any interested person who has no previous detailed knowledge of the subject.

Each volume has a clearly defined theme made plain in its title, and is for that reason self-sufficient: there is no jargon, and references to other volumes are avoided. Nevertheless, taken together, the eight thematic volumes (and the Index and Ready Reference volume which completes the series) provide a complete and reliable survey of human knowledge and achievement. Within each independent volume, the material is arranged in a large number of relatively brief articles in A–Z sequence, varying in length from fifty to one thousand words. This means that each volume is simple to consult, as valuable information is not buried in long and wide-ranging articles. Cross-references are provided whenever they will be helpful to the reader.

The team allocated to each volume is headed by a volume editor eminent in the field. Over four hundred scholars and teachers drawn from around the globe have contributed a total of 2.4 million words to the Encyclopedia. They have worked closely with a team of editors at Oxford whose job it was to ensure that the coverage and content of each entry form part of a coherent whole. Specially commissioned artwork, diagrams, maps, and photographs convey information to supplement the text and add a lively and colourful dimension to the subject portrayed.

Since publication of the first of its volumes in the mid-1980s, the *Oxford Illustrated Encyclopedia* has built up a reputation for usefulness throughout the world. The number of languages into which it has been translated continues to grow. In compiling the volumes, the editors have recognized the new internationalism of its readers who seek to understand the different technological, cultural, political, religious, and commercial factors which together shape the world-view of nations. Their aim has been to present a balanced picture of the forces that influence peoples in all corners of the globe.

I am grateful alike to the volume editors, contributors, consultants, editors, and illustrators whose common aim has been to provide, within the space available, an Encyclopedia that will enrich the reader's understanding of today's world.

HARRY JUDGE

A User's Guide

This book is designed for easy use, but the following notes may be helpful to the reader.

ALPHABETICAL ARRANGEMENT The entries are arranged in strict A–Z order of their headwords up to the first comma (thus **Hale, George Ellery** comes before **Hale telescope**), and after a comma the alphabetical order begins again (thus **relativity, general theory of** precedes **relativity, special theory of**).

HEADWORDS Entries are usually placed under the principal word (the keyword) in the title (the surname, for instance, in a group of names). In cases where there is no obvious keyword, an entry appears under the word that most users are likely to look up first. Generally the familiar has been preferred to the technical, but a form that is familiar to one generation may not be the most familiar to another. For example, parsecs are now the preferred unit of interstellar distances rather than light-years. If an entry does not appear under the word you expect, therefore, it is worth trying an alternative. Where alternatives are in frequent use, a single-line cross-reference guides the reader to that which heads the entry. This applies also to scientific and specialist terms; for example, **Great Looped Nebula** *Cygnus Loop, and **Northern Hemisphere Observatory** *La Palma Observatory.

CROSS-REFERENCES An asterisk (*) in front of a word denotes a cross-reference and indicates the headword of the other entry to which attention is being drawn. Cross-references in the text appear only in places where references are likely to amplify or increase understanding of the entry you are reading. They are not given automatically in all cases where a separate entry can be found, so if you come across a name or a term about which you would like to know more, it is worth looking for an entry in its alphabetical place even if no cross-reference is marked. Nearly all of the technical terms used in this volume are defined under their own or related headwords; otherwise they are defined where they are used. The cross-references can also be used as an index might be: if, for instance, you seek information about **Cepheid variables** and know what they are but have forgotten their name, you will find a reference to the relevant entry in **variable stars** and **galaxy**.

WEIGHTS AND MEASURES Following current practice in astronomy, SI units are used throughout. However, terrestrial distances are accompanied by their imperial equivalent. For the sizes of telescopes, inches followed by SI units are used when the telescope is commonly known by its size in inches; for example, the 40 inch (1.02m) refractor at the Yerkes Observatory. On the other hand, the sizes of modern telescopes are given in metres followed by inches if it is an optical telescope, or in feet if it is a radio telescope. A table at the end of the volume shows how the principal SI and imperial units are related and defines all of the units used in the entries. It is worth bearing in mind that many astronomical quantities, such as luminosity, mass, and the distances to stars and galaxies, are not known accurately and must therefore be taken to be estimates in some cases.

TERMINOLOGY Technical and professional terms that do not appear in standard dictionaries are defined and, where necessary, explained. For example, the location of a celestial body in the night sky is described by a system of coordinates on the celestial sphere; each of these coordinates is explicitly described. Terms whose meanings differ from their everyday usage, such as space-time curvature, are also explained.

ILLUSTRATIONS Pictures and diagrams almost invariably occur on the same page as the entries to which they relate or on a facing page, and where this is not the case the position of an illustration is indicated in the text. The picture captions supplement the information given in the text and indicate in bold type the title of the relevant entry. Major tables of general interest and of relevance to several entries on, for example, Astronomers Royal, asteroids, comets, constellations, meteor showers, planets, satellites, and stars appear at the end of the volume.

A table of mathematical and physical constants and units is also provided, and a concise history of astronomy is to be found in the front endpapers. The back endpapers contain star charts of the northern and southern sky.

RELATIONSHIP TO OTHER VOLUMES This survey of astronomy is part of a series, the *Oxford Illustrated Encyclopedia*, comprising eight thematic volumes and an index and ready reference volume.

1 *The Physical World*
2 *The Natural World*
3 *World History: from earliest times to 1800*
4 *World History: from 1800 to the present day*
5 *The Arts*
6 *Invention and Technology*
7 *Peoples and Cultures*
8 *The Universe*
9 *Index and Ready Reference*

The volume is self-contained, having no cross-references to any of its companions, and is therefore entirely usable on its own. Some aspects of certain topics are, however, the subjects of other volumes. Some of the scientific topics described here will, for example, be covered in Volumes 1 and 6. Such items appearing in more than one volume will be linked by a general index in Volume 9.

aberration, the difference between the observed direction of a star and its true direction. Aberration is due to the combined effect of the Earth's orbital motion and the finite velocity of light. The effect was discovered by *Bradley, third Astronomer Royal, in 1725. To correct for aberration, a telescope has to be tilted in the direction of the Earth's motion in its orbit about the Sun; otherwise, by the time the light from the star has passed through the telescope, the Earth will have moved sufficiently in its orbit to make the star appear to be at a different position in the sky. A star's apparent direction can be as much as 20.5 arc seconds from its true position. The effect is analogous to the passenger's view of rain falling at an angle past the side window of a moving vehicle. A smaller effect, **diurnal aberration**, is due to the Earth's rotation about its axis and amounts to at most 0.32 arc seconds.

Chromatic aberration is the production, by a simple lens, of an image with coloured edges. These coloured fringes arise because the focal length of the lens is different at different wavelengths of light. The effect can be reduced by combining a concave and a convex lens whose chromatic aberrations tend to cancel each other.

absolute magnitude *magnitude.

accretion disc, a disc of hot gas sometimes found surrounding the smaller member of a binary star. Between the two members of a binary star there is a *Lagrangian point of zero gravity where the stars' gravitational attractions pull

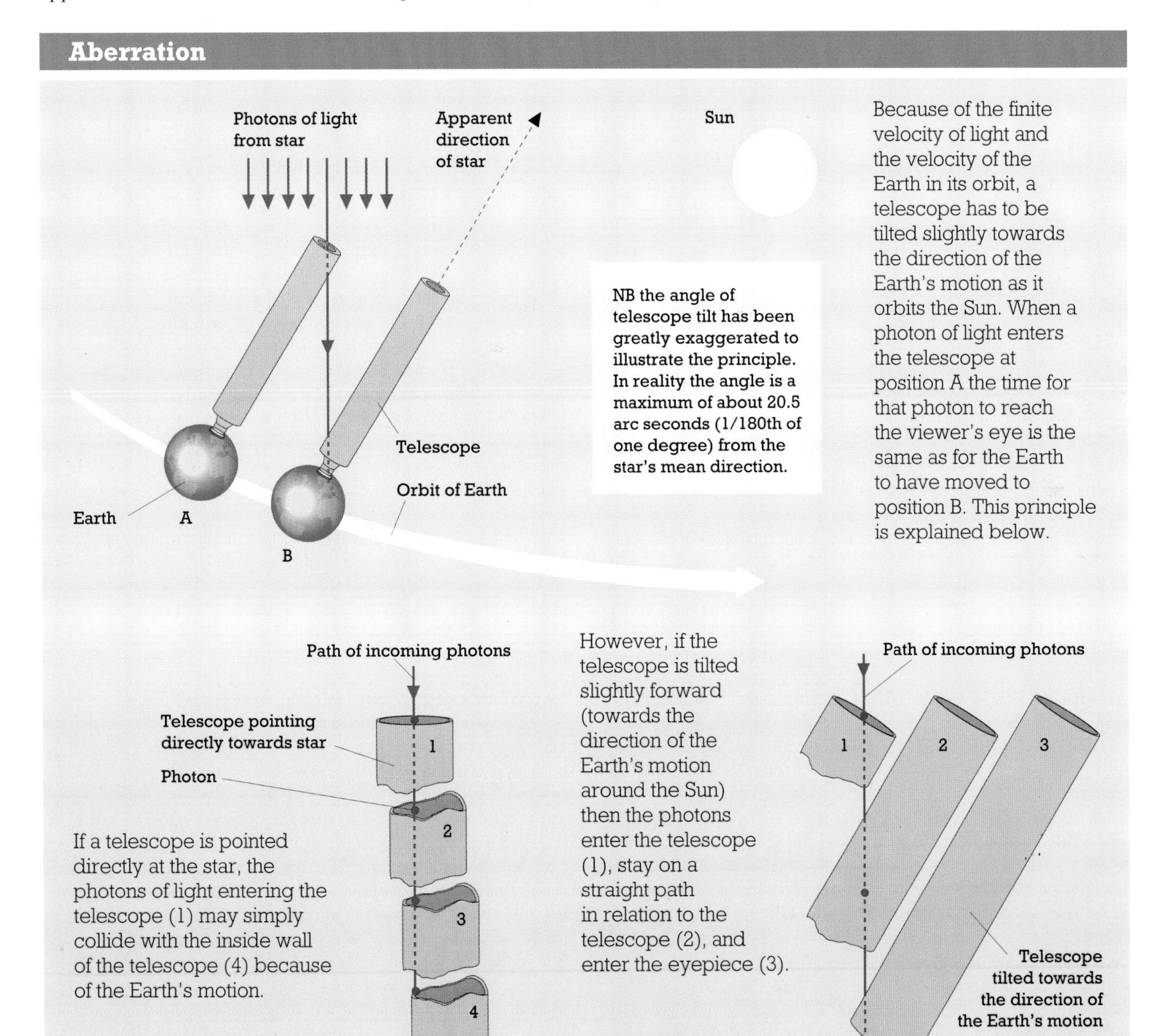

Aberration

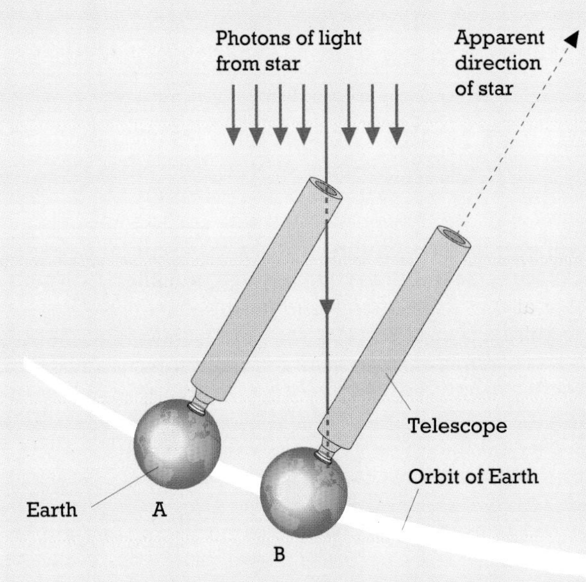

Sun

NB the angle of telescope tilt has been greatly exaggerated to illustrate the principle. In reality the angle is a maximum of about 20.5 arc seconds (1/180th of one degree) from the star's mean direction.

Because of the finite velocity of light and the velocity of the Earth in its orbit, a telescope has to be tilted slightly towards the direction of the Earth's motion as it orbits the Sun. When a photon of light enters the telescope at position A the time for that photon to reach the viewer's eye is the same as for the Earth to have moved to position B. This principle is explained below.

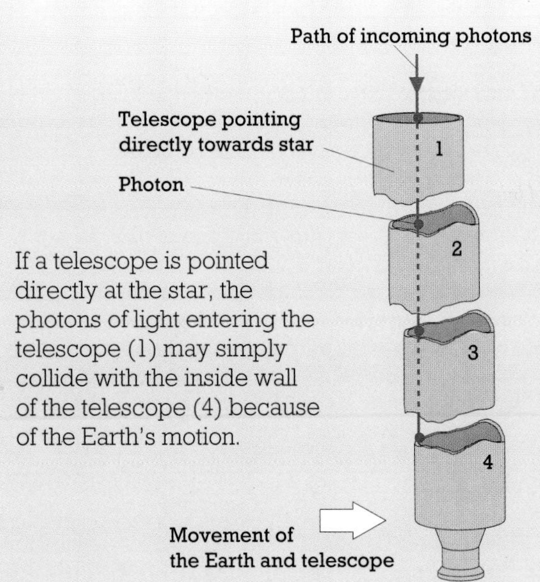

However, if the telescope is tilted slightly forward (towards the direction of the Earth's motion around the Sun) then the photons enter the telescope (1), stay on a straight path in relation to the telescope (2), and enter the eyepiece (3).

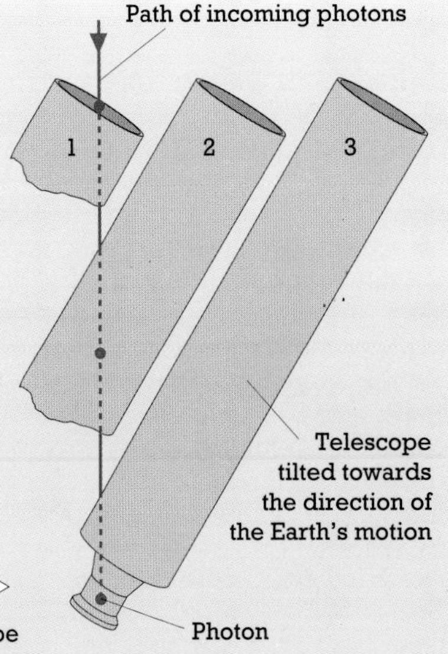

equally in both directions. If one member is very much larger than the other and extends beyond this point of zero gravity so that it overfills its *Roche lobe, matter is drawn away from the larger companion. This material is attracted by the smaller star at supersonic velocity and grows into a disc which surrounds the smaller star. The disc forms under the combined effects of gravitational forces, the orbital motion of the system, and the rotation of the two stars. As matter is drawn from the larger star it falls on to the disc where it produces a hot region of turbulence. Inside the disc, friction causes the material to spiral on to the small star at its centre. Enormous amounts of energy are released at ultra-violet and sometimes X-ray wavelengths. Accretion discs are formed in *cataclysmic variables, *Beta Lyrae stars, and *X-ray binaries. An accretion disc is sometimes also found surrounding the black hole at the centre of a galaxy, or the white dwarf component of a *nova.

Achernar, a star in the constellation of Eridanus, and also known as Alpha Eridani, with an apparent magnitude of 0.46. It is a hot *dwarf star at a distance of 38 parsecs, with about 300 times the luminosity of the Sun. At declination −57 degrees, it does not rise at latitudes north of +33 degrees.

Achilles, a member of the *Trojan group of asteroids discovered in 1906 by the German astronomer Max Wolf. It is 70 km in diameter and was the first of the Trojan group to be discovered. It keeps a position near the *Lagrangian point at the apex of the equilateral triangle formed by the Sun, Jupiter, and those Trojans that precede Jupiter in its orbit round the Sun.

achondrite *chondrite.

Acrux, a first-magnitude *binary star in the constellation of Crux and also known as Alpha Crucis. It comprises a hot subgiant and a hot dwarf star; the orbital period is very long. The brighter component is a spectroscopic binary whose components orbit each other every 76 days. The combined luminosity of Acrux is about 6,000 times that of the Sun, and its distance is 160 parsecs. It is a member of the Scorpius–Centaurus *stellar association, a large moving group which also includes Antares and Becrux. At declination −63 degrees, Acrux does not rise at latitudes north of +27 degrees.

Adams, John Couch (1819–92), British astronomer who discovered the planet Neptune independently of *Le Verrier. He studied the irregularities in the orbital motion of the planet Uranus and thus predicted the position of a planet beyond the orbit of Uranus, later called Neptune. His predictions were given to *Airy in October 1845, but Airy did not institute a search until July 1846. The more assertive French astronomer Le Verrier conducted a similar analysis but published his results sooner. Le Verrier's calculations gave a position which was used by the German astronomers Johann Galle and Heinrich d'Arrest in Berlin to discover Neptune in September 1846. Nevertheless Adams was awarded the Copley Medal of the Royal Society for his work. In 1851 Adams began work on lunar theory, particularly the Moon's motion and distance. In 1861 he became director of the Cambridge Observatory. He studied the orbits of the meteoroids responsible for the 1866 Leonid meteor swarm and used the Cambridge Observatory to map part of the northern sky.

Adonis, an *Apollo–Amor object discovered by the Belgian astronomer Eugene Delporte in 1936. That year it passed within 2 million kilometres of the Earth, a near miss by astronomical standards. Unfortunately it was lost soon afterwards, but was recovered in 1977 by the US astronomer Charles Kowal. The orbit of Adonis has a semi-major axis of 1.87278 astronomical units and an eccentricity of 0.7639.

Adrastea, a small satellite of *Jupiter, discovered by the Voyager science team in 1979. It orbits at about 129,000 km from the centre of the planet, just outside Jupiter's rings and close to the orbit of *Metis. It has an irregular shape with a mean diameter of about 20 km, an albedo of about 0.1, and a surface which is probably of rocky composition.

Agena *Beta Centauri.

age of stellar clusters Open, galactic, or moving star clusters have ages ranging from one million to 15 billion years while all globular clusters are thought to be about 15 billion years old. The age of a cluster is deduced from the turn-off point in the cluster's *Hertzsprung–Russell diagram. The difference between the ages of open and globular clusters enables scientists to deduce how the Galaxy has evolved.

age of the Earth From radioisotope studies, the Earth's oldest crustal rocks must be at least 3.8 billion years old. There is evidence that the Earth formed with the other members of the *solar system some 4.6 billion years ago. The oldest features on the Earth's surface are no longer visible, having been removed by erosion and plate *tectonics.

age of the Galaxy From the age of globular clusters the Galaxy is thought to be about 15 billion years old. These clusters, forming a spherical shell about the galactic equator, contain the oldest stars and show by their absence of dust and gas that they must have formed in the earliest phase of the Galaxy's existence.

age of the solar system Studies of rock samples from the Earth, Moon, and meteorites, together with conclusions drawn from the dynamics of the solar system's bodies, suggest that the solar system formed some 4.6 billion years ago. It is believed that the orbits of most of the major bodies have not varied much in that time.

age of the stars Astrophysical theory and observational evidence give widely differing ages for different types of stars ranging from one million years to 15 billion years. The youngest ages suggest that stars are still forming out of the interstellar medium. In general the mass of a star significantly affects its structure and rate of evolution. The more massive a star is the faster it evolves to its final state of *white dwarf, *neutron star, or *black hole.

age of the universe If the *Big Bang theory of the origin of the universe is accepted then the universe had a beginning. Theories of cosmology and the observed value of the *Hubble constant (at present estimated to be 55 km/s per megaparsec) gives the age of the universe as about 15 billion years.

Airy, Sir George Biddell (1801–92), British scientist and seventh *Astronomer Royal. After he became Astronomer Royal in 1835 he spent the rest of his life at the *Royal

As well as being Astronomer Royal, George **Airy** had many other scientific interests. He calculated the density of the Earth by swinging a pendulum at the top and bottom of a deep mine, and his name is given to the Airy disc, the circle of light in the diffraction pattern of a point source.

Greenwich Observatory. Airy was an eminently practical man who disliked theoretical problems. He transformed the observatory into a highly efficient institution creating a great sense of order and discipline in British observational astronomy. Unfortunately this left little time for astrophysics. His hesitation to start a search for the planet Neptune based on the calculations of *Adams in 1845 lost the initiative to *Le Verrier. However, his observations of Venus and the Moon improved the understanding of their orbits. He investigated the *aberration of starlight and the theory of rainbows and interference fringes.

albedo, the efficiency with which a celestial body such as a planet, satellite, or asteroid reflects light. Unlike stars, planets and satellites have no radiant energy sources of their own and can be seen only by light reflected from them. Albedo is usually expressed as a percentage of the incident sunlight that is reflected from the planet's surface. Thus the albedo of a perfect reflector is unity while an albedo of zero would refer to an object that absorbed all radiation falling upon it.

Aldebaran, a first-magnitude variable star in the constellation of Taurus, and also known as Alpha Tauri. Located in the direction of the Hyades, it is actually a foreground object not directly associated with that cluster. It is a *red giant star at a distance of 18 parsecs, with about 100 times the luminosity of the Sun. A thirteenth-magnitude red dwarf companion has the same *proper motion, and the pair may form a binary star of very long period.

Alfvén's theory, the theory of ionized particles in space and in the interiors of the Sun and stars. It was developed by the Swedish astronomer Hannes Olaf Gösta Alfvén (1908–) in an attempt to account for the formation of the planetary system. He also introduced the concept of frozen lines of magnetic force in describing a *plasma of particles. Alfvén was awarded the Nobel Prize in 1970.

Algol, also known as Beta Persei, a variable star at a distance of 29 parsecs in the constellation of Perseus. Its magnitude varies between 2.1 and 3.4 every 2.87 days. It is an *eclipsing binary star, in which the brighter primary component, a hot dwarf star of about 100 times the luminosity of the Sun, is periodically eclipsed by its fainter secondary companion, a cool subgiant. The brighter star does not fill its *Roche lobe and is therefore approximately spherical; the larger companion does fill its Roche lobe. The magnitude does not vary much except during the eclipses which last nearly ten hours. A third star orbits the eclipsing pair every 1.86 years. The variations in brightness of Algol were discovered by Geminiano Montanari in 1667; the period of these variations was measured in 1782 by *Goodricke who also offered an explanation for them. Algol gives its name to the class of eclipsing binaries that behaves in this way.

Almagest (Arabic, 'The Greatest'), the most important astronomical work of the 2nd-century Greek astronomer,

Algol

Primary component
Secondary component
Orbit of secondary component
Sun to scale

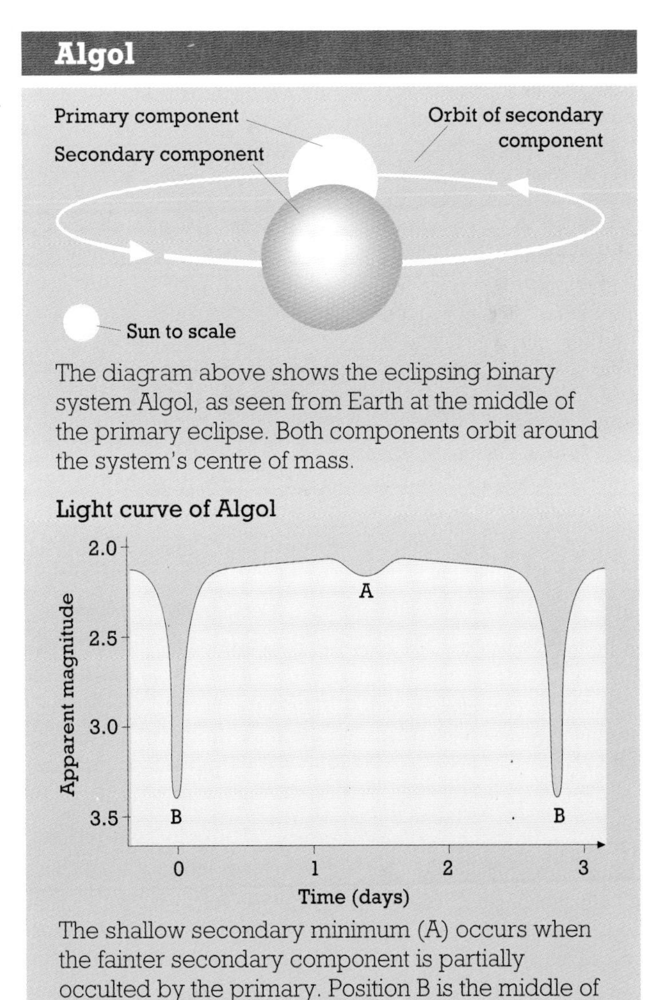

The diagram above shows the eclipsing binary system Algol, as seen from Earth at the middle of the primary eclipse. Both components orbit around the system's centre of mass.

Light curve of Algol

Apparent magnitude
2.0
2.5
3.0
3.5
A
B
B
0
1
2
3
Time (days)

The shallow secondary minimum (A) occurs when the fainter secondary component is partially occulted by the primary. Position B is the middle of the primary eclipse when the greatest amount of the primary component is obscured by the secondary.

geographer, and mathematician Ptolemy (Claudius Ptolemaeus). The original title of the work was *The Mathematical Collection*, but it became known as *The Great Astronomer*, from which, via Arabic, comes the name *Almagest*. Most of the work's thirteen books are devoted to an exposition of his *geocentric system of the universe, now known as the *Ptolemaic system, and to the development of the methods of trigonometry and spherical geometry needed for predicting the motions and apparent changes in size of the Sun, Moon, and planets. It also includes a catalogue of over 1,000 stars, apparently based on an earlier one by *Hipparchus.

almanac, a compilation of observational data for a particular interval of time, usually a year. The data, such as directions, distances, and brightnesses of astronomical objects, are listed at dates throughout the year. Almanacs are published by national observatories often several years in advance. Examples are the *Astronomical Almanac* published jointly by the *Royal Greenwich Observatory and the Washington Naval Observatory, and the *Connaissance des Temps*, published by the *Paris Observatory.

Alpha Centauri, also known as **Rigil Kentaurus**, a nearby first-magnitude multiple star in the constellation of Centaurus. It is 1.3 parsecs away and comprises two *dwarf stars, similar to the Sun, which orbit each other every eighty years. A third member of the system, *Proxima Centauri, is the Sun's nearest known stellar neighbour; it takes about one million years to orbit the other two stars. In 1839, Thomas Henderson, director of the Cape Observatory, measured Alpha Centauri's distance by observing its annual change in position (parallax) as the Earth orbited the Sun. Apart from this seasonal variation, it has a large *proper motion across the sky relative to the background of more distant stars.

Altair, a first-magnitude star in the constellation of Aquila, and also known as Alpha Aquilae. It is a hot dwarf star at a distance of 5 parsecs, with about ten times the luminosity of the Sun. It rotates very rapidly, the speed at its equator being some 250 km/s, and as a result its equatorial diameter may be nearly twice as large as its polar diameter.

Amalthea a satellite of *Jupiter, discovered by the US astronomer Edward Barnard in 1892. Amalthea orbits at 181,300 km from the centre of Jupiter. It has an irregular shape with a mean diameter of about 200 km. The albedo is low (0.06) and the composition is probably rocky. Its reddish colour may be due to contamination by sulphur compounds coming from *Io.

Ambartzumyan, Viktor Amazaspovich (1908–), Armenian astrophysicist who investigated the origin and evolution of stars and galaxies. After teaching at the University of Leningrad, Ambartzumyan became head of the Byurakan Observatory and in 1955 he put forward his theory that the centres of galaxies were the scenes of enormous explosions which would account for many extragalactic radio waves. He also worked on a method for determining the masses of nebulae, and discovered a new type of star system, known as a star association, from which individual stars can escape.

AM Herculis stars, a recently discovered class of variable stars that exhibit periodic variations in luminosity and *polarization, both linear and circular. They appear to have a source of *synchrotron radiation which is swept around the star as it rotates. This radiation is detectable by X-ray observations. AM Herculis stars are sometimes referred to as **polars** because of the large polarization changes that are observed.

Amor, an *asteroid discovered in 1932 by the Belgian astronomer Eugene Delporte. At the time of its discovery it had an apparent magnitude of 12. The absolute magnitude, the magnitude it would have if seen at full *phase at one astronomical unit (a.u.) from both the Sun and the Earth, is 19.1. This asteroid almost reaches the Earth's orbit, having a perihelion distance of 1.0864 a.u. The orbit has an eccentricity of 0.4345, an inclination to the ecliptic of 11.37 degrees, and a semi-major axis of 1.9212 a.u. Amor has a diameter of about 1 km. The closeness of this asteroid to the Earth makes it a potential target for spacecraft exploration. Most asteroids like Amor were found accidentally, their considerable relative velocity producing trailed paths on astronomical photographs taken for other purposes.

Amor group, a collection of *asteroids which have orbits that cross the orbit of Mars but not that of the Earth. The exact characteristics are that the *semi-major axis is greater than 1.0 astronomical unit (a.u.), but the perihelion distance lies between 1.017 and 1.3 a.u. Apart from the asteroid Amor, the group contains Alinda, Betulia, Ivar, Quetzalcóatl, Cuyo, and Anza. The distinction between Amor and *Apollo–Amor objects is somewhat arbitrary because perturbations to their orbits can easily convert an asteroid from one group into the other group. When close to the Earth these small faint asteroids are moving across the sky at around one degree per hour. Thus even favourable and accurately predicted appearances are difficult to observe and often these asteroids can be 'lost' for many decades.

In this woodcut of the constellation of **Andromeda**, published in Johann Bayer's *Uranometria* in 1603, Princess Andromeda offers herself as a sacrifice to the sea monster which is threatening her country. The legend goes on to tell of how she was discovered by Perseus who immediately fell in love with her and saved her by slaying the monster.

Because the **Andromeda Galaxy** has an inclination of about 13 degrees to the line of sight, its spiral structure cannot be clearly seen. The galaxy has two dwarf elliptical companion galaxies: NGC 205, *top*, and NGC 221 (M32), *bottom*.

Andromeda, a *constellation of the northern hemisphere, one of the forty-eight constellations known to the Greeks. It represents the mythical Princess Andromeda, daughter of King Cepheus and Queen Cassiopeia of Ethiopia, who was chained to a rock in sacrifice to a sea monster, represented by *Cetus. The brightest stars in Andromeda are of second magnitude, and the most celebrated feature in the constellation is the *Andromeda Galaxy. The star Alpha Andromedae, known as Sirrah or Alpheratz, forms one corner of the Square of *Pegasus. Gamma Andromedae is an attractive double star for small telescopes, consisting of yellow and blue components of second and fifth magnitude.

Andromeda Galaxy, one of the few external galaxies visible to the naked eye under the right conditions. First shown by *Hubble in the 1920s to be an external star system, it is now known to be a spiral galaxy containing about 4×10^{11} stars. It is larger than our own galaxy and is a member of the *Local Group of galaxies. Because of its relative closeness (about 700,000 parsecs), it has been extensively observed in all parts of the spectrum. It has a bright optical nucleus with spiral arms containing stars, an interstellar medium of dust and gas, numerous HII regions, and an abundance of neutral hydrogen. The central regions rotate so that the orbital periods of revolution of the stars in that region are very similar, while the outer regions have stellar orbits more in accordance with *Kepler's Laws. Analysis of the galaxy's spectrum using the *Doppler effect shows that it is moving towards us.

Andromedids (or Bielids), a *meteor shower associated with *Biela's Comet, which used to return every 6.6 years. However, the comet split apart in 1846, and in 1872, 1885, 1892, and 1899 there were impressive displays of meteors when the Earth crossed the orbit of the vanished comet. On the night of 27 November 1885 the rate was estimated to be 75,000 meteors per hour. The shower is now very poor.

Anglo-Australian telescope *Siding Spring Observatory.

angular diameter *angular measure.

angular measure, as part of a *coordinate system, used to measure the angles between the directions of celestial objects as well as their **angular diameter**, as shown in the figure. It is also used to measure changes in the directions and sizes of objects. If the angle is measured around a great circle, as for example, *declination, the circle is usually divided into 360 degrees each of which is subdivided into 60 arc minutes. Each arc minute is further subdivided into 60 arc seconds. Degrees, arc minutes, and arc seconds are also used for *right ascension, *ecliptic, and other longitudes. However, it is sometimes convenient for a full circle of right ascension to be divided into 24 hours, where each hour equals 60 minutes, or 3,600 seconds of time.

angular momentum, the *momentum a body has by virtue of its rotation. For example, the angular momentum of a particle moving in a circle is the product of the particle's momentum and the radius of the circle; however, angular momentum can be defined for any type of rotation, including spinning bodies. Angular momentum is a vector quantity whose direction is along the axis of rotation perpendicular to the plane of rotation. Unless angular momentum is transferred to or lost by the body, it is conserved, that is, it does not change when the structure of the rotating body changes. For example, if a spinning star

Angular measure

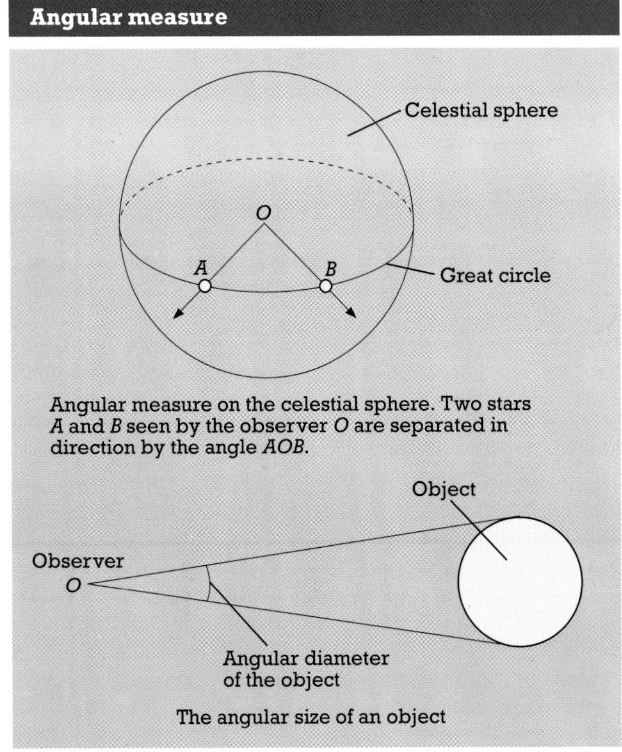

Angular measure on the celestial sphere. Two stars *A* and *B* seen by the observer *O* are separated in direction by the angle *AOB*.

The angular size of an object

Anomaly

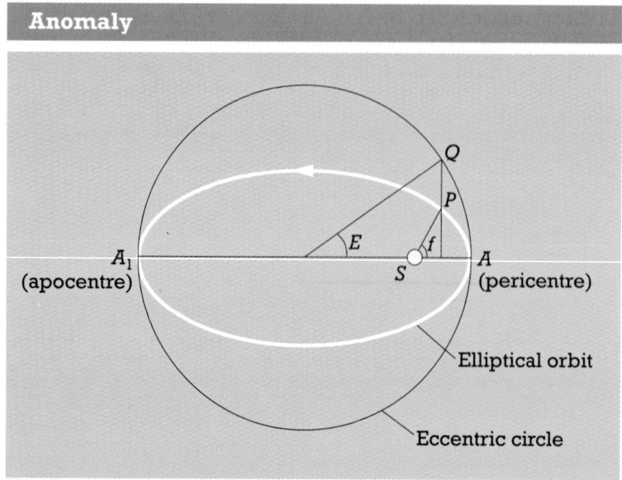

shrinks in size after a *supernova explosion, the conservation of its angular momentum results in a very much smaller *neutron star which spins much more rapidly than the original star.

angular velocity, the rate of rotation of a body about an axis. If the axis is within the body then the body is spinning; for example, the Earth spins on its axis once every day. If the axis is outside the body then the whole of the body revolves about the axis; for example, the Earth moving in its orbit about the Sun takes a year to complete one revolution. This second sense of angular velocity is also used in connection with the motion of a star through space as observed from the Earth. A star's velocity can be made up of motion along the line of sight and motion across the line of sight. The latter is the cause of the star's *proper motion and shows up as a movement across the sky against the background of more distant stars. In this case the proper motion is projected on to the *celestial sphere that is centred on the Sun and is measured with respect to a particular coordinate system associated with the celestial sphere.

anomaly, one of three angles used in calculating the position of a body P in an elliptical orbit about another body S. The **true anomaly** f and the **eccentric anomaly** E are related by the equation

$$\tan f/2 = \sqrt{((1+e)/(1-e))}\tan E/2$$

where e is the orbital *eccentricity. The angles f and E are shown in the figure. The **mean anomaly** M is the angle between SA and the direction of a fictitious body Q moving at a constant angular speed equal to the mean motion of the body P. The angles M and E are related by *Kepler's equation

$$E - e\sin E = M.$$

Antares, a variable first-magnitude *binary star in the constellation of Scorpius, and also known as Alpha Scorpii. The bright component is a red *supergiant star with about 10,000 times the luminosity of the Sun. Measurements of its disc by interferometry suggest that its diameter is about a billion kilometres, two-thirds the size of *Jupiter's orbit. It has a hot fifth-magnitude companion in a 900-year orbit, and is surrounded by a large *reflection nebula. The estimated distance of Antares is 160 parsecs. It is a member of the Scorpius–Centaurus *stellar association, a large moving group which also includes Acrux and Becrux.

Antlia, a faint *constellation of the southern hemisphere, representing an air pump. It was introduced by the French astronomer Nicolas Louis de Lacaille in the 1750s to commemorate the invention of the Irish scientist Robert Boyle. Its brightest star is of only fourth magnitude, though the constellation does contain one spiral galaxy worthy of observation.

aperture synthesis, in radio astronomy, a technique for achieving high angular resolution (image clarity) using *interferometry. In its basic form, it requires two radio telescopes, one in a fixed location, and the other placed successively at a series of positions. In each of these positions, observations and measurements are made with both telescopes. The position of the movable telescope is changed every twenty-four hours. Computer analysis of the complete series of observations synthesizes essentially the same results as those that would have been obtained from a single telescope whose size is equal to the whole of the area covered by the movable telescope.

Large regions of the sky can be very accurately mapped using arrays of pairs of radio telescopes with intercontinental baselines (*VLBI) and recording the response of each constituent pair. As the Earth rotates the effective baselines change, so effectively changing the relative positions of the telescopes, and it is possible to synthesize a radio beam which is comparable to that of a dish whose diameter is equal to the largest spacing of the array; this can be a significant fraction of the diameter of the Earth. This technique of Earth-rotation aperture synthesis has been exploited successfully by radio astronomers at Cambridge, UK, and at Westerbork in The Netherlands.

apex, the point on the *celestial sphere towards which the Sun is moving relative to the 30,000 or so stars in the solar neighbourhood. The velocity and direction of the solar motion can be found from an analysis of the *proper motions and *radial velocities of these stars. The apex lies in the constellation Hercules at right ascension 18 hours, declination 30 degrees north. The point diametrically opposite is referred to as the **antapex**.

apex of the Sun's way *apex.

aphelion *ap(o)-.

ap(o)-, a prefix used in astronomy to denote the point in an orbit furthest from a centre of attraction. Thus:

aphelion is that point in the elliptical orbit of a planet or other body about the Sun where that body is furthest from the Sun;

apogee is that point in the orbit of the Moon or of an artificial Earth satellite when it is furthest from the Earth's centre;

apogalacticon is that point in a star's orbit within the Galaxy furthest from the galactic centre;

apoastron is that point in the orbit of one component of a *binary star about the other when the two components are furthest apart.

The prefix used to denote the point in an orbit nearest to a centre of attraction is *peri-.

apogee *ap(o)-.

Apollo, the first *asteroid found to cross the Earth's orbit, discovered by the German astronomer Karl Reinmuth in 1932. That year it passed within 11 million kilometres of the Earth. This irregularly shaped S-type asteroid has a mean diameter of 1.4 km.

Apollo (space vehicle) *Project Apollo.

Apollo–Amor objects, *asteroids that cross the orbits of Mars and, in some cases, the Earth. They follow highly elliptical orbits which take them closer to the Sun than other asteroids in the main asteroid belt. As they orbit they have a reasonable chance of colliding with the Earth and Mars and it has been concluded that they cannot have existed in their present orbits since the origin of the solar system. Apollo–Amor asteroids are small and there has been considerable debate as to whether they are fragments produced by collisions between large main-belt asteroids, or are the remnants of cometary nuclei after all the ice has disappeared. A difficulty with the latter theory is that the orbits of Apollo–Amor asteroids are smaller than that of Encke's Comet, the smallest cometary orbit known. There is also a strong possibility that comets decay completely and crumble to form meteor streams.

apparent magnitude *magnitude.

apse, apsis, apsides In an elliptical orbit the major axis is often referred to as the *line of apsides or the apse line. The centre of force being at one of the foci of the *ellipse by the first of *Kepler's Laws, the ends of the major axis will be the nearest and farthest points in the orbit from the centre of force. They are referred to as **periapse** (periapsis) and **apoapse** (apoapsis), respectively. The orbital positions of perihelion and aphelion can be referred to as the apsides.

Apus, a faint *constellation near the southern celestial pole, representing a bird of paradise. It was one of the constellations introduced at the end of the 16th century by the Dutch navigators Pieter Dirckszoon Keyser and Frederick de Houtman. It contains no objects of particular interest.

Aquarids, a set of three *meteor showers that have their *radiants in the constellation of Aquarius. The most famous is the Eta Aquarids, a shower that reaches its peak on about 5 May each year and produces meteors for about three weeks around this date. The associated *meteor stream has been created by Halley's Comet and a detailed analysis of the shower has enabled the mass and age of Halley's Comet to be calculated.

Aquarius, the Waterbearer, a *constellation of the zodiac, representing a youth pouring water from an urn. In mythology the youth is said to be Ganymede, a Trojan shepherd boy who was abducted by Zeus to be the cup-bearer of the gods on Olympus. The Sun passes through the constellation from late February to early March. Its brightest stars are of third magnitude. Aquarius contains two celebrated *planetary nebulae: NGC 7293, popularly known as the Helix Nebula, and NGC 7009, known as the Saturn Nebula because of its resemblance to that planet when seen through a telescope. The globular *cluster M2 (NGC 7089) is visible through binoculars.

Aquila, the Eagle, a *constellation of the equatorial region of the sky, one of the forty-eight original constellations known to the Greeks. In mythology Aquila represents the eagle of Zeus. Its brightest star is *Altair, a name that comes from the Arabic meaning 'flying eagle'. Altair is flanked by the stars Beta and Gamma Aquilae which represent the eagle's outstretched wings. Aquila lies in a rich area of the Milky Way.

Ara, the Altar, a *constellation of the southern hemisphere, known to the Greeks. In mythology Ara represented the altar on which the gods of Olympus swore a vow of allegiance before their fight with the Titans. The star fields of the Milky Way represent the smoke rising from the altar. Ara's brightest stars are of third magnitude, but it contains no objects of particular note.

Archer *Sagittarius.

Arcturus, a star in the constellation of Boötes, and also known as Alpha Boötis, with an apparent magnitude of −0.04. It is a red giant star at a distance of 10 parsecs, with about seventy times the luminosity of the Sun. Its large *proper motion, 2.3 arc seconds per year, was first detected by *Halley in 1718.

Although the dish of the **Arecibo Radio Observatory** is fixed, the telescope can scan the sky by relying on the rotation of the Earth and by moving the detectors on the cables suspended high above the dish. The Arecibo telescope has been used to transmit powerful radio signals to any extraterrestrial civilizations which might inhabit our region of the Galaxy.

Arecibo Radio Observatory, an *observatory on the Caribbean island of Puerto Rico. It has the world's largest *radio telescope, a radio *dish 305 m (1,000 feet) in diameter. The instrument is operated by the National Astronomy and Ionosphere Center (NAIC) at Cornell University, USA. The dish hangs on cables inside the crater of an extinct volcano. Radio receivers are mounted in a cable car which is suspended above the dish at its focus. The dish is not steerable but can only point straight up. However, movement of the cable car enables radio sources to be tracked within 20 degrees of the zenith. The dish is used mainly for *radio astronomy and also for atmospheric studies and for *radar astronomy. The enormous sensitivity of the system has enabled radar echoes to be detected from many bodies in the solar system, not only planets but also asteroids, the Galilean satellites, and the nuclei of comets, including Halley's Comet.

Arend–Roland Comet, a very bright long-period *comet which had one of the most impressive dust tails of any comet of the 20th century when it was seen in 1957. Michael L. Finston and Ronald F. Probstein analysed the tail by considering both the action of *radiation pressure on fine dust and the propulsion of dust by sonic and supersonic gas flows. The dust particles responsible for the visual dust tail were found to have diameters of about one-quarter the wavelength of red light. To produce the observed tail the comet must have been ejecting about 10 tonnes of gas per second and this gas was carrying along with it an even greater amount of dust. Arend–Roland also had a thin spiky tail pointing towards the Sun and was seen when the Earth crossed the comet's orbit.

Argyre Planitia, a huge 900 km impact basin on the southern hemisphere of Mars at longitude 42 degrees west and latitude 50 degrees south. The floor of this basin has been flooded with lava and it is surrounded by rugged mountains. These mountains are uplifted blocks of the planetary crust and similar massifs are seen on Mercury and the Moon. Argyre, like most of the large Martian basins, does not have a radial *ejecta blanket, probably because of the high gravitational field on the planetary surface.

In addition to the tail commonly seen in comets, **Arend–Roland**, when it appeared in 1957, seemed to develop a spike pointing towards the Sun. This was due to dust ejected and left behind by the comet some weeks earlier. The spike is the result of the chance alignment of the dust trail and the comet with the Earth.

Ariel, a satellite of Uranus, discovered by the British amateur astronomer William Lassell in 1851. It has a nearly circular orbit above the equator of Uranus at 191,200 km from the planet's centre. It is about 1,158 km in diameter and its density of 1,600 kg/m^3 suggests that the interior contains a mixture of ice and rock. Photographed by *Voyager 2, its surface appears to be young and geologically complex, with cratered regions broken by a global system of fractures and faults. Smooth plains have been formed by flows of a highly viscous fluid that overlapped and buried older craters.

Aries, the Ram, a *constellation of the zodiac. In Greek mythology it represented the ram whose golden fleece was sought by Jason and the Argonauts. Over 2,000 years ago this constellation contained the vernal *equinox, known also as the *First Point of Aries, but this point has since moved into neighbouring *Pisces because of the effect of *precession. Alpha Arietis is the brightest star in the constellation, at second magnitude; it is also known as Hamal, an Arabic name meaning 'sheep'. Gamma Arietis is an attractive pair of fifth-magnitude stars for small telescopes.

Aristarchus of Samos (*c*.310–*c*.230 BC), Greek astronomer. Aristarchus is the first known astronomer to have suggested that the stars and the Sun remain fixed in position, while the Earth moves about the Sun in a circular orbit whose size is infinitesimal in comparison with the distance of the fixed stars. He also attempted to estimate the sizes and distances of the Sun and Moon by geometrical methods described in his only surviving treatise *On the Sizes and Distances of the Sun and Moon.* His work was largely ignored by his contemporaries, but was known to *Copernicus who revived the *heliocentric model. A prominent lunar crater, 40 km (23 miles) in diameter is named after Aristarchus. The brightest feature on the Moon, it is a site of suspected *transient lunar phenomena.

armillary sphere, a hollow three-dimensional model of the solar system and the *celestial sphere. It usually consists of two circular hoops that link the celestial poles, one representing the *prime meridian and the other the *prime vertical. These are intersected by an equatorial and an ecliptic hoop and smaller ones representing the two tropics. The sphere contains a series of rings or arms that show the positions of the planets. There are two common types, one representing the *Ptolemaic system with the Earth at the centre and the Sun and planets in orbit around it. The other represents the correct Sun-centred *Copernican system. Armillary spheres range from a few centimetres to 6 m in diameter. The early ones were fitted with sights but this went out of fashion in the 16th century.

artificial satellite *satellite, artificial.

ashen light, a faint glow seen at times in the unlit part of the disc of Venus when the planet appears as a crescent. The origin of the effect is uncertain but is likely to be from atoms and molecules in Venus's atmosphere emitting light after absorbing energy from fast-moving particles from the Sun.

asterism, a distinctive star pattern that forms part of a *constellation. The Plough is an example of an asterism, its seven stars being part of *Ursa Major, the Great Bear. Asterism is also used to describe a group of stars which gives the appearance of a sparse cluster although these stars are not part of a true association.

The **astrolabe** was used to measure the positions of stars in the night sky. Knowing the time and date of the observation, navigators could determine their location on the Earth's surface provided they were in the latitudes for which the instrument was designed. The astrolabe illustrated here was made by George Hartman of Nuremberg, Germany, in 1548.

asteroid, a minor planet which in general has a near-circular orbit close to the plane of the *ecliptic and which lies between the orbits of Mars and Jupiter. Asteroids are numbered in approximate order of the date of discovery and the establishment of an orbit accurate enough for them to be recovered. In 1796 the French astronomer Joseph Jerome le Français de Lalande had instigated a search for the 'missing' planet between Mars and Jupiter, this planet being predicted by *Bode's Law. The Italian astronomer Giuseppe Piazzi found *Ceres on New Year's Day 1801, although he was not looking for a planet, but was compiling an accurate star catalogue. The German scientist Karl Gauss calculated the orbit. Soon four more asteroids were discovered: *Pallas in 1802, *Juno in 1804, *Vesta in 1807, and Astraea in 1845. A hundred asteroids had been found by 1868, 200 by 1879, and now the list contains around 3,500. Ceres, Pallas, and Vesta have radii of 512, 304, and 290 km, respectively; the rest are smaller. It is estimated that there are about 100 million asteroids in the main belt and the total mass is 1.1×10^{22} kg, about 15 per cent the mass of the Moon. This mass is thought to be about 1/2,200 of the mass that was initially present in this region. The present-day asteroids are fragments of what was a much larger group of Moon-sized minor planets. These initial bodies had metal cores surrounded by more rocky mantles and crusts. Today's asteroids have surfaces that range from rocky to metallic. Asteroids are classified as follows:

C-type have blue colours and featureless spectra, similar to those of carbonaceous chondrite *meteorites;

S-type have a reddish colour and spectra similar to stony-iron siderolite meteorites;

M-type have metallic spectra;

U-type are unclassified.

There are groups of asteroids with similar elements of *orbit. These **Hirayama** families are thought to have resulted from the break-up of a single larger asteroid in the distant past. Most asteroids have spin periods of around 8 hours. All asteroids smaller than about 120 km radius are irregular in shape. The orbits of the asteroids have been affected by the gravitational influence of Jupiter. This has introduced gaps into the distribution of the orbital semi-major axes. Any asteroids in these *Kirkwood gaps would have orbital periods related in a simple way to the period of Jupiter, for example, $\frac{1}{3}$ and $\frac{2}{5}$ of the orbital period of Jupiter. The inner and outer edges of the belt are near the regions where the orbital periods would be $\frac{1}{4}$ and $\frac{1}{2}$ that of Jupiter. The *Trojan group of asteroids have similar orbital periods to Jupiter but lie near the L4 and L5 *Lagrangian points. For tables of the biggest asteroids see page 193.

astrograph, a *telescope that is specifically designed for the accurate photography of areas of the sky. The instrument is often a highly stable long-focal-length refractor which can produce a large photographic plate with an attached measurement scale such that a small area of the sky is magnified to spread over a large region of the plate. This enables the relative positions of the stars to be measured accurately. Photographs of the same region of the sky are taken season by season and year by year and the plate measurements are then used to measure the distances between the Sun and the stars, and the stars' proper motion, that is, their angular velocity across the sky.

astrolabe, a flat circular brass instrument which embodies a stereographic projection of a hemisphere of the heavens. The point of projection is usually the south pole and the plane of projection is the equator. Around the outer edge is a scale reading from 0 degrees to 360 degrees. The instrument is crossed by a sighting arm which is pivoted at the centre of the circle. This arm is used to measure the altitude of a star. A fretted disc (the **rete**) is free to rotate about the central pivot. Inside the rete are a number of points, each positioned to represent a specific star. The rete is turned until the pointer that represents the star being observed rests on the circle of the stereographic projection that represents the measured altitude. If the day is known, the time of the observation can then be read off. Astrolabes were first produced shortly before the beginning of the Christian era but most of them are Islamic. The astrolabe formed the basis of the *planisphere introduced in the early 19th century.

astrology, a pseudo-science which attempts to make predictions about worldly events and people's lives by describing supposed influences of celestial bodies, principally the Sun, Moon, and planets. Developed in Mesopotamia in the 2nd millennium BC, it interpreted astronomical and meteorological phenomena as astral omens, the phenomena being indications of the gods' intentions for kings and kingdoms. In time the nature of astrology changed and in Hellenistic times the horoscope was devised, which gives the fortune of an individual from the positions of celestial bodies at the moment of birth. The ancient division of the *ecliptic into the twelve signs of the *zodiac is basic to astrology and it is claimed that the configurations of Sun, Moon, and planets within those signs are crucial in forecasting influences and trends occurring at any moment in the individual's life. With our increased understanding of the size of the universe and its nature, and of human psychology, the non-scientific claims of astrology have been recognized as baseless though many today still remain devoted to it, even to the complete nonsense of daily newspaper astrology.

astrometry, the high-precision measurement of the positions of stars and other celestial bodies on the *celestial sphere. The fundamental instrument of optical astrometry is the meridian circle, which is a telescope that is free to move only in the *meridian, or north–south plane. The elevation of the telescope above the horizon as a star passes through the meridian can be related to the star's declination, and the time of passage is linked to the star's right ascension. In addition, astrometric measurements in limited areas of the sky are made with more conventional telescopes, especially long-focal-length refractors, such as the one at the *Yerkes Observatory. In recent times the use of photography has largely replaced the meridian circle in astrometry. High-quality positional information is also produced by certain radio telescopes. Astrometric measurements underpin studies of numerous astronomical phenomena, including precession, nutation, refraction, the aberration of starlight, parallax, planetary and stellar orbits, the distribution of velocities within star clusters, and quasar expansion velocities. The launch into Earth orbit of the satellite *Hipparcos has recently obtained highly accurate astrometric measurements for 120,000 stars.

astronautics, the science of manned space flight, in contrast to aeronautics. Pioneers of astronautics include Robert H. Goddard (1882–1945), a US physicist; Robert Esnault-Pelterie (1881–1957), a French engineer; and Constantin Tsiolkovsky (1857–1935), a Russian school teacher. Goddard and Esnault-Pelterie were experimenters and theoreticians while Tsiolkovsky published a dozen influential theoretical works on astronautics. Rapid progress in rocketry took place during and after World War II when rocket and space flight enthusiasts such as the German Wernher von Braun developed powerful intercontinental ballistic missiles enabling artificial satellites to be orbited. Sputnik 1, the first artificial Earth satellite, was launched by the Soviet Union on 4 October 1957. A month later, Sputnik 2, with the dog Laika on board, was put into orbit. The first human to orbit the Earth was Yuri Gagarin, of the Soviet Air Force; he was launched on 12 April 1961. John H. Glenn, launched on 20 February 1962, was the first American to orbit the Earth though two Americans had made suborbital flights before him. Many men and women have now been placed in orbit for long periods of time. The successful US *Project Apollo took astronauts to the Moon. The Moon-landing missions each carried three men, two of whom landed on the Moon's surface in the lunar excursion module while the third man orbited the Moon in the command module. On each manned landing a group of experiments was set up and samples of lunar dust and rocks collected. The experiments continued to operate for several years after the astronauts left the Moon. In the last thirty years, progress in astronautics has continued, culminating in the establishment in Earth orbit of prototype space stations such as the US Skylab and the Soviet *Mir in which crews of scientists can remain for months at a time carrying out a wide range of scientific, technological, and biological experiments. It is likely that by the end of the 20th century there will be a number of permanently crewed space stations in Earth orbit. Additionally, plans to set up a permanent base on the Moon and to send a manned mission to the planet Mars will be well advanced.

astronomer, a scientist who studies the extraterrestrial universe. Astronomers use both observational and theoretical techniques to study celestial bodies such as planets, stars, and galaxies, the material in the spaces between them, and how all these things interact. The study of *astronomy is as old as civilization but it has significantly advanced recently in line with the development of mathematics, physics, and chemistry, and with the introduction of space science. Observational astronomy now encompasses all the wavelengths of the electromagnetic spectrum and not just light. Early astronomers would only have been concerned with the movement of astronomical objects; today most scientists in the field regard themselves as astrophysicists and are more interested in the physics and chemistry of celestial objects and processes.

Astronomer Royal, originally, the chief astronomer in the UK. The post of Astronomer Royal has been filled by eminent astronomers since 1675 when *Flamsteed was appointed first Astronomer Royal by King Charles II. Before 1971, the post of director of the *Royal Greenwich Observatory was also held by the Astronomer Royal but since that year they have been separate appointments. All Astronomers Royal from Airy onwards have been knighted. A special post of Astronomer Royal for Scotland has existed since 1834, being held by the director of the *Royal Observatory, Edinburgh. For a table of Astronomers Royal see page 193.

astronomical clock, a highly accurate clock keeping either *sidereal time or *Greenwich Mean Time (Universal Time). The former is based on the time between passages across the observer's *meridian of the vernal *equinox (First Point of Aries); the latter is essentially based on the time between passages of the *mean sun across the observer's meridian. Both are therefore related to the rotation of the Earth. In former centuries such clocks were checked by timing the passage of *clock stars across the meridian using meridian or transit telescopes and a book of clock errors; the rate of change of error was kept for each clock. Nowadays the Earth is seen to be an irregular timekeeper and atomic clocks are used.

astronomical units In common with most branches of science, astronomy has some units that have convenient values for the measurements being made. The distances between the Sun and planets can be calculated as multiples of the distance between the Sun and Earth, the **astronomical unit** (a.u.), and was first done by *Kepler in the 17th century, but the a.u. was not accurately measured until 1930–31, following a world-wide concerted effort to measure the *parallax of *Eros. Its currently accepted value is 149,600,000 km.

The distances of stars are usually given in *parsecs. One parsec is the distance at which 1 a.u. subtends an angle of 1 arc second (1/3,600 degree). In other words, imagine a very tall thin triangle with a base 1 a.u. long and a height of 1 parsec (206,265 a.u. or 3×10^{13} km). The angle at the top is then 1 arc second. A parsec is a very large distance but even so the distances to galaxies are so large that they are usually expressed in terms of megaparsecs, that is millions of parsecs. The *light-year, the distance that light travels in one year (9.5×10^{12} km, about $\frac{1}{3}$ parsec), is used in most popular articles about astronomy but is rarely used by astronomers. Another unit used in astronomy is the mass of the Sun, the masses of stars being often given in solar mass units. In *radio astronomy the amount of power received at the Earth's surface from a radio source is measured in Janskys, where one Jansky is 10^{-26} W m^{-2} Hz^{-1}.

astronomy, the study and understanding of the universe beyond the Earth. Man has evolved on a planet from which he could see celestial phenomena: the motions of the Sun, Moon, planets, and stars; the dramatic appearance of comets and meteors; and the occasional eclipsing of the Sun or Moon. As early as the 3rd millennium BC, *prehistoric astronomy recognized the connection between the yearly cycle of the positions of the rising and setting of the Sun and the cycle of the seasons; in *Egyptian astronomy the coincidence of the time of the *heliacal rising of the bright star Sirius and the annual inundation of the Nile Valley with its life-giving waters was observed. The Chinese people had astrological schools as early as 2000 BC which contributed to the development of *Chinese astronomy. The astronomer–priests of the Sumero–Akkadian civilizations of the Middle East spent many centuries recording the movements of the Sun, Moon, and planets against the rotating stellar background, attaching astrological significance to their patterns and to the occurrence from time to time of eclipses of the Sun and Moon. Many of the hundreds of clay tablets recovered from the ruins of Nineveh and other once powerful Babylonian cities are astronomical *ephemerides of the Sun, Moon, and planets strongly reminiscent of the astronomical almanacs published today. If we today really understand the universe better than our ancestors, it is because of the discovery and use of the scientific method, providing a civilization in which groups of people can devote their lives to astronomy, utilizing the scientific and technological tools provided by that civilization.

The Greeks, who possessed little in the way of instruments and depended upon the unaided eye, demonstrated the power of the mind. In *Greek astronomy a much clearer understanding of the nature of the universe arose during the flowering of Greek genius in the 1st millennium BC. *Aristarchus of Samos taught a heliocentric theory of the solar system with the planets, including the Earth, orbiting the Sun. Philolaus also believed that the Earth moved through space. *Eratosthenes in *c.*250 BC accurately measured the circumference of the Earth while other Greeks harnessed the brilliance of their minds to simple observations, not only demonstrating that the Moon and Sun lay at enormous distances compared to the size of the Earth but also obtaining a reasonable estimate of the distances and sizes of these bodies. *Hipparchus, probably the most outstanding of Greek astronomers, discovered the precession of the equinoxes, measured the Sun's distance, and provided theories of the motion of the Sun and Moon. Ptolemy, in his great work, the **Almagest*, published *c.*140 AD, collected the whole corpus of Greek astronomical knowledge. This work survived the Dark Ages of Europe and influenced astronomical thought right up to the 17th century. During these Dark Ages, *Islamic astronomy flourished with Ptolemy's *Almagest* being held in high esteem. The influence of Muslim astronomers on astronomical thought was profound. Many terms and names used in the science are of Arabic origin, for example zenith, nadir, and almanac. Algol, Aldebaran, and other star names are also a legacy from this great school of astronomy. The awakening of Europe from its sleep of ignorance brought a new revolution in astronomy. *Tycho Brahe's highly accurate observations of the planetary positions, coupled with *Copernicus's heliocentric theory of the solar system, provided *Kepler with the data and perspective necessary to formulate his three laws of planetary motion. The invention of the telescope and its use by *Galileo to observe the heavens provided crucial discoveries regarding the enormous scale of the universe, as well as the nature of planets. In particular the discoveries of the Galilean satellites of Jupiter and the lunar phases of Venus demonstrated the inapplicability of the Ptolemaic geocentric theory of the solar system, lending support to the *Copernican system. The final step was taken by *Newton. The application of his law of universal gravitation and his three laws of motion to the dynamics of the Moon and the planets integrated the works of Copernicus, Kepler, and Galileo. During the next two centuries there flowed two main streams of development in astronomy. The first was the exploration of wider and wider regions of the universe by means of the bigger and better telescopes built in that era. The second was the successful application of Newtonian dynamics to the motions of the planets and their satellites, to comets and asteroids, and to the increasing number of binary systems of stars discovered. Everything obeyed Newton's Law of Gravitation, its culminating triumph being the discovery of a new planet by *Adams and *Le Verrier from their mathematical investigations of the orbit of Uranus.

The last half of the 19th century saw three further torrents of discovery with the application of the camera and spectroscope to astronomy and the measurement of the distances of an increasing number of stars. The camera, with its ability to make long time exposures, enabled telescopes to probe deeper than ever into space and allowed the images not only of stars and nebulae but also of their spectra to be permanently recorded for subsequent analysis in the laboratory. Spectroscopic analysis of celestial objects such as the Sun, stars, and nebulae revealed their composition and the conditions within them, and the measurement of stellar distances confirmed the immense scale of the universe. The early 20th century saw further spectacular progress in the methods by which astronomers studied the universe and increased their understanding of its structure. By now the main branches of the science were traditional spherical or positional astronomy, celestial mechanics, stellar kinematics and stellar dynamics, astrophysics, and cosmology. Observations of the solar system bodies, the Sun and the stars of our own galaxy, the nebulae, and a myriad other galaxies were being carried out in a large number of national and university observatories. Developments in atomic physics enabled the US physicist Hans Bethe and the German physicist Carl Friedrich von Weizacker to produce a breakthrough in astronomers' understanding of the major problems of the source of stellar energy, the ages of stars and star clusters, their creation and evolution, and the observed relationships between stellar luminosities, sizes, and masses. The development of radar and rockets during and after World War II provided two new major branches of astronomy: radio astronomy and space research. Radio astronomy's discovery that celestial objects such as supernovae and galaxies give out radiation in the radio region of the electromagnetic spectrum is matched in its importance by the discovery of hitherto unsuspected objects such as pulsars and quasars. The greater penetrating power of the most powerful radio telescopes and their high resolution produced by interferometric arrays has been found to be of relevance in distinguishing between cosmological models such as the Big Bang and the Steady State theories. The ability of the rocket

Atmosphere, planetary

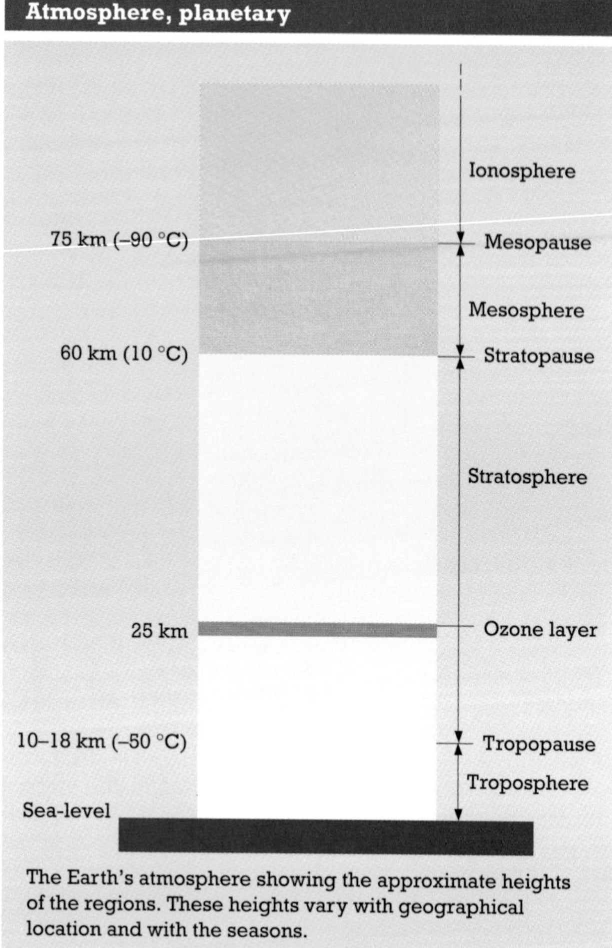

The Earth's atmosphere showing the approximate heights of the regions. These heights vary with geographical location and with the seasons.

to place packages of scientific instruments in orbit above the atmosphere surrounding our planet has opened the whole of the electromagnetic spectrum to man's scrutiny, not just the narrow optical and short-wave radio regions which were the only wavelengths capable of penetrating the atmosphere. Thus gamma-ray, X-ray, ultraviolet, infra-red, and millimetre astronomy has been born and the exciting discovery has been made that there exist in space many objects radiating in these regions. Exotic objects such as black holes are being looked for; these are thought to be regions of matter so dense that not even light itself is fast enough to achieve escape velocity in their gravitational fields. Packages of instruments have also been sent to fly past, orbit, or land upon the planets, in particular Mars, where the search for extraterrestrial life goes on. Modern astronomy is then a growth science, exploding outwards as spectacularly as the universe itself. Man's increasing understanding of this enormous and ancient universe he inhabits, and his growing appreciation that it is either overwhelmingly probable that life is common throughout the cosmos or occurs only on Earth, cannot but profoundly influence his philosophical attitudes and his fight to achieve maturity, a maturity no one would deny that he desperately needs.

astrophysics, the study of the physical nature of astronomical objects, especially stars. Astrophysics has developed mainly in the 20th century and complements the traditional domains of astronomy such as *astrometry, *stellar dynamics and kinematics, and *celestial mechanics. Observational astrophysics interprets the electromagnetic radiation emitted by celestial objects. For example, analysis of the *spectral lines and *polarization of light from galaxies, stars, and planets enables their properties, such as temperature, density, and magnetic field, to be inferred. This is sometimes complemented by analysis of radiation at other wavelengths to reveal the mechanism responsible for its emission. Theoretical astrophysics is concerned with understanding these observations in terms of mathematical models. These models are constructed from the known laws of physics, and may occasionally result in the discovery of new laws under extreme astrophysical conditions. Physics is applied to astronomy in areas such as the understanding of the spectra of stellar atmospheres; the nuclear energy processes which fuel the stars; and the structure of *pulsars and *neutron stars.

Atlas, a small *shepherding satellite of Saturn. Atlas is approximately spherical with a mean diameter of 30 km, moving in a circular orbit of radius 137,670 km from Saturn in a period of orbital revolution of 0.602 days.

An atlas is also a variety of *catalogue or map giving plotted or photographic information about astronomical objects such as stars, galaxies, and nebulae.

atmosphere, planetary, the gaseous blanket that envelops a planet. For the *inner planets most of the gaseous molecules have been produced by the thermal break-up of the mantle rocks. Whereas Mercury has no atmosphere, the atmospheres of Venus and Mars consist mainly of carbon dioxide. The Earth's atmosphere is 78 per cent nitrogen and 21 per cent oxygen. The *outer planets have atmospheres that are mainly hydrogen and helium and these gases were probably accreted in the early stages of the formation of these planets. The retention of an atmosphere depends on the *escape velocity from the planetary surface. This is at least five times greater than the mean thermal velocity of the atmospheric molecules. The weather in an inner planetary atmosphere is driven by solar heating. In the case of the outer planets the atmospheric mass accounts for a large percentage of the total planetary mass, and internal energy sources and the rapid rotation of the planets play a significant role in atmospheric dynamics.

Auriga, the Charioteer, a *constellation of the northern sky, one of the forty-eight constellations known to the Greeks. It is usually said to represent Erichthonius, a legendary king of Athens who was a champion charioteer, but sometimes it is identified as Myrtilus, charioteer of King Oenomaus of Elis. The brightest star in Auriga is *Capella, representing a goat carried by the charioteer. Its two kids are represented by the stars Eta and Zeta Aurigae. Epsilon Aurigae is a supergiant star that is eclipsed every twenty-seven years by a dark companion; this companion is now thought to be a close binary surrounded by a disc of dust. The last eclipse was in 1983. Auriga contains three star *clusters visible through binoculars: M36, M37, and M38.

aurora, or polar lights, a display of visible light seen in the upper atmosphere at night. The name Aurora Borealis or Northern Lights is attributed to the French philosopher Pierre Gassendi (1621), though descriptions of this beautiful natural phenomenon in the northern hemisphere go back to Aristotle and Pliny. The aurorae are described in the mythology of Scandinavia and of the Inuit, where they were attributed to supernatural manifestations. The southern hemisphere equivalent is the Aurora Australis first named by

The **aurora** known as the Aurora Borealis, or Northern Lights, is seen here near Fairbanks, Alaska. These luminous displays appear in the night sky at high altitudes over the poles and form rapidly changing shapes. Bands and arcs in a variety of colours can extend over 1,000 km, and luminous rays follow the lines of the Earth's magnetic field. Although they are a spectacular sight to the naked eye, they emit much more radiation in the invisible parts of the spectrum.

the British explorer James Cook in 1773, following his observation of auroral displays in the southern Indian Ocean. Aurorae are most frequently observed in two oval belts (the auroral ovals) surrounding the Earth's magnetic poles at a magnetic latitude averaging about 70 degrees. In regions under the auroral ovals the displays are to be seen overhead at heights of about 100 km on most clear nights. At middle latitudes they are usually observed only as a glow resembling dawn on the horizon in the direction of the pole but occasionally, following large solar *flares, the auroral ovals move to lower latitudes and then they are observed overhead. Auroral displays take many forms from luminous arcs, bands and coronae, to diffuse patches resembling clouds. Amongst the most dramatic are folding bands which resemble a great moving curtain in the sky. Faint aurorae are often white but the more vivid displays are greenish yellow or red in colour. They are produced when upper atmospheric atoms (mainly atomic oxygen) fluoresce on being struck by fast-moving charged particles. These particles originate in the Sun and are guided into the auroral regions by the Earth's magnetic field. Although great auroral displays have been associated with magnetic storms and solar flares for well over a century, only since the discovery of the *solar wind and its influence on the *magnetosphere in the early 1960s has it been possible to begin to understand the complex relationships between these phenomena.

Australia Telescope National Facility, a radio *observatory based at Narrabri in New South Wales. Officially opened in 1988 as part of the Australian bicentenary celebrations, it has the most powerful radio *interferometry system in the southern hemisphere. The instrument combines the features of several different *radio telescopes. The Compact Array consists of five telescopes which can be moved along a 3 km (1.9 miles) railway track, and a sixth telescope which is fixed. The dishes are 22 m (72 feet) in diameter. Their high surface accuracy enables measurements to be made at radio wavelengths of less than 1 cm. The Long Baseline Array links the Compact Array with other radio telescopes in Australia by the satellite AUSSAT for *VLBI observations.

azimuth *celestial sphere.

B

Baade, Walter (1893–1960), German-born US astronomer who proposed the existence of two *stellar populations. He deduced this from photographs taken of the satellites of the Andromeda Galaxy (M31). Young stars belonging to the spiral arms of galaxies he called Population I stars and the old stars associated with the more central regions he called Population II stars. Baade reinvestigated the *period–luminosity law for variable stars which can be used to measure the distances of nearby galaxies. In 1952 he proposed a revision of the law which meant that the estimated distances of these galaxies had to be doubled.

Barnard's Star, a tenth-magnitude *binary star in the constellation of Ophiuchus. It was discovered by the US astronomer Edward Barnard in 1916 and is noted for its large *proper motion of 10.3 arc seconds per year with respect to the Sun. It is the second-nearest star to the Sun, only 1.8 parsecs distant, and it moves so rapidly that in under 200 years its position against the stellar background changes by half a degree, the angular diameter of the full moon. Its luminosity is 2,000 times less than that of the Sun and it may be orbited by planets.

barred spiral galaxy, a *galaxy in which the nucleus is crossed by a bar of dust, gas, and stars with the spiral arms emerging from the ends of the bar. These types of galaxies have the *Hubble classifications SBa to SBc. The reason for the existence of the bar is not fully understood.

Bayer letters, a sequence of Greek letters used to identify the brightest stars in a constellation. The system was introduced by Johann Bayer (1572–1625), a German astronomer, on his star atlas *Uranometria* published in 1603. Bayer assigned the Greek letters to stars usually in order of brightness, although in some cases where there were several stars of similar apparent brightness, such as in Ursa Major, the letters were assigned in order of *right ascension.

Barred spiral galaxies were once thought to be intermediate in the evolution of galaxies from elliptical to fully spiral galaxies, whose spiral arms grew out of the elliptical body. It is now thought they are a totally distinct type of galaxy.

Becrux, a first-magnitude *variable star in the constellation of *Crux and also known as Beta Crucis. It is a hot giant star at a distance of 140 parsecs, with about 6,000 times the luminosity of the Sun. It is a member of the Scorpius–Centaurus *stellar association, a large moving group which also includes Acrux and Antares. At declination −60 degrees, Becrux does not rise at latitudes north of +30 degrees.

Bessel, Friedrich Wilhelm (1784–1864), German astronomer and applied mathematician. His first major work, *Fundamenta Astronomiae* (1818), described techniques which were used to correct the observations of *Bradley enabling the positions of approximately 4,000 stars to be determined. From observations of the binary star 61 Cygni, Bessel published a value for its distance in 1838. In around 1834 Bessel commenced a series of observations of the small wave-like irregularities in the proper motion of Sirius and he concluded that Sirius had a dark companion. In 1862 the companion star was observed by the US telescope maker Alvan Clark. Bessel was the first to measure the apparent diameter of the Sun using a *heliometer. In studying the *three-body problem he investigated the properties of a set of mathematical functions (Bessel functions); these are now fundamental to much of mathematical physics.

Beta Centauri, a first-magnitude *binary star in the constellation of Centaurus, and also known as **Agena** or Hadar. It comprises a hot bright giant star with about 5,000 times the luminosity of the Sun, with a fourth-magnitude companion, in an orbit of very long period. The distance of Beta Centauri is 100 parsecs. At declination −60 degrees, it does not rise at latitudes north of +30 degrees.

Beta Lyrae stars, a class of *eclipsing binary in which, unlike *Algol, the magnitude varies continuously even when eclipses are not taking place. Usually the component of greater surface brightness, the primary, fills its *Roche lobe and is not spherical, so that the visible surface area changes as the system revolves. Primary and secondary minima in the luminosity are of unequal depth when plotted as a light curve as illustrated in the figure. Beta Lyrae itself varies between magnitudes 3.3 and 4.4 in a period of 12.94 days; its changes were discovered by *Goodricke in 1784. Matter is rapidly being transferred from the bright component to the smaller but more massive faint one via an *accretion disc, and the period is slowly increasing.

Betelgeuse, a first-magnitude red *supergiant semi-regular *variable star in the constellation of Orion, and also known as Alpha Orionis. Its magnitude changes between 0.0 and 1.3 in a varying period of about six years, with shorter changes of between seven and thirteen months. The variations were discovered by John *Herschel in 1836. The distance of Betelgeuse is not known precisely. If it is a member of the Orion *stellar association at a distance of 400 parsecs, its mean luminosity is some 60,000 times that of the Sun. Measurements of its disc by *interferometry indicate a diameter of 1.9×10^9 km, larger than the orbit of Jupiter. Other estimates give a lower distance, and consequently a lower luminosity and smaller diameter.

Bethe cycle *carbon–nitrogen cycle.

Beta Lyrae

The diagram below shows the eclipsing binary system Beta Lyrae. The secondary component is concealed from view by a surrounding disc of matter accreted from the less massive but brighter star.

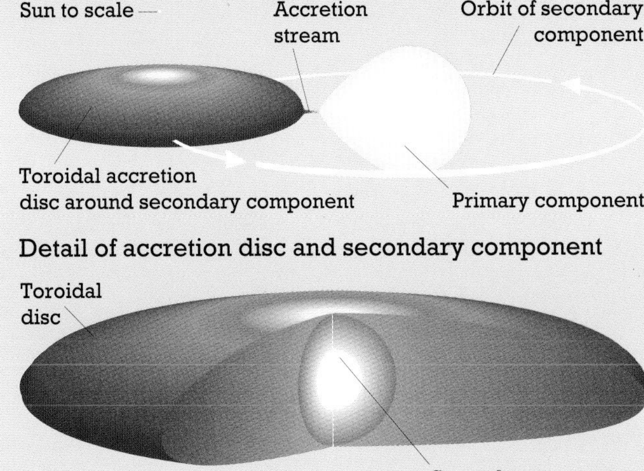

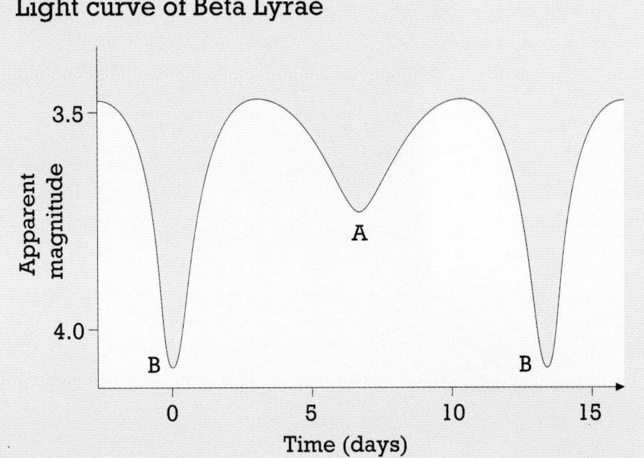

The shallow secondary minimum (A) occurs when the fainter secondary component is partially occulted by the primary. Position B is the middle of the primary eclipse when the greatest amount of the primary component is obscured by the secondary.

Biela's Comet, a *comet first seen by the French astronomer Jacques Montaigne in 1772 and then by Jean-Louis Pons in 1805 but whose orbit was calculated accurately only after it was rediscovered by the Austrian Wilhelm von Biela in 1826. After calculating its orbit, he showed that it was similar to that of the comets visible in 1772 and 1805. The comet returned every 6.6 years and its orbit intersected the Earth's orbit. When the comet returned in 1846 it was found to have split in two. Sometimes one and sometimes the other portion was brighter. By 1852 the two components were separated by 2.4 million km and the comet has not been seen since. However, the *Andromedid meteor shower was observed to be associated with the cometary remnants. This shower is now very poor.

Big Bang theory, the most generally accepted theory in *cosmology which states that the universe began in a primordial explosion about 15 billion years ago. The *microwave background radiation found in 1965 by the US physicists Arno Penzias and Robert Wilson is a remnant of that earliest phase. Various versions of the Big Bang theory, such as the inflation theory, have been proposed recently to account for a number of observational features of the universe. These include the distribution of matter, the types of celestial object, and the microwave background radiation at various distances. Physicists are now beginning to simulate, on a subatomic scale, the conditions that prevailed during the Big Bang. However, a complete understanding of the origin and evolution of the universe is still some way off. The *Steady State theory, once the principal alternative to the Big Bang theory, does not agree with current observational results.

Big Crunch, in some cosmological models, the end of the universe when the present expansion is reversed and the universe contracts under its own gravitation. *Oscillating universe models suggest that the Big Crunch is followed by a new *Big Bang producing another expanding universe.

Big Dipper *Plough.

binary pulsar, a *pulsar that is one component of a *binary star. The best known is PSR 1913+16 in the constellation of Aquila, discovered in 1974, whose pulse period is 0.059 seconds. It is in an elliptical orbit with an unseen companion thought to be a *white dwarf or a *neutron star that emits no radio signals. The pulsar's orbital motion can be measured by the *Doppler effect on the arrival time of its radio pulses. It completes an orbit every 7 hours 40 minutes. The strong gravitational field provides a sensitive test of the general theory of *relativity, which predicts that the periastron (see *peri-) should advance at almost exactly the observed rate of 4.2 degrees per year. This is stronger confirmation of the theory than that given by the advance of the perihelion of Mercury.

binary star, a pair of stars in orbit about their common centre of gravity under their mutual gravitational attraction. Binary stars are sometimes called **double stars** and the individual stars in the system are called its **components**. Such pairs of stars, of which there are several types, are exceedingly common. A **visual binary** appears as a distinct pair of stars when viewed through a telescope with their separation and orientation changing slowly over the years. In an *eclipsing binary each star regularly eclipses the other resulting in periodic variations in the total brightness. A **spectroscopic binary** contains members that are too close together to be seen as separate stars. As a star orbits its companion, it will periodically be moving towards and away from the Earth. By examining its spectrum, the *Doppler effect can be used to measure the changes in its motion. A **polarization binary** displays periodic changes in the *polarization of its light. In such binaries, as the stars orbit they illuminate gas and dust in the space between them, and the angle at which their light strikes this matter changes periodically. In this way the scattered light is polarized. Accurate measurement of these effects enables the orbits to be calculated, giving information about the relationship between the stars' masses, sizes, velocities, and separation. For example, if a star is both an eclipsing and spectroscopic binary, the

Binary star (spectroscopic)

The top diagrams show the positions of two stars A and B as they orbit their centre of mass C. As a result of the Doppler effect, the wavelengths of light from each star are displaced toward the blue or red end of the spectrum, depending on whether the star is approaching or receding from the Earth. Any spectral line in the spectrum of each star is therefore seen to oscillate back and forth across its normal wavelength position, as shown in the row of spectra.

masses of each star may be determined, together with the inclination of the orbit. Details of the way the light changes at the times of the eclipses give the relative sizes of the stars and allow the structure of their atmospheres to be studied. A binary star that emits X-rays is known as an *X-ray binary.

In a number of cases a binary star reveals itself under extended observation to be a triple star with the third star in orbit about the centre of mass of the binary components, or it may be discovered that one of the components of the binary is itself a double star as in the case of *Acrux. Whereas the close binary components of the triple star may have a period of a few days, the more distant third component may orbit the centre of mass of the closer pair in a period of time measured in hundreds or even thousands of years.

Birr Castle, the home of Lord Rosse in Ireland, where he built his giant 72-inch (1.8 m) *reflecting telescope between 1842 and 1845. He mounted it between massive masonry walls. With it he was able to detect the spiral nature of a number of nebulae now known to be spiral galaxies. In 1844 he also named the famous nebula in Taurus the *Crab Nebula because of its filamentary structure.

black body, a hypothetical entity that perfectly absorbs all *electromagnetic radiation (of whatever wavelength) that falls on it. A black body is also the most efficient emitter of electromagnetic radiation that is theoretically possible, and, as such, the radiation emitted is characterized by the temperature, T, of the black body, and no other property. The spectral distribution of black body radiation can be calculated using quantum mechanics, and is called *Planck's Law. For a black body of a particular temperature, the emission peaks at a particular wavelength. This peak moves to shorter wavelengths for hotter temperatures, as given by *Wien's Displacement Law. The total power emitted per unit surface area of a black body is given by *Stefan's Law, and increases exceedingly rapidly with increasing temperature. For many purposes the electromagnetic emission from so-called 'thermal' astronomical sources can usefully be approximated by black body radiation. Examples include the *microwave background radiation (characterized by T = 2.7 K), the infra-red emission from dust clouds in the interstellar medium (T = 10–300 K) or surrounding certain types of star (T up to 1,500 K), the infra-red–optical–ultraviolet emission from normal stars (T = 3,000–30,000 K),

Black body

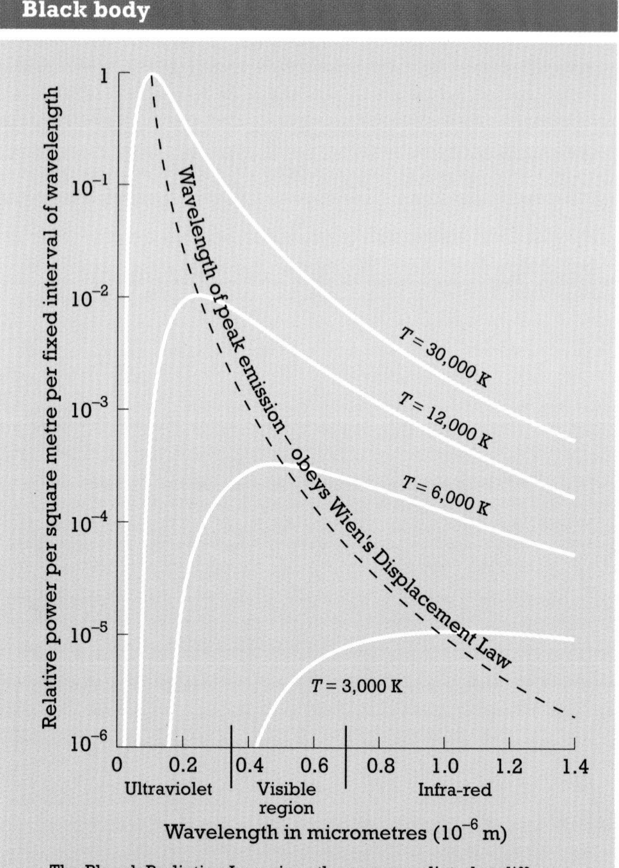

The Planck Radiation Law gives the power radiated at different wavelengths by black bodies at various temperatures.
The position of the peak is given by Wien's Displacement Law.
Stefan's Law shows that a 30,000 K black body radiates 10,000 times as much total power as a 3,000 K black body.

and the extreme-ultraviolet emission from proto-white dwarf stars and planetary nebula nuclei (T = 100,000 K).

black hole, a possible final state of a star, when its mass, and hence its gravitational pull, is so large that the star undergoes catastrophic gravitational collapse inwards, completely overwhelming the stabilizing forces within it. The matter becomes more and more compressed during this process which ultimately results in an object of zero size and infinite density. Indeed according to Einstein's theory of general *relativity a space–time *singularity is formed where the gravitational forces become infinite. As this inward collapse proceeds, the gravitational field on the surface of the star increases so much that it becomes more and more difficult for particles and light emitted from the star to escape. Eventually all of the star is hidden inside a one-way membrane or *event horizon through which no light can escape. The collapsing star has now become a black hole and all that can be felt from the outside is the gravitational pull of the original star; the star is otherwise completely invisible. The radius of the event horizon is called the *Schwarzschild radius after the German astronomer Karl Schwarzschild, who, in 1916, discovered the correct solution of Einstein's equations. This solution describes an isolated non-rotating black hole. In 1963 the corresponding solution for a rotating black hole was discovered by the New Zealand physicist Roy Kerr. The Schwarzschild radius of a black hole can be very small. A star three times the Sun's mass will probably form a black hole with a Schwarzschild radius of only 9 km. Astronomers believe that many black holes are to be found in the Milky Way. Indeed many believe that X-rays coming from *Cygnus X-1, a binary star system in the constellation of Cygnus, are due to the presence of a black hole which forms one component of the binary system. Gas drawn from the surface of the ordinary visible star in the system is swept around into an *accretion disc surrounding the apparently invisible member of the system, that is, the black hole. This swirling disc of gas is believed to be heated by friction to temperatures of tens of millions of degrees. Such a hot gas would emit X-rays which are probably those seen during satellite observations of this system.

Some astronomers have suggested that very large black holes may also be found at the centres of some galaxies including, perhaps, our own. It has also been suggested that very small black holes may have been formed during the highly compressed initial phase of the universe. If these black holes are within a Schwarzschild radius comparable to the size of an atom then quantum mechanics must be considered. From the point of view of Einstein's theory of gravitation, the event horizon represents an absolutely impenetrable barrier. However, when we try to harmonize quantum mechanics with gravitation theory, we discover that the matter inside black holes can 'tunnel' out, rather like a kind of radioactive decay. The quantum decay of black holes takes an enormously long time, unless the black hole is very small. Indeed a tiny black hole, the mass of an average mountain, but whose Schwarzschild radius is as small as an atomic nucleus, would decay explosively. It is possible that such an explosion could give rise to gamma rays.

The possible existence of black holes was foreseen by *Laplace in 1798, basing his ideas on Newton's theory of gravitation. He conceived of light as a stream of tiny particles and calculated that with a body of sufficient density, these particles would not be able to escape from its surface. His black hole is different from the one based on Einstein's theory of relativity, even though the end result is similar.

Bode's Law, or the Titius–Bode Law, a mathematical sequence that described the relative sizes of the orbits of the planets known in the 18th century. *Kepler had tried to explain planetary spacings using geometry, but it was Johann Titius von Wittenberg who in 1766 found such a sequence. His publication was not widely read and it was the better-known German astronomer Johann Bode (1747–1826) who made the law famous. In the two centuries since its formation, opinion has varied as to whether the law is a numerological coincidence, or whether it reflects aspects of whatever processes shaped the solar system.

The average distances of the first six planets from the Sun are proportional to $r = 4 + 3 \times 2^n$. The table shows how the sequence progresses. Belief in a physical basis for the law was strengthened by the discovery in 1781 of the planet Uranus at approximately the $n = 6$ position in the sequence, corresponding to $r = 196$, and by the discovery in 1801 of the asteroid Ceres at $n = 3$, corresponding to $r = 28$, thus filling a gap in the sequence. However in 1846 Neptune was discovered far short of the $n = 7$ position (despite *Le Verrier's having used Bode's Law in his prediction of Neptune's orbit) and there was also the puzzle that Mercury occupies the $n = -\infty$, rather than the $n = -1$ position. In 1913 the British astronomer Mary Blagg derived a modified formula which fits not only planetary distances but also most of the major moons of the giant planets. Although tidal and other interactions can certainly affect orbits, it is nevertheless still unclear whether there is any physical significance in the precise spacing of planets and moons in the solar system.

	n	r
Mercury	$-\infty$	4
Venus	0	7
Earth	1	10
Mars	2	16
Jupiter	4	52
Saturn	5	100

bolide *meteor.

bolometer, an electrical instrument for measuring heat radiation. In one form, radiant energy is concentrated on a tiny metal strip which forms one branch of an electrical circuit known as a Wheatstone Bridge. The intensity of the energy can then be transformed into an electrical reading.

A **bolometric magnitude** scale is a standard scale of stellar brightness. The sensitivity of most optical measuring instruments varies with the wavelength of the incident light. By using the bolometric magnitude scale a star's *magnitude can be measured independently of these variations. To obtain the bolometric magnitude from a measurement made by any detector an adjustment called the bolometric correction is made. Measurements made by different detector systems can then be adjusted to the bolometric magnitude scale.

Boötes, the Herdsman, a *constellation of the northern hemisphere, one of the forty-eight constellations known to the Greeks. In mythology it represents Arcas, son of the god Zeus and the nymph Callisto. Boötes is often depicted as a man driving the great bear around the pole. The constellation's brightest star is the red giant *Arcturus. A celebrated double star is Epsilon Boötis, also known as Izar or Pulcherrima, consisting of a close pair of orange and blue stars of third and fifth magnitude that make a beautiful colour contrast in telescopes. There are several other interesting double stars in Boötes. Every January the *Quadrantid meteors radiate from a point in the northern part of Boötes, an area that was once occupied by the now defunct constellation of Quadrans Muralis, the mural quadrant.

Bradley, James (1693–1762), British astronomer. He became a Fellow of the Royal Society in 1718, Savilian Professor of Astronomy at Oxford in 1721, and *Astronomer Royal at the *Royal Greenwich Observatory in 1742. Bradley was an accurate and industrious observer. He attempted to measure the distance between the Earth and Gamma Draconis using stellar *parallax. This star was chosen because it passes almost through the zenith at Greenwich and thus the problems of refraction and bending of the telescope tube were avoided. However, the annual movement of Gamma Draconis could not be accounted for by parallax alone and he was led to the discovery of the *aberration of starlight. Bradley had provided the first direct proof that the Earth moved round the Sun as in the *Copernican system. He discovered *nutation, and also derived practical rules showing how the refraction angle of an astronomical object depended on its *zenith distance and the atmospheric temperature and pressure. By 1750 Bradley had re-equipped the Greenwich Observatory and in the next twelve years obtained 60,000 accurate stellar positions that are still used as a starting point for the study of stellar movement.

Brahe, Tycho *Tycho Brahe.

bremsstrahlung (German 'braking radiation'), electromagnetic radiation produced when high-energy electrons are accelerated by a magnetic field. The observed bremsstrahlung yield of radiation from the vicinity of a solar *flare results from the injection of a beam of energetic electrons over the flare's area. Bremsstrahlung that originates in remote parts of the Galaxy can be observed as a significant constituent of *cosmic rays. High-energy bremsstrahlung radiation is often observed at gamma-ray frequencies.

brightness *luminosity.

Bruno, Giordano (1548–1600), Italian philosopher and astronomer. He rejected the *Ptolemaic system which put the Earth at the centre of the universe and was the first to see the logical consequences of the removal of the Earth from this position. He held that the universe was infinite, with countless individual worlds similar to the Earth, rather than a fixed sphere of stars as in the *Copernican system. He was arrested by the Inquisition of the Roman Catholic Church for his unorthodox beliefs and was burnt at the stake.

Bull *Taurus.

burster, a stellar source of X-rays exhibiting a sudden fluctuation in the strength of its X-ray emission several times a day. The X-rays may be emitted by gas being transferred to a *neutron star from a companion star. The short bursts are thought to be created by the enormous accelerations and collisions of the gas as it is transferred.

butterfly diagram, a diagram showing how the number and solar latitude of *sunspots change during their eleven-year cycle (see page 164). The first spots in a new cycle appear in latitudes far from the solar equator with later spots appearing nearer and nearer to the equator as the cycle continues. The diagram is an illustration of *Spörer's Law.

C

Caelum, the Chisel, a small, faint, southern hemisphere *constellation, introduced in the 1750s by the French astronomer Nicolas Louis de Lacaille. Located near *Orion, it represents a sculptor's chisel. There are no objects of note in Caelum, the brightest stars being only of fourth magnitude.

calendar, any system for fixing the beginning, length, order, and subdivisions of the year. Calendrical systems have been used by societies since the earliest times, nearly all of them based on one of two astronomical cycles: the cycle of the phases of the Moon (the *synodic month or lunation), often of major ritual and religious significance, and the cycle of the *seasons (the period of the Earth's orbit around the Sun), of importance in agriculture. The two cycles are incompatible in that the synodic month has a period of about 29.5 days, giving a lunar year (12 months) of just over 354 days, over 11 days shorter than the mean solar year of 365.2422 days. In most societies the *lunar calendar, in which the month, not the year, is the basic interval, was the first to be used, and different systems were developed to reconcile this cycle with that of the seasons. The early Egyptians had two completely separate calendars running concurrently, one for religious purposes and one for agricultural use. For some early calendars the correspondence between the seasons and the phases of the Moon was secondary and an uninterrupted ritual cycle was of prime importance. For example, the Mayan civilization of Central America had two calendars: a ritual cycle of 260 days, formed by combining the numbers 1–13 with 20 day names, ran concurrently with a yearly cycle of 365 days consisting of 18 months of 20 days each and 5 additional days. The beginnings of the two cycles came into coincidence every 52 years, this being exactly 72 ritual cycles. By specifying a date in both systems it was possible to identify that date uniquely within this 52-year period. The Mayan civilization maintained this calendar over several millennia. For the Babylonians the first appearance of the new crescent moon in the western evening sky fixed the beginning of a new month, days being reckoned to begin at sunset. The average length of these months was therefore the synodic month, and the normal year had 12 months, that is, about 354 days. In order to maintain the correspondence of the calendar to the seasons an additional month was **intercalated** (added) as necessary. Usually the last month of the year (Adaru) was simply repeated. At first these intercalations were carried out empirically but by the early 4th century BC definite rules had been established based on the *Metonic cycle in which 19 years is equated to 235 months and to 6,940 days.

The *Julian calendar was introduced to the Roman Empire by Julius Caesar in 46 BC. It was developed from a lunar calendar, as is evident from its division into 12 months. However, the months no longer correspond to lunations, as days have been added to give a total year length of 365 days. Almost exact correspondence with the mean solar year is maintained by the intercalation of a **leap year** containing an extra day, on 29 February, every four years. The average length of the year was therefore 365.25 days which is close to the length of the mean solar year. The discrepancy however amounts to one day in 128 years, and by the 16th century it was found that the vernal *equinox was occurring about ten days early. The *Gregorian calendar, first introduced in 1582 by Pope Gregory XIII and in almost universal civil use today, superseded, with only slight modification, the Julian calendar. The Gregorian reform of 1582 omitted ten days from the calendar that year, the day after 4 October becoming 15 October. This restored the vernal equinox to 21 March and, to maintain this, three leap years are now suppressed every 400 years, centurial years ceasing to be leap years unless they are divisible by 400. For example, 1900 was not a leap year but 2000 is. The average length of the calendar year is now reduced to 365.2425 days, so close to the mean solar year that no adjustment will be required before AD 5000. The Gregorian calendar was devised principally as a basis for fixing the date of Easter and therefore the whole ecclesiastical calendar. As such it was adopted at once in predominantly Roman Catholic countries but only gradually over the next two centuries in the Protestant countries of Western Europe. This calendar only became established for international transactions early in the 20th century.

The system of numbering years from the beginning of the Christian era was introduced by a Roman monk, Dionysius Exiguus, in AD 525. This system was soon generally adopted in Western Europe. In different countries, however, a variety of dates was adopted as the first day of the New Year, the most common choices, apart from 1 January, being 25 December, 1 March, and 25 March, and these differences persisted as late as the 18th century. The names of the months are of Roman origin. In fact the ancient Roman calendar had March as its first month, although in 153 BC the calendar New Year was transferred to 1 January. Originally, therefore, September to December were months seven to ten as their names suggest. The two previous months were originally named Quintilis and Sextilis but were renamed July and August in honour of Julius Caesar and Augustus Caesar.

Other calendrical systems continue to be used, particularly for religious purposes, alongside the Gregorian system. The present Jewish calendar uses the 19-year Metonic cycle made up of 12 common years and 7 leap years. The common years have 12 months, each of 29 or 30 days, while the leap years have an additional month. The rules governing the detailed construction of the calendar are very complicated but the year begins on the first day of Tishri, an autumn month. Years are reckoned from the era of creation (*anno mundi*) for which the epoch adopted is 7 October 3761 BC. Thus the year beginning in autumn 1988 was AM 5749. The Islamic calendar is wholly lunar, the year always containing 12 months without intercalation. This means that the Muslim New Year occurs seasonally about 11 days earlier each year. The months have alternately 30 and 29 days and are fixed in length, except for the twelfth month (Dulheggia) which has one intercalatory day in 11 years out of a cycle of 30 calendar years. This period of 360 months amounts to 10,631 days which differs from 360 lunations by only 17.3 minutes. So, although the Islamic calendar bears no relation to the solar year, it is, within its own lunar terms of reference, of an accuracy comparable to the Gregorian calendar. The years in the Islamic calendar are reckoned from the Hegira, the flight of Mohammed, which was 16 July 622 (Julian calendar).

The week, unlike the day, month, or year, is an artificial device with no astronomical origin. First introduced by the Babylonians, it became established in the Christian era only when it was linked to the Julian calendar by Constantine in AD 321. As a continuously uninterrupted cycle it can be useful in removing ambiguities. The ancient Chinese had a similar system of day naming but in cycles of 60 that were unaffected by any intercalation. As the day name is usually included with the date in Chinese records, it is possible to verify the exact epoch of early Chinese chronology.

Callisto, a *Galilean satellite of Jupiter, discovered by Galileo in 1610. Callisto has a nearly circular orbit above Jupiter's equator at 1.883 million km from the planet's centre. Its diameter is 4,800 km, and its density is 1,860 kg/m^3. Its composition is probably a mixture of water ice and rocks. *Voyager images have shown that its icy surface is heavily cratered, including a 600 km wide impact basin, **Valhalla**, which is surrounded by concentric rings. The craters are brighter than the surrounding terrain. There is no sign on Callisto of any activity in the interior of the satellite or of *tectonic patterns.

Caloris Basin, an impact crater with a diameter of 1,300 km, the largest feature on *Mercury. It was formed relatively recently and is named Caloris because at local noon it is one of the hottest regions on the planet (600 K), the Sun being overhead there when the planet is at the perihelion of its orbit. The impact produced a series of mountain ranges, the 3 km high, 40 km wide Calores Montes being a typical example, and also a region known as the weird terrain. This terrain is at the antipodes of the basin and the hilly landscape was produced by the focusing of seismic activity. The floor of the Caloris Basin has been filled in by lava flows and then broken up by a series of radial and concentric fractures and ridges which were probably produced when the infill material settled.

Camelopardalis, the Giraffe, a *constellation in the north polar region of the sky, introduced in 1613 by the Dutchman Petrus Plancius. It represents the animal on which Rebecca rode into Canaan for her marriage to Isaac. Its brightest stars are of only fourth magnitude.

Canada–France–Hawaii telescope, a 3.6 m (142-inch) aperture *reflecting telescope for optical and infra-red observations installed at an altitude of 4,200 m (13,800 feet) on Mauna Kea in Hawaii. At such a height it operates under very favourable atmospheric conditions which, for example, enable it to be used to measure the angular diameters of Pluto and its satellite Charon.

canals, in planetology, the fine, straight, artificial-looking lines that were seen crossing the deserts of Mars. They were initially observed telescopically by the Italian astronomer Giovanni Schiaparelli in 1877 and he named them 'canali'. This word was translated into English as 'canals' and soon the markings became associated with a global network of artificial waterways and irrigation systems built by the Martian inhabitants. Henri Perrotin and Louis Thollon, of the Nice Observatory, confirmed their existence in 1886 and *Lowell founded an observatory at Flagstaff, Arizona, primarily to study the form and development of the canal system. Other observers, using refracting telescopes of a similar size to Lowell's, either failed to see the canals or drew them as broad diffuse streaks. Some thought the canals were the result of a physiological effect by which the eye, working at the limits of its resolution, has the tendency to interpret a series of dark unrelated patches as a line. The controversy over canals on Mars continued into the space age and was in part responsible for stimulating public support for planetary exploration. The series of *Mariner spacecraft that flew past and photographed the surface of Mars found that most canals did not actually exist. However a few were found to be just *tectonic features and bright and dark patterns.

Cancer, the Crab, a *constellation of the zodiac. In mythology it represents the crab that attacked Hercules during his fight with the Hydra, but which was crushed underfoot. It is the faintest of the zodiacal constellations, with no star brighter than fourth magnitude. Its most important feature is a large star cluster called *Praesepe (M44). The Sun lies within the boundaries of the constellation from late July to early August.

Canes Venatici, the Hunting Dogs, a *constellation of the northern celestial hemisphere, introduced by the Polish astronomer Johannes Hevelius (1611–87). It represents a pair of dogs held on a leash by Boötes. Alpha Canum Venaticorum, of third magnitude, is also known as *Cor Caroli.

This photograph of the **Caloris Basin** on Mercury was taken by Mariner 10. The basin is the large area at the left, half in daylight. Mariner 10 was the first spacecraft to fly past the planet, making three separate encounters during 1974–5. This image is a mosaic constructed from the best photographs of the area taken during all three encounters.

A globular *cluster, M3 (NGC 5272), is visible through binoculars. The most famous feature in Canes Venatici is the *Whirlpool Galaxy (M51).

Canis Major, the Greater Dog, a prominent *constellation of the southern hemisphere of the sky, one of the forty-eight constellations known to the Greeks. It represents one of the two dogs of Orion and contains *Sirius, the brightest star in the sky, popularly known as the Dog Star. A few degrees below Sirius is M41 (NGC 2287), a large star *cluster visible through binoculars.

Canis Minor, the Lesser Dog, a northern hemisphere *constellation on the celestial equator. It is one of the forty-eight constellations known to the Greeks and represents the smaller of the two dogs following Orion. Its only prominent feature is the bright star *Procyon, a visual binary star.

Canopus, a first-magnitude star in the constellation of Carina and also known as Alpha Carinae. It is the second-brightest star in the sky after Sirius. Of whitish hue, Canopus is a hot luminous *supergiant or *giant star twenty-five times the Sun's diameter and 1,200 times its luminosity. It is a young *Population I star in the disc of the Galaxy, and lies some 23 parsecs from the Sun. Because of its great brightness, satellite star trackers often use Canopus for orientation.

Capella, a *binary star in the constellation of Auriga, and also known as Alpha Aurigae, with an apparent magnitude of 0.08. Two *giant stars with temperatures similar to that of the Sun move in a 104-day orbit. A distant twelfth-magnitude companion, itself a binary comprising two red *dwarf stars, has the same *proper motion, and the two pairs may form a binary of very long period. Capella's distance is 13 parsecs, and its combined luminosity is about sixty times that of the Sun.

Capricornus, the Sea Goat, a *constellation of the zodiac. In Greek mythology Capricornus represents the god Pan who jumped into a river on the approach of the monster Typhon and was turned into a fish. The star Alpha Capricorni, sometimes called Algedi or Giedi, both names from the Arabic meaning 'kid', consists of a pair of fourth-magnitude stars separable by the naked eye or through binoculars. The Sun crosses in front of the constellation from late January to mid-February.

carbon–nitrogen cycle, the dominant *nuclear fusion reaction in *main-sequence stars whose central temperature is above 1.8×10^7K. The cycle, also known as the **Bethe cycle**, is a six-stage reaction in which hydrogen is converted to helium via the nuclei of carbon, nitrogen, and oxygen which exist in small quantities in stellar interiors. Gamma rays and *neutrinos are also produced in this cycle. Stars such as the Sun whose central temperature is less than 1.8×10^7K are dominated by the *proton–proton reaction.

Carina, the Keel, a *constellation of the southern skies. It is one of the three parts into which the ancient Greek constellation of Argo Navis was dismembered by the 18th-century French astronomer Nicolas Louis de Lacaille. Its brightest star is *Canopus. Its most celebrated feature is the nebula NGC 3372, known as the *Keyhole Nebula, visible to the naked eye, which contains the unusual variable star Eta Carinae. In 1843 Eta Carinae temporarily became the second brightest star in the sky but now lies around sixth mag-

Cataclysmic variable

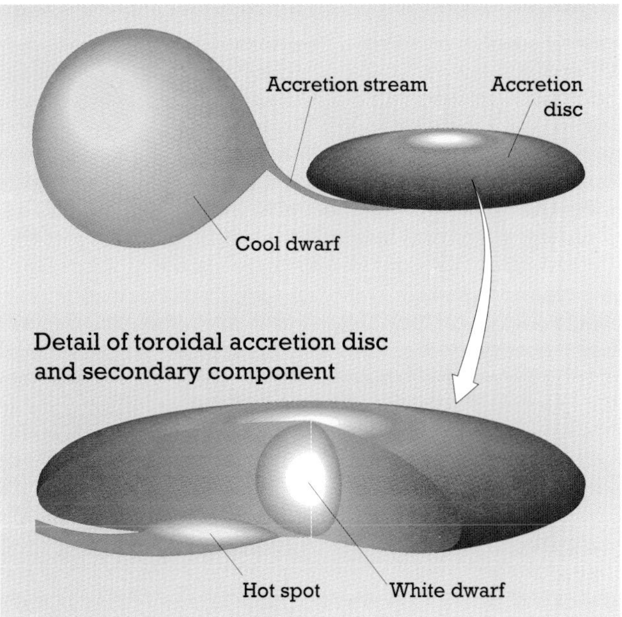

The structure of a typical cataclysmic variable. Material from the larger cool dwarf star is drawn towards the smaller white dwarf star forming an accretion disc around the white dwarf. The hot spot on the accretion disc is caused by the friction of the incoming material colliding with the disc.

nitude. Eta Carinae is thought to be an unstable *supergiant star that may one day explode as a *supernova.

Cassegrain telescope *reflecting telescope.

Cassini family, the four generations of Cassini who managed and directed the *Paris Observatory from its inception in 1667 until around 1793. **Giovanni Domenico Cassini** (1625–1712) (also known as Cassini I) was an Italian-born French astronomer and a precise and systematic observer. He used long-focus telescopes to observe the surfaces of planets and made accurate measurements of the spin periods of Jupiter and Mars. He discovered four satellites of Saturn and the division in *Saturn's rings that is named after him. He was the first to maintain records of the *zodiacal light. His son **Jacques Cassini** (1677–1756) (Cassini II) managed the observatory from 1710. Under his directorship the observatory began measuring the shape of the Earth, and the determination of the arc of the meridian between Dunkirk and Perpignan was completed in 1718. **César-François Cassini** (1714–84) (Cassini III) continued the work of his father. He was also a cartographer and was instrumental in producing the first modern map of France. **Jacques-Dominique Cassini** (1748–1845) (Cassini IV) directed the observatory from 1784 but had to abandon his plans of restoring and equipping it after his imprisonment during the French Revolution (1789).

Cassini's division *Saturn's rings.

Cassiopeia, a *constellation of the northern sky, one of the forty-eight constellations known to the ancient Greeks. It represents Queen Cassiopeia, the vain wife of King Cepheus of Ethiopia; she is depicted sitting in a chair, circling the north celestial pole. The five brightest stars of the constellation form a distinctive W-shape. The centre star of the W, called Gamma Cassiopeiae, is an unpredictable variable known as a *shell star. In 1572 the Danish astronomer Tycho Brahe observed a brilliant supernova in Cassiopeia, now known as *Tycho's Star. Also in Cassiopeia lies the strong radio source Cassiopeia A, the remains of a supernova that erupted around the year 1600 but which went unnoticed at the time. Lying within the rich star fields of the Milky Way, Cassiopeia contains a number of open star *clusters including M52 and M103.

Castor, a second-magnitude *multiple star in the constellation of Gemini, and also known as Alpha Geminorum. Two hot *dwarf stars move in a 500-year orbit, and each component is a spectroscopic *binary star. A third, distant member of the system is the ninth-magnitude *eclipsing binary YY Geminorum, in which both components are *flare stars. Thus Castor is a system of six stars. Castor's distance is 15 parsecs, and its combined luminosity is about fifty times that of the Sun.

cataclysmic variable, a *variable star that undergoes outbursts when it may brighten by several magnitudes. Most such objects are very close *binary stars in which a cool star fills its *Roche lobe and loses material in the direction of a *white dwarf companion. The mass may form a bright *accretion disc or, if the white dwarf has a strong magnetic field, as in the *AM Herculis stars, may land directly on one or both of its magnetic poles. In *novae, the outbursts are due to the nuclear fusion of hydrogen on the surface of the white dwarf. In **dwarf novae**, which may have outbursts every few weeks, the outbursts are the release of gravitational energy in the form of light and heat, triggered by an instability in the accretion disc. Binary stars with a similar structure may be included among the cataclysmic variables even though no outbursts have been detected; they resemble ex-novae and are called **nova-like variables**. Some authorities also include the *symbiotic stars and *supernovae as cataclysmic variables.

catalogue, in astronomy, a collection of information about stars, nebulae, or galaxies. The brightest stars have proper names of Greek, Roman, or Arabic origin, though in practice astronomers prefer the system of Greek *Bayer letters introduced by the German astronomer Johann Bayer in 1603, whereby the stars in each constellation were designated, usually in order of brightness, by α, β, γ, and so on. On exhausting the Greek alphabet, recourse was made to the Roman alphabet, namely a, b, c, ... and after these were used capital letters A, B, C, ... There are so many stars, however, that the modern procedure is to designate them by their number in a catalogue. Thus Lalande 21185 is the 21,185th star in Lalande's catalogue. In a catalogue, against each star number is listed such information as the star's coordinates on the *celestial sphere, its *proper motion, its *magnitude, *parallax, whether it is a *variable star or a component of a *binary star, its *radial velocity, *colour index, and *spectral class. Other information, such as its mass, may be listed if it is known. Because the stars' coordinates change with time due to *precession and *nutation, the catalogue is usually made for a particular epoch, say AD 2000.0, that is, the beginning of the year AD 2000.

Over twenty well-known and useful catalogues have been compiled, among them Flamsteed's *British Catalogue* (published in 1725), the *Henry Draper Catalogue* (1900.0), the *Bonner Durchmusterung* (1855.0), compiled by the Prussian astronomer Friedrich Wilhelm Argelander, the *General Catalogue of Double Stars* (1950.0) by P. N. Kholopov, and *Norton's Star Atlas* (2000.0) by I. Ridpath *et al.* **Star atlases** are to some extent a type of catalogue in which the stellar positions are plotted or photographed. During the past century a number of sky surveys have been made resulting in, for example, the *True Visual Magnitude Photographic Star Atlas* (1950.0) by Christos Papadopoulos and the *Palomar Sky Survey* (1950.0) by the National Geographic Society and the Mount Palomar Observatory. Photographic atlases such as these catalogue photographs of nebulae, galaxies, star clusters, and so on.

catastrophe theory, the mathematical description of dynamical systems that undergo abrupt change. Originally developed by the French mathematician René Thom in 1972, it can be used to describe sudden discontinuous changes in a system as it evolves in time. Such behaviour is well known in many mathematical theories of astrophysical phenomena, for example, in the dynamics of stars under the gravitational attraction of a galaxy.

A quite unrelated theory of catastrophes attempts to describe the effect of violent astronomical events on the biosphere of the Earth. For example, the impact of an enormous meteorite, either an *Apollo–Amor object or a comet, has been proposed as the cause of the extinction of many species, including all the dinosaurs, at the end of the Cretaceous geological period.

CCD camera, in astronomy, an instrument that creates an electronic image of a celestial object using a charge-coupled device (CCD). A CCD contains a light-sensitive semiconductor chip which is divided up into an array of elements called **pixels**. For example, a typical astronomical CCD has a rectangular array of 576 × 384 pixels and which measures 13 mm × 9 mm (0.5 inches × 0.33 inches). The light from a celestial object is focused by a telescope into an optical image on the chip. Electrical charge accumulates at each pixel in proportion to the amount of light that falls on it so that the optical image is represented by the distribution of charge across the array. At the end of the exposure the charge at each pixel is transferred from the chip into subsidiary electronic equipment where it is measured and its value stored in a computer for display and analysis. CCD cameras are now widely used in optical and near infra-red astronomical work because of their high sensitivity and their great accuracy in measuring the intensity of light. Extremely faint objects require exposures of several hours' duration, so CCD cameras are usually cooled with liquid nitrogen in order to minimize unwanted thermal generation of charge in the chip. A CCD camera is often used as the detector in a *photometer.

celestial axis *celestial sphere.

celestial equator *celestial sphere.

celestial latitude *celestial sphere.

celestial longitude *celestial sphere.

celestial mechanics, the branch of astronomy dealing with the orbits of planets, natural and artificial satellites, comets, meteors, asteroids, and binary or multiple stars under the action of *gravitation, and, in the case of some satellites, atmospheric drag and *radiation pressure. These orbits are often subject to small changes, known as **perturbations**, caused by the gravitational influence of nearby celestial bodies. Begun by *Newton and based on his Law of Gravitation and three Laws of Motion, celestial mechanics attempts to produce analytical theories to account for and to predict the movements of celestial objects. Many important sections of applied mathematics have stemmed from such researches, the subject receiving its fullest classical development during the 18th and 19th centuries at the hands of outstanding theoretical astronomers such as *Laplace, *Le Verrier, and *Poincaré. Problems such as the Moon's orbital stability and evolution, the long-term stability of the solar system, the origin of Saturn's hundreds of rings, and the general *three-body problem, are among those which in recent years have benefited from the use of computers. In addition, the success of projects such as the landing of men on the Moon or the *Voyager missions to Jupiter, Saturn, Uranus, and Neptune depend upon correct use of celestial mechanics to plan and modify the trajectories of spacecraft in flight.

celestial meridian *meridian.

celestial poles *celestial sphere.

celestial sphere It is convenient for many purposes to consider the sky as represented by a sphere of arbitrarily large radius centred on the observer, the celestial objects being projected on to this sphere. The position of any celestial object on this sphere at any moment will be defined by two angular arcs. For this purpose a fixed reference plane passing through the centre of the celestial sphere is chosen, intersecting the sphere in a **great circle**. The two points 90 degrees away from this great circle are its *poles (analogous to the equator and the north and south poles on Earth). The great circle through the poles and a chosen point on the reference plane's intersection with the sphere's surface provides the necessary second reference circle.

Various reference or *coordinate systems are used based on *angular measure and these are illustrated in the figure. The **alt-azimuth system** uses the *horizon and vertical great circles from the zenith to the nadir, the poles of the horizon circle. A celestial object's **altitude** is the arc from horizon to object measured along the vertical great circle through the object. Its **azimuth** is the arc measured from the most northerly point of the horizon along the horizon in an easterly direction to the foot of the vertical through the object. The **equatorial system** uses the **celestial equator**, the great circle formed by the intersection of the celestial sphere by the extended plane of the Earth's equator. The **north** and **south celestial poles**, formed by extending the Earth's axis of rotation to pierce the celestial sphere, are also used, *meridians being drawn on the sphere from the north to the south pole. The angular distance of an object from the north celestial pole measured along the meridian from the pole to the object is known as the **north polar distance**. The **zenith distance** of an object is its angular distance from the zenith and is equal to 90 degrees minus the object's altitude. One version of the equatorial system uses as a reference meridian the half great circle from the pole above the horizon through the zenith to the pole below the horizon. This particular meridian is termed the **observer's meridian**. The angle between the observer's meridian and the meridian drawn through the celestial object, the angle being measured westwards from the observer's meridian, is the **hour angle** of the object and is shown in the figure. Since the celestial sphere appears to rotate westwards about the **celestial axis** (because of the Earth's rotation in the opposite direction), a star's hour angle in the course of one *sidereal day increases from zero (when the star is on the observer's meridian) to 360 degrees (when it returns to the meridian). To get rid of this variable it is customary to choose a point on the celestial equator, rotating with the celestial sphere, so that the angle between the meridians through the point and the star is constant. The point chosen is the vernal equinox or *First Point of Aries. The angular distance between the meridian through Aries and that through the star, measured eastwards, is called the **right ascension** of the star. The second coordinate in both these versions of the equatorial system is the **declination**. The declination of a celestial object is its angular distance north or south of the celestial equator, measured along the meridian through the object.

A third commonly used coordinate system illustrated in the figure uses the *ecliptic as reference circle. The vernal equinox also lies on the ecliptic, being one of the two points of intersection of the ecliptic and the equator. The half great circle through the poles of the ecliptic and the vernal equinox forms the other necessary circle. Then the **celestial** or **ecliptic latitude** of a celestial object is its angular distance north or south of the ecliptic, measured along the half circle through the object and the poles of the ecliptic. The **celestial** or **ecliptic longitude** of the object is the angular distance between Aries and the point of intersection of the latitude half circle through the object with the ecliptic. This distance is measured eastwards. Transformation from one coordinate system to another can be effected by the use of the formulae of spherical trigonometry.

Celestial sphere/Coordinate system

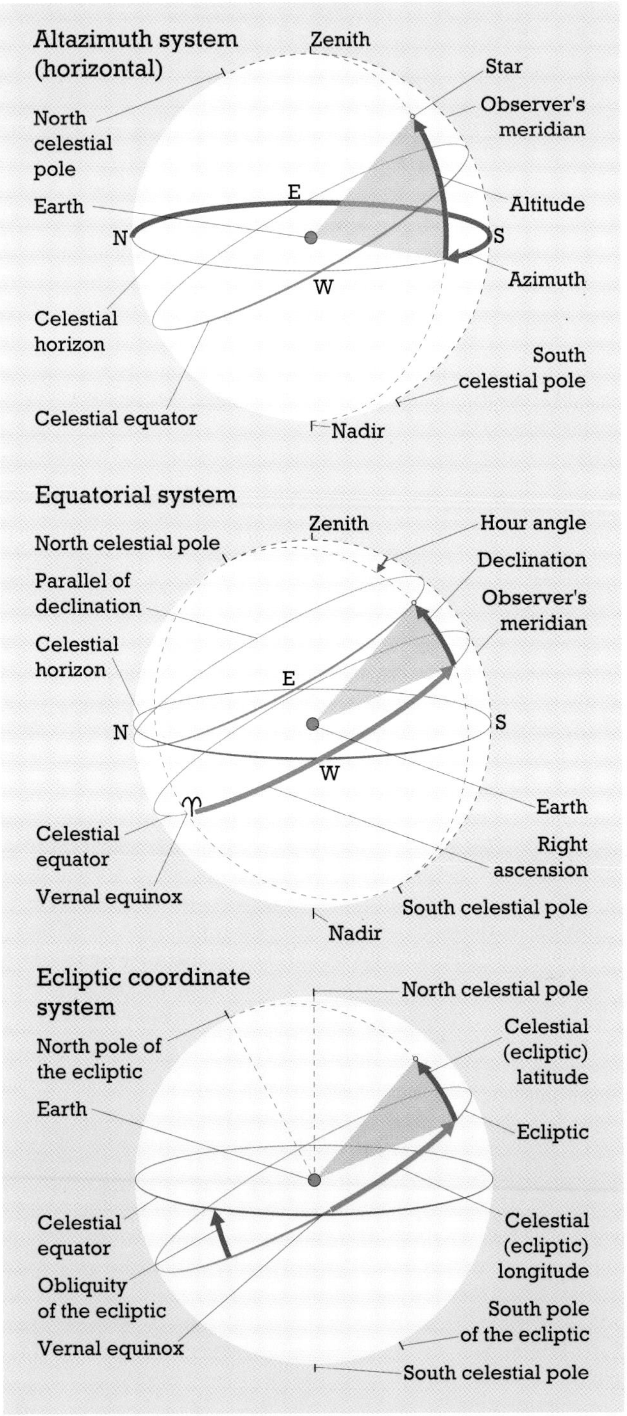

Centaurus, the Centaur, a prominent *constellation of the southern sky, one of the forty-eight constellations known to the ancient Greeks. It represents Chiron the centaur, a mythical beast with four legs like a horse but the torso, head, and arms of a human. Its two brightest stars are *Alpha Centauri (Rigil Kentaurus) and *Beta Centauri (Hadar or Agena), representing the creature's front legs; they act as pointers to the Southern Cross, *Crux. Centaurus contains the largest and brightest globular *cluster in the sky, Omega Centauri, which appears like a fourth-magnitude fuzzy star covering an area greater than the full moon; it lies approximately 5,000 parsecs away.

Centaurus A, a large elliptical *radio galaxy in the constellation of Centaurus at a distance of 4 megaparsecs. It is a strong radio source which also emits infra-red and gamma radiation. Its regions of radio emission have a complicated structure and it is likely that the source of radiation is the visible galaxy NGC 5128. This has rotating gas and dust belts extending far beyond the visible galaxy.

Cepheid variable stars, an important class of highly luminous *variable stars named after Delta Cephei which was the first of their type to be discovered, by *Goodricke in 1784. They typically exhibit variations in luminosity, temperature, and radius with well-defined periods in the range 1–50 days. During their pulsation periods they rapidly rise to maximum brightness and slowly fade during expansion. The pulsations are caused by successive loss and recapture of electrons by helium. Observing Cepheids in the *Magellanic Clouds in 1912, the US astronomer Henrietta Leavitt discovered that their periods were uniquely related to their apparent magnitudes. Since all stars in the Clouds are at essentially the same distance from the Earth, it follows that a relationship must exist between the period and the true luminosity (absolute magnitude) of a Cepheid. It then follows that if, by finding the actual luminosity for a single Cepheid, we can obtain an accurate absolute calibration of this *period–luminosity law, then we have a very powerful method of measuring astronomical distances, as follows. Firstly, measure the period of a Cepheid in the system observed; then determine its luminosity from the period–luminosity law; and finally compare this with its apparent brightness to deduce its distance. If necessary the distance can be compared with that obtained from the spec-

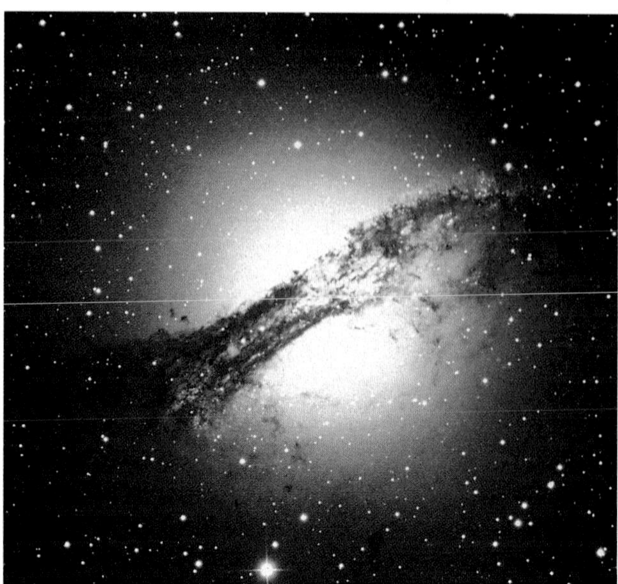

NGC 5128 is the probable source of the intense radio signals from **Centaurus A**. There are two radio sources within the visible galaxy and two more spread out from the galaxy by about a million parsecs.

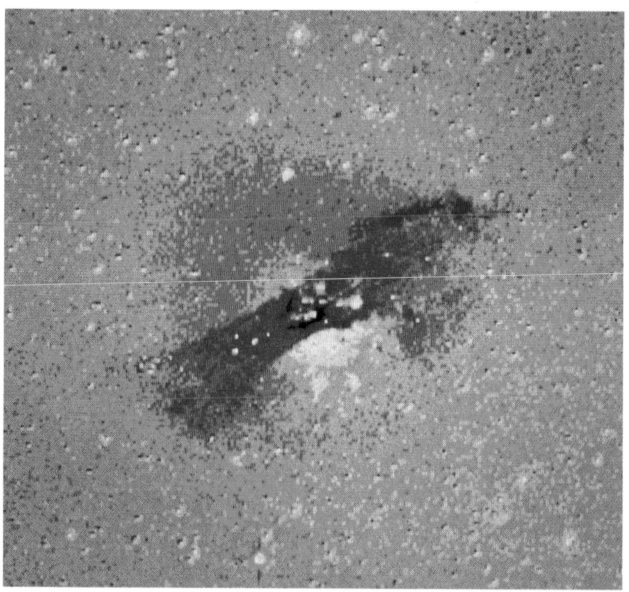

This image-processed optical photograph of **Centaurus A** was taken with the 3.9-metre Anglo–Australian telescope. The most distinctive feature is the great lane of dust (here coloured red) that bisects the galaxy. Astronomers now believe the dust lane is a great disc of dust, yet to be formed into stars, which surrounds the galaxy's core and which we see edge-on.

troscopic *parallax. The power of the method lies in the fact that Cepheid periods and the period–luminosity law are well defined and that the high luminosity of Cepheids permits application of the method to quite distant galaxies. In 1952 it emerged that distances so determined had been too small by a factor of two owing to an erroneous assumption made in calibrating the period–luminosity law. The error lay in assuming that Cepheids in the globular *clusters of the Galaxy were identical to those in the Magellanic Clouds. It is now recognized, however, that Cepheids fall into two types according to the *stellar population in which they reside. Type I, the classical Cepheid of Population I, is more luminous than Type II, the W Virginis or cluster Cepheid, of the same period. The difference is attributed to the effect of their different chemical compositions on their oscillations.

Cepheus, a *constellation of the north polar region of the sky, one of the forty-eight constellations known to the ancient Greeks. It represents King Cepheus of Ethiopia, husband of Cassiopeia and father of Andromeda. Its most celebrated star is Delta Cephei, prototype of the *Cepheid variable stars. Delta Cephei is also a double star for small telescopes, consisting of a fourth-magnitude yellow star (the variable) and a sixth-magnitude blue-white companion. Mu Cephei is a red supergiant variable, called the Garnet Star by the British astronomer William *Herschel.

Cerenkov radiation, radiation emitted by atomic particles when they pass through a transparent medium at a velocity greater than the *velocity of light in that medium. It is impossible for any radiation or particle to travel faster than the velocity of light in a vacuum. However, when light enters a transparent medium its velocity is sometimes reduced and it is then possible for a particle to travel faster than the velocity of light in that medium but still slower than the velocity of light in a vacuum.

Ceres, the largest *asteroid and the first to be discovered (by the Italian astronomer Giuseppe Piazzi in 1801). Its orbit has a semi-major axis of 2.768 astronomical units, an eccentricity of 0.077, and an inclination of 10.6 degrees to the ecliptic. About 1,023 km across, nearly spherical in shape, with an albedo of about 0.06 and a spin period of 9.075 hours, Ceres was probably a growing *planetesimal when accretion was stopped in the primordial asteroid belt. Like the other large asteroids, it is likely to have retained its original size and rotational state, while its surface is probably covered by a thick *regolith layer of fragmented material.

Cerro Tololo Inter-American Observatory, an *observatory at La Serena, Chile, at an altitude of 2,160 m (7,080 feet). The observatory is run by the Association of Universities for Research in Astronomy, a group of seventeen US university astronomy departments. The main instrument is a 4 m (157-inch) *reflecting telescope, commissioned in 1976, which is twinned with the Mayall reflector at the *Kitt Peak National Observatory in Arizona, USA. These telescopes have *Ritchey–Chrétien optics of fused quartz with a *focal ratio of 2.8, and the coudé focus has a focal ratio of 190, giving an effective focal length of 700 m (2,300 feet). The Cerro Tololo telescope has been used for such projects as the mapping of giant luminous arcs between galaxies and the investigation of the high velocities of the hydrogen gas near the centre of our own galaxy; these velocities may be caused by a central black hole.

Cetus, the Whale, the fourth-largest *constellation. It lies across the celestial equator and was one of the forty-eight constellations known to the ancient Greeks. Cetus represents the sea monster to which Andromeda was to be sacrificed before her rescue by Perseus. It contains the long-period variable star *Mira Ceti, also known as Omicron Ceti. *Tau Ceti is a yellow dwarf star similar to the Sun. M77 (NGC 1068), at ninth magnitude, is the brightest example of a *Seyfert galaxy.

Chamaeleon, a small and faint *constellation near the south celestial pole, representing a chameleon. It was introduced by the Dutch navigators Pieter Dirkszoon Keyser and Frederick de Houtman at the end of the 16th century. Its brightest stars are of only fourth magnitude. It lies near the conspicuous southern constellation *Crux (the Southern Cross).

Chandler wobble, the *precession of the Earth's axis of rotation about its axis of symmetry. To a first approximation the Earth is symmetrical about the polar axis and slightly flattened at the poles, and the rotation axis is inclined at a small angle of a few tenths of an arc second to the symmetry axis. As a consequence the rotation axis wobbles, that is, describes a cone around the symmetry axis and makes the latitude of a given location on the surface of the Earth change accordingly. This phenomenon is known in mechanics as free precession; it is called free to distinguish it from forced luni-solar precession by the Moon and the Sun. By approximating the Earth to a rigid body the period of the Earth's wobble was first predicted by the Swiss mathematician Leonhard Euler in 1765 to be 300 days. In 1891 it was found by US geophysicist Seth Chandler to be about 428 days on the basis of his analysis of a long-term record of latitude measurements. This free precession, or Chandler wobble, can be distinguished from luni-solar precession by its effects: it changes the latitude of the observer, while luni-solar precession changes the declination of stars.

Chandrasekhar, Subrahmanyan (1910–), Indian-born US astronomer who developed the theory of stellar structure and evolution. His theories on the later stages of stellar evolution postulated that a star whose mass is greater than 1.44 times the mass of the Sun (the *Chandrasekhar limit) will not form a *white dwarf but will continue to collapse into a *neutron star or *black hole. In 1983 he was awarded the Nobel Prize for Physics. Chandrasekhar has also made major contributions to many other areas of theoretical astronomy including the internal constitution of stars, the theory of how energy is radiated within stellar atmospheres, and general relativity.

Chandrasekhar limit, the maximum possible mass for a *white dwarf, 1.44 times the Sun's mass. If a white dwarf's mass were to exceed this limit, the weight of the layers overlying its core would be too great to be supported by the pressure of the *degenerate matter in the core. Consequently, stars with a mass greater than the Chandrasekhar limit must either lose enough mass to bring them below it, or ultimately collapse to form a *neutron star, or even a *black hole. The Chandrasekhar limit is named after the Indian astrophysicist whose calculations identified it; it is distinct from the *Schönberg–Chandrasekhar limit.

chaos, a state of apparent randomness and unpredictability which can be observed in any dynamical system that is highly sensitive to small changes in external conditions. Chaos theory was first discovered by *Poincaré in his investigation of the *N-body problem, although he did not use the term 'chaos'. The serious study of chaos began in the late 1960s but it was only in the 1980s that the study of these phenomena came to be called chaos theory. Chaos occurs when the system has many degrees of freedom and not enough symmetry. The long-term behaviour of the system after a small initial disturbance is unpredictable and cannot even be described qualitatively because too much information is needed to describe accurately how the system responds to the disturbance. Thus the system is in principle deterministic, that is, subject to the laws of physics, but is apparently random and complex.

An often-quoted terrestrial example is the **butterfly effect** in weather systems. These systems are so sensitive to small disturbances that it is said that a butterfly fluttering its wings on one side of the world could determine whether or not a tornado occurs on the other side. In astronomy, chaos-like phenomena include the orbits of asteroids, comets, and satellites, where small disturbances can have dramatic and complex effects. Turbulent flows in hydrodynamics is another example. By the beginning of the 1990s the study of chaotic systems was already providing important insights into all areas of science, partly because chaos theory views systems as dynamic and developing rather than static. The understanding of chaos is one of the main challenges of modern science, with profound implications for all branches of astrophysics.

Charon, the only known satellite of Pluto, discovered by the US astronomers James Christy and Robert Harrington in 1978. It is the only natural satellite in the solar system whose orbital period is *synchronous with the rotation period of its planet. It is also the largest satellite in the solar system relative to its planet: its diameter of 1,200 km is about half that of Pluto. Its surface is probably covered by water ice. In 1985–90 its mutual eclipses with Pluto were observed by *photometry, yielding the orbital elements of the system and a rough map of Pluto's surface.

chemistry, cosmological, the study of the chemical reactions taking place in the various bodies in the universe and in the interplanetary, interstellar, and intergalactic mediums between them. Such studies provide information about the evolution and structure of planets, stars, and galaxies and demonstrate that chemical and physical laws operate universally. Many different types of molecules have been detected in interstellar space by the radio and infra-red radiation they emit.

Chinese astronomy Almost certainly Chinese astronomy has had as long a history as astronomy in the West. Because of China's isolation, it developed independently, producing original methods of solving various astronomical problems. It is probable that ancient astronomical schools existed in the 2nd and 3rd millennia BC. As in Babylonian astronomy, astronomers were also astrologers, teaching that the movements, positions, and appearances of heavenly bodies, particularly the Moon, gave information about terrestrial events and lives. Unusual events such as eclipses or the appearance of comets were indications of calamities caused by bad government. Astrologer–astronomers were certainly playing important roles in governmental policy-making by the 4th century BC. A lunar *calendar was in operation and the seasons were considered to begin by the positions of certain constellations. The astronomers were responsible for timekeeping and geographical measurements. Catalogues and maps list some 122 constellations including over 800 stars. There were twenty-eight moon stations or directions, planetary phenomena were observed, and the *synodic periods of the planets and the Moon were very accurately computed.

In the first few centuries AD further progress took place. It was appreciated that the Moon's motion includes various irregularities, the seasons' unequal lengths were discovered, and a measurement of the *precession of the equinoxes was carried out by the astronomer Yu-Hsi. By now various trade routes to the West were opening up so that a two-way flow of knowledge was set in motion. Some Chinese sources mention how, in AD 164, people with astronomical knowledge came from western Asia to China. It is also said that in AD 440 Ho-Tsheng-Tien was taught astronomy by an Indian priest. From then on, Chinese astronomy, while developing vigorously in ways moulded by Chinese philosophy and religion, was also influenced by Arab sources and, in the 16th century, by Jesuits visiting China. During those centuries, the Moon's orbit was studied and better understood, text books were written, many tables of the positions of the Sun and Moon appeared, and latitudes and the obliquity of the *ecliptic were measured. In their observatories the Chinese astronomers used shadow poles, *quadrants, and *armillary spheres. In June 1054, the Chinese were the only people to record the appearance of a 'guest star' in the constellation Taurus. It was the supernova, now known to astronomers as the *Crab Nebula.

The progress of Chinese astronomy, as in other parts of the world, has been subject to conditions of national stability and to periods of government and religious control. In recent years it has made a strong recovery from the effects of the Cultural Revolution (1966–76), with a growing number of astronomers playing an increasingly important part in the international community's astronomical effort. Many Chinese universities teach and conduct research in astronomy and observatories such as Purple Mountain Observatory of the Chinese Academy of Sciences, Shanghai Observatory, and Beijing Astronomical Observatory are carrying out valuable observational programmes.

Chiron, the most distant of the objects classified as *asteroids, discovered by the US astronomer Charles Kowal in 1977. Its orbit has a semi-major axis of 13.69 astronomical units and an eccentricity of 0.379, implying that the orbit lies mostly between Saturn and Uranus, resembling those of short-period comets. Because of strong perturbations and relatively frequent close encounters with planets, the orbit is chaotic and is rapidly evolving. The discovery, in 1989, of a faint *coma around Chiron confirms that this object experiences comet-like emissions of volatile materials; it probably originated in the outer solar system and perhaps came in recent times from the *Oort–Öpik Cloud.

chondrite, a stony *meteorite containing granules. The granules, known as **chondrules**, are solidified droplets of rock that was once molten. Chondrites were formed at the birth of the solar system 4.6 billion years ago and constitute the most ancient and chemically primitive material in the solar system. Though they have suffered little, if any, chemical change since their formation, there is evidence that they have been modified physically by heat. There are three basic classes of chondrite: **carbonaceous**, which are thought to have a high carbon content; **enstatites**, which have an abundance of $MgSiO_3$; and ordinary chondrites, which are the most common. These three types were formed at progressively higher temperatures. Stony meteorites that have no chondrules are known as **achondrites**.

chromosphere, the lowest layer of the *Sun's tenuous atmosphere above the **reversing layer** and the *photosphere. The chromosphere can be seen as a thin red band of light surrounding the Moon's disc during a total solar *eclipse. The strong magnetic fields and convection currents in the chromosphere create *spicules. The reversing layer is a few hundred kilometres thick and is composed of the gases of many of the familiar terrestrial elements.

Chryse Planitia, a multiple-ring impact basin in the northern hemisphere of Mars at longitude 47 degrees west and latitude 19 degrees north. The southern edge consists of enormous channels 2.5 km deep and in excess of 150 km in width, together with canyons and depressions that are associated with *Valles Marineris. Some of these features extend across the basin and have been produced by vast floods of water that have occurred many times in Martian history. Teardrop-shaped islands can also be seen in the basin.

Circinus, the Compasses, a *constellation of the south celestial hemisphere, introduced by the 18th-century French astronomer Nicolas Louis de Lacaille. It represents a pair of drawing compasses or dividers used by draughtsmen and navigators. Circinus contains no objects of particular interest, the brightest star within it being of third magnitude.

circumpolar star, a star whose proximity to the *north celestial pole ensures that it never rises or sets but remains above the observer's horizon, circling the pole once every 24 *sidereal hours. Stars so near the south celestial pole that they remain below the observer's horizon form a second group of circumpolar stars. From a given site, a star is circumpolar if its declination plus the latitude of the site is greater than or equal to 90 degrees.

clock star, a star of precisely known right ascension used to measure the error of an observatory's *sidereal clock. When the star is on the observer's *meridian the local sidereal time equals the star's right ascension. The advent of highly accurate atomic clocks has diminished the need to use clock stars because atomic clocks are more accurate time-keepers than the Earth whose rotation rate is subject to small but detectable changes.

The globular **cluster** M13 in the constellation of Hercules is just visible to the naked eye as an indistinct patch of light. It contains several hundred thousand stars held together by their mutual gravitational attraction which dominates the gravitational effects of the centre of the Galaxy, 9,000 parsecs away.

cluster, star, a group of stars which are relatively close together and which move through space with the same velocity as each other. *Open clusters or galactic clusters, such as the *Pleiades, typically contain between ten and a thousand young *Population I stars and lie in the galactic disc. Open clusters are subject to disruption from the gravitational attractions of the galactic centre and interstellar clouds, and are generally short-lived entities (10^7–10^9 years). **Globular clusters** contain 10^5–10^6 densely packed Population II stars and are located in the *galactic halo. These clusters all have ages of around 10^{10} years as indicated by the low stellar mass at the turn-off point where the giant branch leaves the main sequence of the *Hertzsprung–Russell diagram. Because of their greater populations and position in the galactic halo, globular clusters are stable groups. The centres of some globular clusters are sources of X-rays, possibly due to the gravitational collapse of the inner core.

Coalsack *Southern Coalsack.

coelostat, a flat mirror fitted in a mount that is mechanically driven so that, as the Earth rotates on its axis, light from a chosen area of the sky can be continuously directed into an optical analysing instrument. Whereas a telescope is rotated by its *mounting to keep the image within its field of view, this is not always practical for heavy optical analysing equipment. Therefore it is often simpler to keep the equipment fixed and to use a coelostat to follow the rotation of the celestial sphere, constantly guiding light into the main optical system. The *heliostat performs a similar function for keeping an image of the Sun fixed.

colour index, a comparison of the *magnitude of a star at two different wavelengths (colours) of light. In the widely used Johnson system of photometry, a star's brightness is measured through a 'U' filter which transmits ultraviolet and violet light, a 'B' filter which transmits blue light, and a 'V' filter which transmits green-yellow light. If a star's observed magnitudes (corrected for the absorption of light in the Earth's atmosphere) are U, B, and V, then the colour indices are the differences $U - B$ and $B - V$. The great utility of colour indices is that they relate to the relative amounts of light radiated at different wavelengths, not the actual light levels, because a difference of logarithmically defined magnitudes corresponds to a ratio of brightnesses. Furthermore, because it is a ratio, the colour index does not depend on the distance to the object. Colour indices can thus indicate whether a star emits proportionately more violet or red light and hence tell whether the stellar temperature is high or low. Colour indices can also be defined which are sensitive to *spectral lines or bands. By comparing several colour indices and magnitudes it is usually possible to determine many important details about a star, such as its intrinsic luminosity, temperature, and distance, and the extent to which its light has suffered absorption by any intervening interstellar dust. The colour indices of galaxies are useful in understanding their stellar content.

colour–luminosity diagram *Hertzsprung–Russell diagram.

Columba, the Dove, a *constellation in the southern half of the sky, introduced by the Dutchman Petrus Plancius (1552–1622) from stars that the 2nd-century Greek astronomer Ptolemy had catalogued south of *Canis Major. It represents Noah's dove, sent out to find dry land. Mu

Coelostat

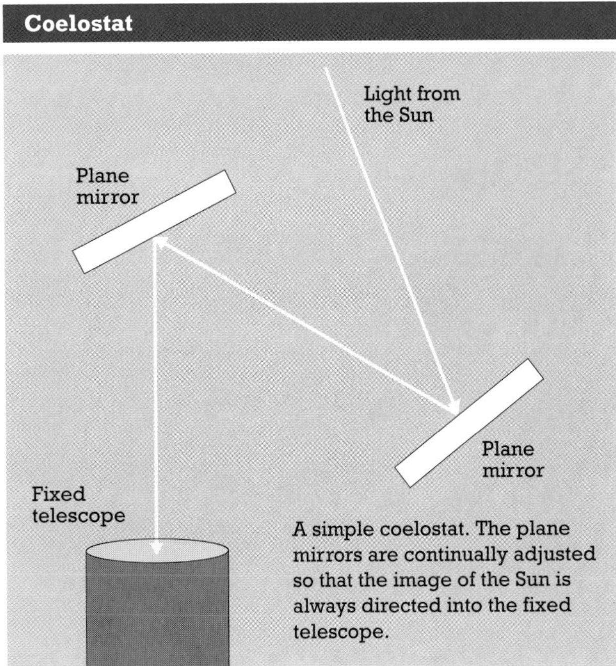

A simple coelostat. The plane mirrors are continually adjusted so that the image of the Sun is always directed into the fixed telescope.

Columbae is one of three so-called runaway stars that appear to be diverging from a point in *Orion; the other two stars are 53 Arietis and AE Aurigae. Possibly they were once members of a multiple star system that was disrupted when a fourth companion exploded as a supernova. As seen from Earth, the site of this explosion lies in Orion.

coma, in astronomy, the almost spherical halo of gas and dust that surrounds a *comet when it is in the vicinity of the Sun. This dusty 'atmosphere' is continually being swept away by the *solar wind and is replenished by gas and dust emitted by the nucleus. The coma can be about 100,000 km in diameter. The velocity of the escaping gas and dust is several kilometres per second relative to the nucleus and it disperses into the near vacuum of interplanetary space, some of it via the cometary *tail.

Coma Berenices, a *constellation of the northern hemisphere of the sky, representing the hair of Queen Berenice of Egypt. The constellation consists mostly of a scattered cluster of faint stars. This group of stars was known to the Greeks, but was considered by them to be part of *Leo. It was made into a separate constellation in 1551 by the Dutch cartographer Gerardus Mercator. Part of the *Virgo cluster of galaxies spills over into Coma Berenices. The most famous galaxy in Coma is M64 (NGC 4826), known popularly as the Black Eye Galaxy because of a dark cloud of dust near its centre. M64 is closer to us than the Virgo cluster and is not a member of it.

comet, a ball of ice and dust that orbits the Sun. As the comet nears the Sun, it forms a *coma and a *tail and the ball is referred to as the nucleus. The nucleus is frequently described as a dirty snowball and cometary nuclei are thought to be the remnants of the huge collection of icy *planetesimals that were formed at the birth of the solar system and which now reside in the *Oort–Öpik Cloud. An average cometary nucleus has a mass of $10^9 - 10^{11}$ kg, a diameter of 200–1,200 m, and a density of 200 kg per cubic metre, only one-fifth the density of water. The nucleus contains hollow regions, is fragile, and has about twice as much

Comet, structure of

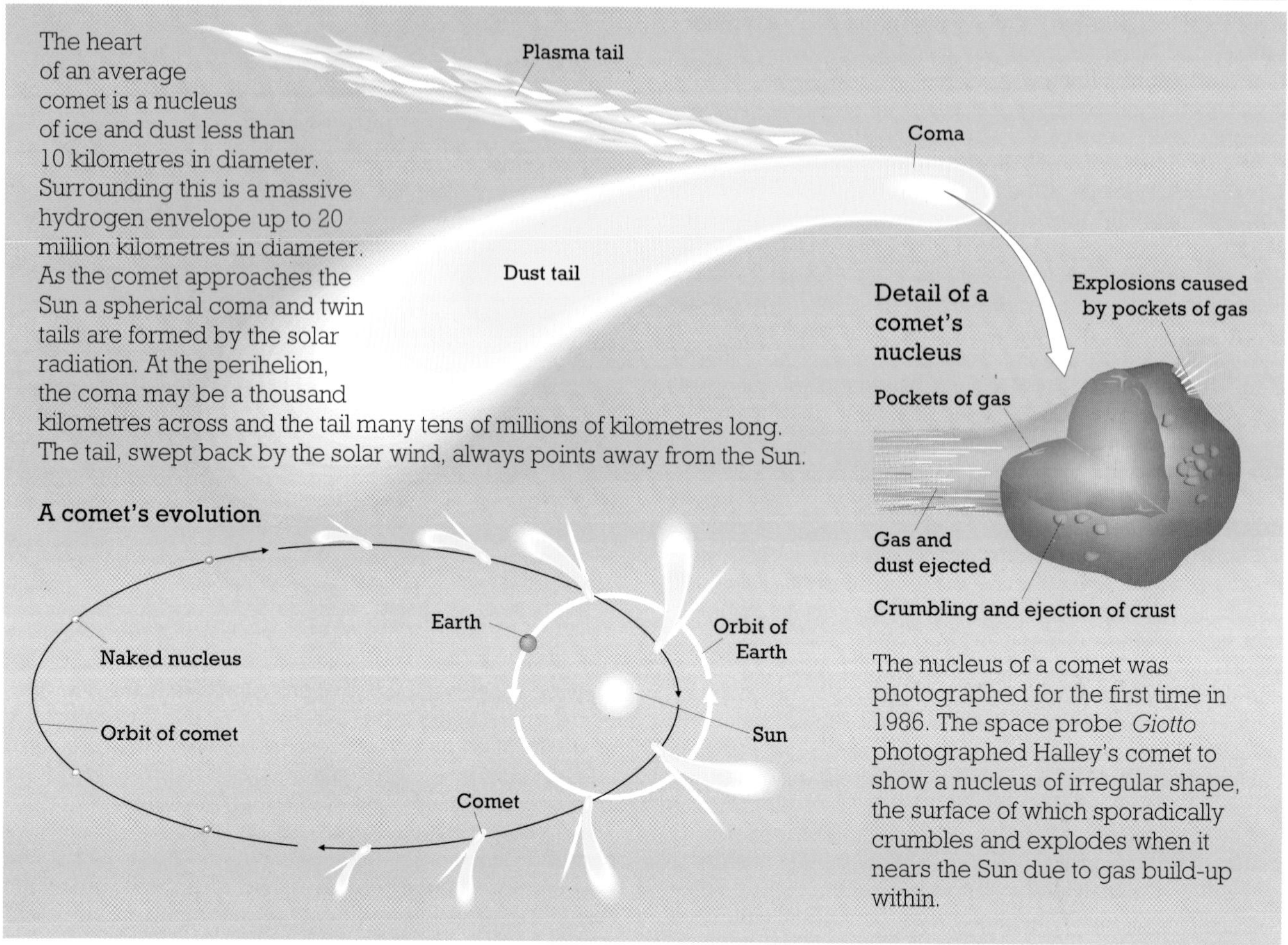
The heart of an average comet is a nucleus of ice and dust less than 10 kilometres in diameter. Surrounding this is a massive hydrogen envelope up to 20 million kilometres in diameter. As the comet approaches the Sun a spherical coma and twin tails are formed by the solar radiation. At the perihelion, the coma may be a thousand kilometres across and the tail many tens of millions of kilometres long. The tail, swept back by the solar wind, always points away from the Sun.
Plasma tail
Coma
Dust tail
Detail of a comet's nucleus
Explosions caused by pockets of gas
Pockets of gas
Gas and dust ejected
Crumbling and ejection of crust
The nucleus of a comet was photographed for the first time in 1986. The space probe *Giotto* photographed Halley's comet to show a nucleus of irregular shape, the surface of which sporadically crumbles and explodes when it nears the Sun due to gas build-up within.
A comet's evolution
Earth
Orbit of Earth
Naked nucleus
Orbit of comet
Sun
Comet

dust as ice. The ice is mainly water ice but is contaminated with other substances. Each time a comet passes close to the Sun the Sun's radiation causes the ice to give off gas which carries with it dust and dirty snowflakes. This material produces the spherical coma around the nucleus, and the long tail. The tail is in two parts, one of dust and the other of *plasma, and generally points away from the Sun. The total quantity of matter lost depends upon the amount of insulating dust covering the nucleus and how close the comet approaches the Sun at its *perihelion. The spacecraft *Giotto has observed *Halley's Comet at very close range, confirming many of the theories about cometary structure.

Comets are usually named after their discoverer and the year in which they were last seen. They divide naturally into short-period and long-period comets. Short-period comets orbit the Sun every few years. The average time is about eight years, and *Encke's Comet, with a period of a little over three years, has the shortest period. Short-period comets have been captured by Jupiter's gravitational field into relatively small orbits. A typical short-period comet has a perihelion distance of 1.5 astronomical units and will decay completely after about 5,000 orbits, creating a *meteor stream. Comet West was observed to break up in 1976, and *Biela's Comet has also been seen to break up. By contrast, long-period comets take between 10,000 and a million years to orbit the Sun and at their aphelion they can be nearly one-third of the way towards the nearest stars. At present about 140 short-period and 800 long-period comets are known and around thirty new comets are being discovered each year. Our knowledge of the cometary population is biased because comets are discovered only when they come within about 2.5 astronomical units of the Sun. It is thought that the Sun actually has about a million million comets in orbit around it. For tables of bright comets and other famous comets see page 194.

components *binary star.

conjunction *planetary configurations.

constellation, one of the eighty-eight areas into which the entire night sky is divided. Originally a constellation was simply a star pattern with no definite boundary, but in 1930 the International Astronomical Union adopted official boundaries to the constellations, drawn along lines of right ascension and declination for the epoch 1875. This date was chosen because it had already been used by the US astronomer Benjamin Apthorp Gould (1824–96) to draw up constellation boundaries for the southern hemisphere. The 1930 IAU boundaries were the work of the Belgian astronomer Eugene Delporte (1882–1955).

Various civilizations have imagined their own patterns in the stars, but the constellations used internationally today stem from a group of forty-eight known to the ancient Greeks and listed by the 2nd-century Greek astronomer Ptolemy in the **Almagest*. At the end of the 16th century two Dutch navigators, Pieter Dirkszoon Keyser (*c.*1540–96) and Frederick de Houtman (1571–1627), added twelve new constellations in the far southern part of the sky that was below the horizon to Greek astronomers. These twelve were: Apus, Chamaeleon, Dorado, Grus, Hydrus, Indus, Musca, Pavo, Phoenix, Triangulum Australe, Tucana, and Volans. Shortly thereafter the Dutch celestial cartographer Petrus Plancius (1552–1622) added Columba, Monoceros, and Camelopardalis in between the constellations known to the Greeks.

The constellations of the northern sky were completed by eleven figures invented by the Polish astronomer Johannes Hevelius (1611–87). These were first shown on his star atlas *Firmamentum Sobiescianum*, published posthumously in 1690: these include Canes Venatici, Lacerta, Leo Minor, Lynx, Scutum, Sextans, and Vulpecula. After a visit to the Cape of Good Hope in 1751–2, the French astronomer Nicolas Louis de Lacaille (1713–62) filled out the southern sky with fourteen new constellations mostly representing instruments used in science and the arts: Antlia, Caelum, Circinus, Fornax, Horologium, Mensa, Microscopium, Norma, Octans, Pictor, Pyxis, Reticulum, Sculptor, and Telescopium. He also split up the Greek constellation of Argo Navis into three parts: Carina, Puppis, and Vela.

Other astronomers invented additional constellations, but they never stood the test of time. The current list of eighty-eight constellations was officially adopted by the International Astronomical Union in 1922, along with the three-letter abbreviations shown in the table on page 195. The Latin genitive case of the constellation name is used when referring to a star within it; for example, Alpha Orionis means the star Alpha in Orion. The principal constellations are shown in the endpapers. See also the illustrations on pages 30–1.

co-orbital satellites, two of the small inner satellites of Saturn, *Janus and **Epimetheus**, which essentially share the same orbit. In fact, the orbits are only 50 km apart and although the two satellites do not collide, they do interact strongly at *conjunction. The diameters of these satellites are about 200 km and 120 km, respectively. Their icy surfaces are rugged and heavily cratered.

coordinate system, a set of numbers (called coordinates) used to fix the position of a point. Right ascension and declination are a coordinate system based on *angular measure for the points on the *celestial sphere; the determination of such coordinates is usually performed by measuring the angular distance to some stars listed in a catalogue. Three perpendicular axes can be used to measure the coordinates for a point in ordinary space. To define such a coordinate system it is necessary to choose an origin (the point with zero coordinates), the direction of the three axes, and a unit of length. For the positions of the planets, the unit of length is usually the *astronomical unit; the directions of the axes are defined by some **reference plane** such as the equator of the Earth or the *ecliptic, and by a reference direction such as the *First Point of Aries, the direction of the ascending intersection of the ecliptic with the equator. *Heliocentric coordinates use the Sun as the origin. In order to compute the orbit of a planet, it can be expedient to use a coordinate system in which *Newton's Laws of Motion directly apply. To this purpose the centre of mass of the entire solar system is used as the origin. In *relativity the coordinate system includes the time coordinate. The change of spatial coordinates (for example, the use of a new set of axes moving with respect to the previous ones) also implies a change in the time-scale.

Copernican system, a *heliocentric model of the solar system introduced by *Copernicus in 1543, which treated the Earth as one of the planets circling the Sun. The Moon revolved about the Earth as in the *Ptolemaic system. Almost every known motion of the planets was accounted for by the Copernican model although he had to keep some epicycles in the system. According to Copernicus the apparent daily rotation of the heavens was due to the real diurnal rotation of the Earth, and the Earth actually moved along its own orbit about the Sun. The retrograde loops, during which the outer planets move backwards in the sky, became mere optical illusions caused by the relative motion of these planets and the Earth as they orbit the Sun. Venus and Mercury were correctly located as inferior planets, orbiting the central Sun, on smaller orbits than the Earth's and at greater speeds; the Sun was no longer categorized as a planet. This heliocentric theory aroused intense opposition. Some opponents maintained that if the Earth moved, the Moon must be left behind, but a more reasonable criticism was that the brighter and nearer stars should show a yearly shift in *parallax against the fainter, more distant stars if the Earth revolved about the Sun. This was subsequently found to be true. The orbits of the planets around the Sun can now be described, to a good approximation, by *Kepler's Laws. The Copernican system had profound consequences, leading to a reassessment of the size of the universe. Also, because objects had previously been assumed to fall to the centre of the universe, that is, the Earth, ideas about gravity needed to be revised. This eventually resulted in *Newton's Law of Gravitation.

Copernicus, Nicolaus (1473–1543), Polish-German astronomer and founder of modern astronomy. In 1543 he published, in his book *De Revolutionibus Orbium Coelestium* ('The Revolution of the Heavenly Orbs'), a *heliocentric model of the solar system, now known as the *Copernican system, whose centre was near the Sun, not the Earth, as in the older *Ptolemaic system. A preface in the book suggests that the system be treated merely as a simple mathematical device but it seems likely that Copernicus, who did not write the preface, believed it to be true. The heliocentric theory, displacing as it did the Earth from the centre of the heavenly stage, aroused fierce religious opposition.

Cor Caroli, a variable third-magnitude *binary star in the constellation of Canes Venatici, and also known as Alpha Canum Venaticorum. It comprises a hot *dwarf star and another star somewhat hotter than the Sun, moving in an orbit of very long period. The distance of Cor Caroli is 40 parsecs, and its combined luminosity is about 110 times that of the Sun. The brighter component is a magnetic star with a variable spectrum.

corona, the tenuous solar atmosphere seen as a ghostly halo of light which surrounds the Moon's disc during a total solar *eclipse. The corona is illuminated by light from the *photosphere being scattered by electrons in the very hot (2×10^6 K) outer atmosphere of the Sun. The corona is now frequently observed using artificial eclipse instruments, known as coronagraphs, and by means of its own high-temperature X-ray emission. Such X-ray photographs reveal coronal holes, low-density regions from which the *solar wind escapes.

Corona Australis, the Southern Crown, a small *constellation of the southern hemisphere of the sky. It was known

Constellations of the northern hemisphere

to the ancient Greeks, to whom it represented a crown or wreath of leaves at the foot of Sagittarius. Its brightest stars are of only fourth magnitude.

Corona Borealis, the Northern Crown, a *constellation of the northern sky, one of the forty-eight constellations known to the Greeks. It represents the jewelled crown worn by Princess Ariadne of Crete when she married the god Dionysus. Fittingly, its brightest star is known as Gemma, Latin for 'jewel'. The constellation contains two bizarre variable stars. *R Coronae Borealis is a supergiant star that normally appears around sixth magnitude, but every few years it fades suddenly to around fourteenth magnitude, evidently a result of the accumulation of carbon particles in its atmosphere. T Coronae Borealis, called the Blaze Star, is a recurrent *nova, normally around tenth magnitude, that occasionally brightens to second or third magnitude.

Corvus, the Crow, a *constellation lying just south of the celestial equator, one of the forty-eight constellations known to the ancient Greeks. In mythology it represents a crow sent by Apollo to fetch water in a cup; the cup is depicted by the neighbouring constellation of *Crater.

cosmic background radiation *microwave background radiation.

cosmic composition, the proportions of material that make up celestial objects. In stars, hydrogen forms about two-thirds of the total mass, with helium making up most of

Constellations of the southern hemisphere

the remaining one-third. About 2 per cent of mass is composed of carbon, oxygen, silicon, and nitrogen with only 0.5 per cent being composed of other elements. These proportions vary somewhat during *stellar evolution. For planets, the cosmic composition can be divided roughly into four components: metal, rock, ice, and gas. The first two form earthy material similar to that which makes up the *inner planets; the icy material is composed of the hydrogen compounds of oxygen, nitrogen, and carbon; and the gaseous material is predominantly hydrogen and helium. The *giant planets contain earthy, icy, and some of the gaseous material in the ratios of 1 to 2.2 to 200 by mass.

cosmic rays Cosmic rays were discovered by Victor Hess in 1912 when he found that the electrical conductivity of the atmosphere increased with altitude, so indicating a 'radiation from above'. Robert Millikan developed the hypothesis in the 1920s that this radiation was cosmic in origin and consisted of gamma rays; he gave it the name cosmic rays. However, with the development of charged particle detectors (in particular the Geiger–Müller counter) at the end of the decade came the discovery in 1930 by German physicists Walther Bothe and Werner Kölhorster that cosmic rays are high-energy charged particles. At the same time Bruno Rossi established that the radiation showed a small excess from the west which indicated that primary cosmic rays are influenced by the Earth's magnetic field; this implied not only that they were charged particles but that the charge was positive. It is now established that primary cosmic rays (those above the Earth's atmosphere) are the nuclei of atoms stripped of their electrons. Because of the preponderance of hydrogen in the universe they are mainly protons. Cosmic rays have a very wide span of energies, the highest being over 100 million times those achieved artificially with particle accelerators. Detailed studies of the composition of the low-energy particles suggest that the majority of cosmic rays are galactic in origin, the most likely sources being supernovae. Studies of the directions of approach of the very highest-energy cosmic rays, however, indicate that these particles may have an extragalactic origin.

When primary cosmic rays interact with nitrogen and oxygen atoms in the Earth's atmosphere they produce cascades of secondary particles including the so-called elementary particles. Their discovery in the 1930s and 1940s led to a new branch of physics, elementary particle physics, and to the development of accelerators which could produce the 'new' particles under more controlled conditions than with cosmic rays. At the Earth's surface the secondary cosmic rays consist mainly of muons, electrons, and photons with an intensity of about one particle per square centimetre per minute. The intensity is constant with time to within a few per cent but because the lowest-energy primary particles are affected by the interplanetary medium as they pass into the solar system there are small solar-induced intensity variations. During very large solar flares the Sun also accelerates particles to cosmic ray energies. These are often referred to as solar cosmic rays.

A by-product of cosmic ray interactions in the atmosphere is the isotope carbon-14 which is absorbed by living organisms through atmospheric carbon dioxide. After death, absorption ceases and the concentration of carbon-14 decreases. This leads to the carbon-14 *radioactive dating technique, developed by Willard Libby in 1949, now widely used for archaeological specimens up to 10,000 years old.

cosmogony, the study of the origin and development of the universe, or of a particular system in the universe, such as a planetary system. When applied to planetary systems, cosmogony is a difficult subject because the solar system is 4.6 billion years old and is the only such system to have been studied in detail. The age of the solar system has allowed sufficient time for all the planets to settle down into orbits that are in the same direction and in nearly the same plane.

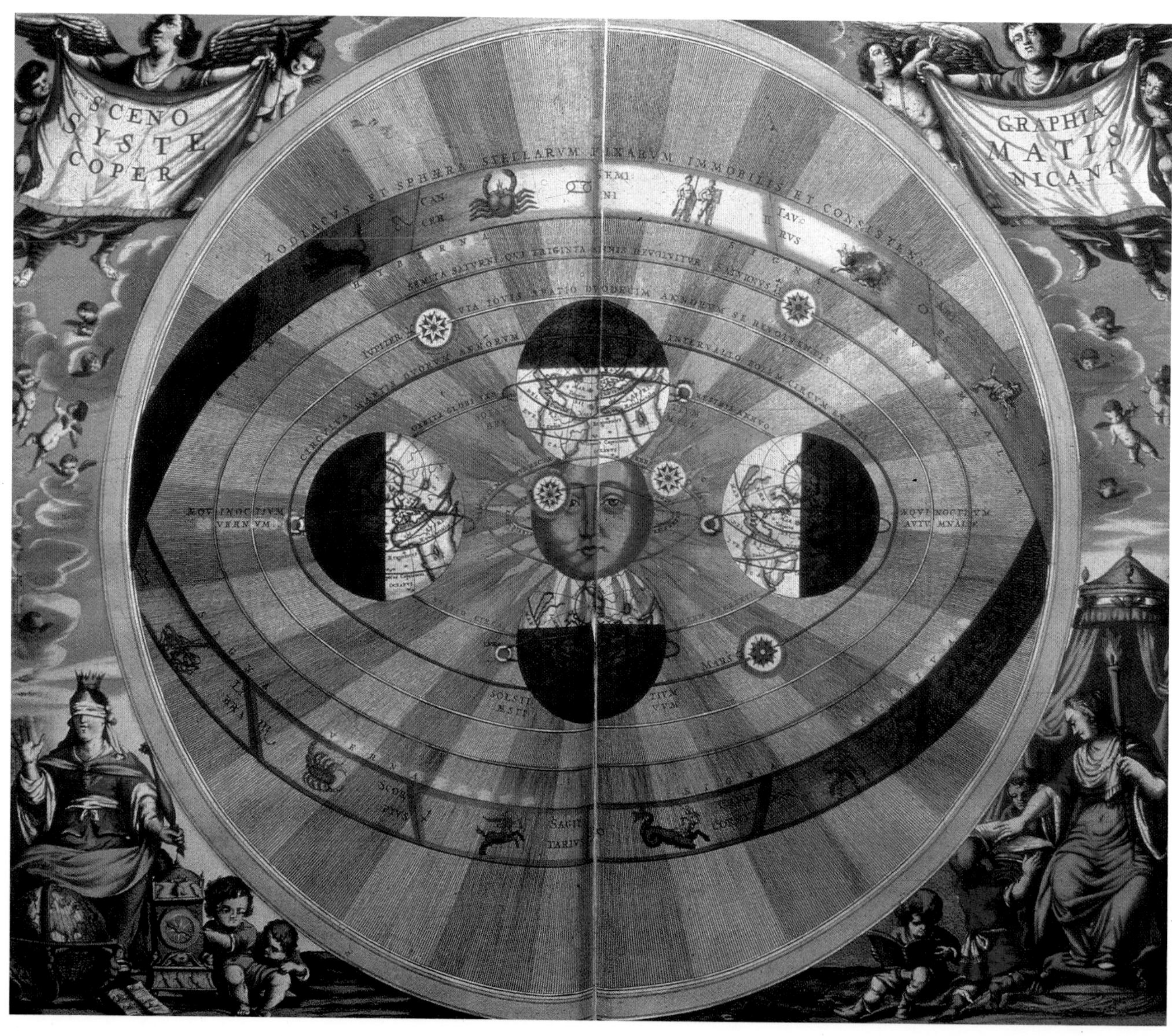

The publication of the **Copernican system** in 1543 was a watershed in the development of astronomy. Although in its simplest form it did not predict the movement of the planets any better than the more complex Ptolemaic system, it was eventually proved to be the correct description of the solar system.

Its size, mass, and speed of rotation suggest that it condensed from a much more massive cloud of gas and dust. Stars are formed from clouds of gas and dust in the spaces between other stars, particularly in nebulae. A large or rapidly rotating cloud can condense into a double star. If the cloud is small or rotates slowly then all the mass is pulled into a single star. In between these extremes the condensing star is surrounded by a nebular cloud, parts of which may subsequently condense to form planets. The *Laplace nebular hypothesis and *Weizsacker's theory propose mechanisms for the creation of planets, although the origin of planetary systems is still not fully understood. It is now thought that the formation of planetary systems is common during the first stage of star formation and that about one in four of the stars in the universe have planetary systems.

cosmological constant, a mathematical constant introduced by *Einstein into the field equations of his general theory of *relativity. It enabled a static non-expanding universe to be obtained as a solution to the equations; it seemed more plausible to him that the universe was static rather than expanding. However, the discovery of the *recession of the galaxies meant that the universe is expanding and so alternative solutions to the equations are required.

cosmology, the study of the structure and evolution of the universe as a complete entity. The term may also be used to describe a particular theory of the universe, such as the *Big Bang theory, and it has also largely replaced the older term *cosmogony. Each civilization has developed its own picture of the universe, the best-known ones of ancient times being those of the Greeks Ptolemy and Aristotle. It is worth noting that prior to this *Aristarchus of Samos had postulated a *heliocentric solar system with the stars as distant suns. The Chinese also favoured this cosmology until the arrival of European culture in the early Middle Ages. Modern cosmology originates from the theory of *Copernicus that the Earth and the other planets orbit the Sun. The dominant force in large-scale interactions is *gravitation and among the triumphs of *Newton's Law of Gravitation is its explanation of the motion of the known planets and the prediction and subsequent discovery of the planet Neptune. It is only during the 20th century, however, that we have reached any understanding of the universe as a whole and the modern concept of an expanding universe.

The first major step in observational cosmology during this century was the realization in the 1920s that the Milky Way is one *galaxy among millions of others, with the Sun just one of millions of stars in the Milky Way. During the ensuing study of the external galaxies, it was found that they are all receding from our own galaxy, with the more distant galaxies having the highest recession velocities (measured by the *red shift of their spectra). We therefore inhabit an **expanding universe**. The *recession of the galaxies was summarized by *Hubble in his law that the red shift of a galaxy is proportional to its distance from us. The constant of proportionality, the *Hubble constant H is very important as most theories give the present age of the universe as approximately equal to its inverse. It is a difficult constant to measure and the first attempts, made in the 1930s, gave a value equivalent to an age of 2 billion years, much less than the age of the Earth. A more modern value of the Hubble constant is about 55 km/s per megaparsec, which may still be in error by up to 20 per cent and is equivalent to an age of 15 billion years. Hubble's Law also tells us that the red-shifted light from more distant objects will have less energy than that from nearby ones. This is the resolution of *Olbers' paradox of 1826 which says that, if the universe is infinite, then an observer will see a star in every direction. Thus, we should never see a dark sky. The darkness of the night sky is due to the expansion of the universe; the simple fact that the sky is dark at night is a very important cosmological observation.

The advent of *radio astronomy in the 1950s brought further results over the next two decades. It was soon realized that most of the radio sources in the sky such as *quasars and radio galaxies, are associated with astronomically distant objects. For example, one of the brightest radio sources, *Cygnus A, is a peculiar galaxy at a distance of 300 million parsecs, sufficiently far away for it to be near the limit of detection of all but the largest optical telescopes. The only way to measure the distance to a radio source is to measure the red shift of its optical counterpart (if any) and apply Hubble's Law. The validity of Hubble's Law, especially when applied to the large red shifts of the quasars which are up to twenty times those of the galaxies, was the subject of considerable debate (the red-shift controversy) during the 1960s and early 1970s, but the law is now believed to apply to both galaxies and quasars. Since the distances calculated from the red shifts are a significant fraction of the size of the universe, the radio and light waves take a period of time approaching the age of the universe to reach us. Therefore by looking at faint radio sources we are looking at earlier epochs in the history of the universe. Counting the number of radio sources in each range of signal intensity (the **radio source counts**) gives results that are in complete disagreement with all theories which assume that the universe has looked the same at all times, a view favoured by the *Steady State theory of cosmology put forward by the British scientists *Hoyle and Hermann Bondi, and the US astronomer Thomas Gold. The only possible explanation is that there was a time when there were no radio sources, followed by an epoch with many sources, and the present time with few sources. This is dramatic evidence for the Big Bang or evolutionary cosmologies.

The most important observation in support of the Big Bang theory was the discovery of the *microwave background radiation by the US physicists Arno Penzias and Robert Wilson in 1965. Looking at a blank area of sky, they found a radiation field equivalent to that emitted by a *black body at a temperature of 2.7 K. This is the relic (or echo) of the original fireball predicted by the US physicist George Gamov in 1948 and proposed by him as a test of the evolutionary (Big Bang) cosmologies. In recent years this background has been studied at all wavelengths from the radio through to the gamma-ray part of the spectrum and a great deal of attention has also been focused on its *isotropy as it gives information about the earliest epochs of the universe. All cosmological theories incorporate the **cosmological principle** which states that we are not located in a special part of the universe; in other words the universe is homogeneous and isotropic and would look the same to an observer in any other galaxy apart from small-scale random variations. It is also usually assumed that the laws of physics and their fundamental constants (particularly the gravitational constant G) do not change with time. There is no evidence to refute this, despite intensive investigation by many scientists. *Einstein's general theory of *relativity and its discussion of gravitation is the starting point for most cosmologies (the classical cosmologies), the principal exceptions being the Brans–Dicke theory which does not use general relativity as its starting point, and the Hoyle–Narlikar theory which is a variation of the Steady State theory. If our universe incorporates the above cosmological principle and assumptions then the interval between two events involves two parameters that change with time; one is the *curvature of space and the second is a scale factor R which describes the separation between two 'fundamental observers', that is, a theoretical ideal observer. The Steady State theory derives a formula for R but it differs from other theories in that it starts from the perfect cosmological principle, that is, that the universe is the same for all observers at all times. The result of this and the observed recession of the galaxies is that matter must be continuously created and the theory is not now generally accepted. Einstein's general theory shows how R varies with time; this introduces two more parameters. The first, A, is the *cosmological constant which was introduced by Einstein in an attempt to obtain a non-expanding universe solution from the field equations. The subsequent discovery of the red shift in the spectra of galaxies seemed to contradict this model of a non-expanding universe. The second parameter, the density parameter D, involves the amount of matter in the universe and the Hubble constant H. The different cosmological models (or **world models**) then differ in their choice of the value of A and D, with the most generally used one, the **Einstein–de Sitter** universe, having $A = 0$ and $D = 1$. This is a model universe which expands for all time. Among the others are the **static universe** which neither expands nor evolves, the pseudostatic **Lemaitre model**, and models which have universes that expand for a limited time and then collapse again. An *oscillating universe even has successive big bangs, each universe ending in the so-called *Big Crunch. The early stages of all the models have a very hot dense **primordial fireball** in which everything is in the form of radiation, but as the fireball cools down matter condenses, mainly in the form of hydrogen and helium. Cosmology is in a very active state at present with both discoveries in particle physics and observations of the universe at large producing new and exciting theories about the origin, evolution, and destination of the universe.

The **Crab Nebula** is the debris from a star that exploded in 1054. The filamentary gaseous remnants of the outer layers of the star are aligned along the lines of the magnetic fields in the nebula and are expanding outwards from the centre at some 1,500 km/s. The Crab Nebula is one of the most extensively studied objects in the sky.

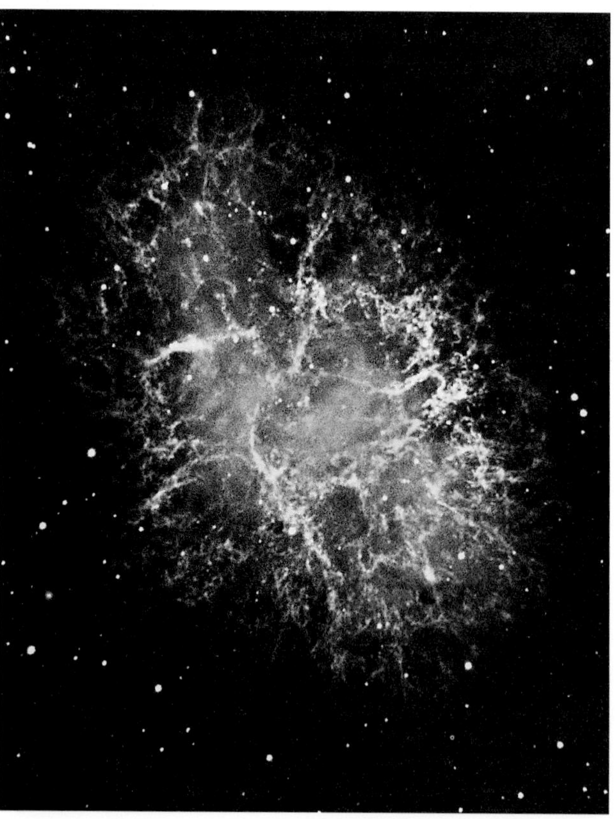

coudé telescope *reflecting telescope.

counterglow *gegenschein.

Crab *Cancer.

Crab Nebula, a luminous gaseous *nebula, in the constellation of Taurus. It has a filamentary structure and is expanding as a remnant of the *supernova of 1054. In 1948 it was found to emit radio waves (*Taurus A) and in 1964 it was found to be an X-ray source (*Taurus X-1). In fact the emission from the Crab Nebula extends throughout the electromagnetic spectrum to gamma rays and is shown by its *polarization to be *synchrotron radiation. The energetic electrons responsible cannot have originated in the supernova itself and must be supplied by the rotating *neutron star within the nebula, now observed directly as a radio, optical, and X-ray *pulsar.

crater, an approximately circular feature found in large numbers on the Moon, Mercury, Mars, Venus, and most of the planetary satellites. Craters are caused either by volcanic activity or by the impact of bodies of various sizes, and they range in diameter from a few centimetres to more than 1,000 km. Many of the older ones have been partially obliterated by more recently formed ones. Most are now thought to be of impact origin rather than volcanic, the vast majority of the former being created in the early days of the solar system when thousands of planetesimals, asteroids, and meteoroids existed and were swept up by the larger planets and satellites. The *kinetic energy of even a small object is large enough to vaporize and shatter much of the material that forms it, together with the ground it impacts. The explosion then creates a crater, shallow relative to its diameter, pushing up one or more concentric rings and sending out *ejecta. Often, in the bigger events, a central mountain is left. While the record of all such collisions from the solar system's early days is plain to see on the faces of the planets and their moons, most craters of this kind on the Earth have been destroyed through erosion by wind, water, and geological processes except for those made in the last few hundred million years such as the Arizona meteor crater. Impact craters are still being formed when the debris remaining in the solar system strikes a planet or satellite. A class of asteroids called *Apollo–Amor objects cross the orbits of the four inner planets and members of this class must occasionally collide with the planets. Volcanic craters such as those found on the Earth also exist on the Moon, Mars, and Venus though all seem to be extinct. The largest, *Nix Olympica, is on Mars and dwarfs anything on our planet.

Crater, the Cup, a *constellation just south of the celestial equator, one of the forty-eight constellations known to the ancient Greeks. In mythology it represents the cup in which Corvus, the Crow, was supposed to bring water to Apollo.

Curvature

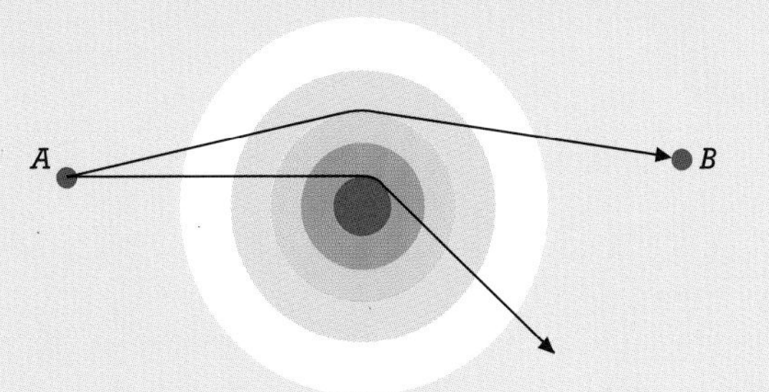

The curvature of space–time is traced by a beam of light. The shading shows the degree of space–time curvature around a massive body. If the transmitter at *A* tries to send a signal directly to *B*, as in ordinary flat space, the beam, following the curvature of space–time, will be deflected away from *B*. The geodesic is the upper path which, though apparently curved when seen by the observer at *B*, is the shortest path in curved space–time. This effect is used in the grazing of starlight past the Sun during a total eclipse as a demonstration of the general theory of relativity.

However, the crow was late with the water because he stopped along the way to eat figs. Apollo condemned the crow to a life of eternal thirst by placing it in the sky just out of reach of the water in the cup. The brightest stars in Crater are of only fourth magnitude and there are no objects of interest.

Crux, the Southern Cross, the smallest *constellation in the sky. The stars of Crux were known to the ancient Greeks, who regarded them as part of the hind legs of Centaurus, the Centaur. Crux was made a separate constellation in the 16th century by European navigators, who used it as a pointer to the south celestial pole. Its brightest star is Alpha Crucis, also known as *Acrux. Crux lies in a rich part of the Milky Way and includes the *Jewel Box Cluster. Most of the *Coalsack Nebula lies in Crux.

culmination, the two points on the *celestial sphere in the 24-hour circuit (the *diurnal motion) of a celestial body when it is on the observer's meridian. For an object that rises and sets, upper culmination is its highest altitude above the horizon, lower culmination being below the horizon and therefore unseen. For a *circumpolar star, both upper and lower culminations are observable.

curvature, in cosmology, the modification of *space–time in the presence of matter. Light will always travel by the shortest route, known as a **geodesic**, between two points. If a transmitter at *A* wishes to send a beam of light to a point *B*, then, when there is no matter present, the path taken by the light is the straight line between *A* and *B*. This corresponds to the experience of everyday life. However, in the presence of a large amount of matter, such as in the vicinity of a very massive star or a *black hole, the light received at *B* will appear to have followed a curved path. In fact the light has still followed a geodesic, but it is space–time that has been curved by the massive body. According to the general theory of *relativity, the gravitational field of a massive body is then an effect of the curvature of space–time. A two-dimensional analogy of curvature is the surface of a hilly putting green. In order to send the golf-ball into the cup it is sometimes necessary to aim not directly at the cup but to one side so that the curvature of the surface will bring the ball to the cup. On this two-dimensional curved surface the geodesic is no longer a straight line as it would be on a flat green.

Cyclops Project, a proposed *radio astronomy project in which a large array of radio telescopes would be built to search for extraterrestrial intelligence (*SETI) by detecting radio signals from them. There have been several other SETI programmes, such as the *Ozma Project, but these have not yet been successful. For this reason, and also because the building of such a large Cyclops array would be extremely expensive, it is unlikely that Cyclops will ever be funded.

61 Cygni (the 'flying star'), at a distance of 3.4 parsecs the first star other than the Sun for which a convincing measurement of distance was made. After several centuries of unsuccessful attempts by others, the German astronomer *Bessel successfully measured its distance in 1838. He succeeded because he used an excellent telescope; because he was able to eliminate interfering effects like the *aberration of starlight; and because he did not assume that the brightest stars were necessarily the nearest. He rightly presumed that fifth-magnitude 61 Cygni would be nearby because it is a *binary star whose components could easily be seen as separate stars through a telescope and because it had a large *proper motion when observed against the background of more distant stars. To determine its distance he made frequent and consistent observations of the minute annual variation (parallax) of the star's position due to the orbiting of the Earth around the Sun.

Cygnus, the Swan, a *constellation of the northern sky, one of the forty-eight constellations known to the ancient Greeks. In mythology it represents the swan as which Zeus disguised himself to seduce Queen Leda of Sparta. Cygnus is popularly known as the Northern Cross because of its distinctive shape. Its brightest star is *Deneb. At the foot of the cross lies Beta Cygni, known as Albireo, a beautiful coloured double star for small telescopes, consisting of orange and green components of third and fifth magnitude. Cygnus lies in a rich part of the Milky Way where it is bisected by a dark dust cloud known as the Cygnus Rift. Near Deneb lies the *North America Nebula.

Cygnus A, one of the brightest radio sources in the sky, and the first *radio galaxy to be discovered. The galaxy was identified as a result of accurate radio position measurements made at Cambridge, UK, in the early 1950s by Francis Graham-Smith, later to become Astronomer Royal. The galaxy is 300 million parsecs away, and its radio output is millions of times greater than that of a normal galaxy. Most of the radio emission comes from two gigantic lobes which lie on either side of the galaxy. A narrow jet of high-energy particles carries energy to the lobes from the mysterious energy source which lies at the heart of the radio galaxy.

Cygnus Loop, a *supernova remnant in the constellation of Cygnus. Also known as the **Great Looped Nebula**, it is an enormous shell of material more than 3 degrees across ejected in a supernova explosion 30,000 years ago. Observations made by the infra-red astronomical satellite IRAS reveal that much of this shell is dust heated by the shock wave from the explosion. An alternative name for part of the

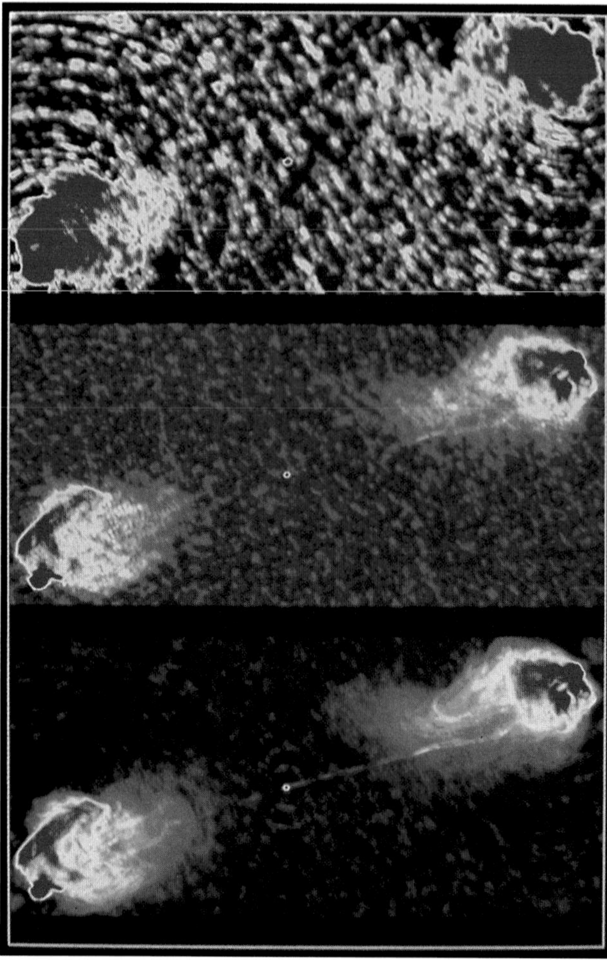

These three images of the radio galaxy **Cygnus A** represent different stages in the processing of an interferometric image produced by a radio telescope. The top image is the raw data as it appears before processing. Most of the structure is not real, but is caused by the process that synthesizes the beam signals. The centre image has been enhanced by computer to remove this structure. The final image has been enhanced again to remove the effects of small-scale irregularities in the Earth's atmosphere. The three images represent some thirty hours of data taken by the Very Large Array radio telescope in New Mexico, USA.

remnant is the *Veil Nebula. The object is visible in high-power binoculars but is best studied in photographs.

Cygnus X-1, an *X-ray binary, identified as a part of the spectroscopic binary star HDE 226868, probably containing a *black hole. The strong X-ray emission fluctuates rapidly suggesting that the source is compact. Theoretical analysis suggests that the X-ray object may have a mass many times larger than the Sun and be accompanied by a supergiant in orbit round it.

dark nebula *nebula.

Dawes limit, an empirical expression describing the ability of an optical telescope used with the eye to separate two stars which are close together in angle. The formula is based on the observational experience of the British amateur astronomer William Dawes and is approximately given by $\alpha = 0.12/D$, where α is the resolved angle and D is the telescope diameter in metres. The limit is rarely used by professional astronomers.

day, in ancient times, the interval of time between sunrise and sunset, but nowadays the time for one rotation of the Earth counted from one midnight to the next. For two thousand years and more, the year has been divided into days and it was known that the proportion of daylight time to darkness in a particular day altered with the *seasons, being smaller in winter than in summer. Nevertheless, the division of daylight time into the same number of hours throughout the year which could be measured by a *sundial was convenient. The apparent solar day is then the time between successive passages of the Sun over the observer's *meridian. However, this varies slightly during the year because the Earth's elliptical orbit about the Sun causes it to travel a little faster in January than in July. Another source of variation is the inclination of the Earth's equator to its orbit which makes the Sun's apparent motion faster at a *solstice. As the necessity grew to have a time interval which did not vary with the seasons, a fictitious body, the **mean sun**, was introduced. It moved above the equator with the Sun's mean angular velocity against the stellar background. A **mean solar day** is the time between successive passages of the mean sun across the observer's meridian and is the same throughout the year. For many astronomical purposes, however, the time between successive passages of the *First Point of Aries or the vernal equinox across the observer's meridian is more useful. This is the **sidereal day** and is essentially the time of one apparent revolution of the celestial sphere or stellar background because of the Earth's rotation.

Motion that is performed in the course of a day is called *diurnal motion.

declination *celestial sphere.

deferent *Ptolemaic system.

degenerate matter, a state of matter which is not gas, liquid, or solid, reached when the central density in a star has become so high (several thousand tonnes per cubic metre) that a state said to be quantum mechanically degenerate is produced. White dwarfs and neutron stars contain degenerate matter and so are often described as **degenerate stars**.

Deimos, the outer satellite of Mars, discovered by the US astronomer Asaph Hall in 1877. It has a nearly circular orbit at 23,460 km from the planet's centre. Markedly irregular in shape, the mean radius of Deimos is about 6 km. The mean density is about 1,700 kg/m^3 and the rotation is synchronous, that is, its rotational period is the same as its orbital period. Its surface is dark, heavily cratered, and covered by a deep *regolith. There are extensive patches of bright material and the rock and dust filling most craters is conspicuous.

Delphinus, the Dolphin, a *constellation just north of the celestial equator. It was one of the forty-eight constellations known to the ancient Greeks, and represents either the messenger of the sea god Poseidon or the dolphin that carried the musician Arion to safety after he was attacked by robbers on board ship. Its most distinctive feature is a rectangle of stars that form an *asterism called Job's Coffin. Two of these stars bear the peculiar names Sualocin and Rotanev. When reversed, these names spell Nicolaus Venator, the Latinized name of the Italian astronomer Niccolo Cacciatore who gave the stars their names.

Deneb, a first-magnitude star in the constellation of Cygnus, and also known as Alpha Cygni. It is a very luminous hot *supergiant star at an estimated distance of 500 parsecs, with about 60,000 times the luminosity of the Sun. It is situated at one end of the long arm of Cygnus.

Denebola, a second-magnitude *variable star, the third-brightest in the constellation of Leo, and also known as Beta Leonis. It is a hot *dwarf star at a distance of 12 parsecs, with about seventeen times the luminosity of the Sun. Because of its high temperature, it appears white.

density, the *mass of a unit volume of an object. The mean density of a celestial object is obtained by dividing its mass by its volume. The Sun has a mean density of 1,400 kg/m^3 but this value can be misleading because the density changes logarithmically as a function of distance from the centre of the Sun, and the density of the material there is about 160,000 kg/m^3, whereas it is around 1 kg/m^3 near the visible surface. The Earth can be thought of as being made up of rock and iron and these have densities of around 3,000 and 8,000 kg/m^3, respectively. The mean density of the Earth is 5,520 kg/m^3. Saturn, with its large gaseous atmosphere, has a mean density of 700 kg/m^3 which is 30 per cent less than the density of water.

This photograph of the Martian satellite **Deimos** was taken by the Viking 1 orbiter. The properties of the surface of Deimos suggest that its composition is probably similar to that of carbonaceous chondrite meteorites and it may be a captured asteroid.

This image of Saturn's moon **Dione**, recorded by Voyager 1 during its encounter with Saturn in November 1980, shows Dione's heavily cratered surface. Different amounts of cratering indicate that there have been episodes of surface renewal since the period of heavy bombardment. Troughs or canyons are also visible at the upper left of the photograph.

de Sitter model, a mathematical model in *cosmology in which the universe contains no matter. It was derived by the Dutch physicist, Willem de Sitter from the field equations of the general theory of *relativity. Although such models may seem unrealistic and trivial when it is known that the universe does contain matter, nevertheless a knowledge of the full range of solutions (models) of the field equations under various conditions gives valuable insight into cosmological questions.

diamond ring effect, a sudden flash of light observed as the Sun begins to reappear after a total solar *eclipse. The Sun's lower atmosphere, that is, the *chromosphere and reversing layer, emerges shining brightly, followed by the lower-lying photosphere, while the outer solar *corona vanishes. The inner corona remains as a thin ring. When the first part of the photosphere emerges, it is intensely bright and appears much larger than it really is; together with the ring it forms the diamond ring effect.

diffraction, the behaviour of light near the edge of an opaque obstacle or when it passes through a narrow slit or small aperture, resulting in the formation of alternate bright and dark bands or rings. Such a formation, known as a **diffraction pattern**, is a function of the *wavelength of the light. In a telescope the image of a star under perfect seeing conditions is a diffraction pattern consisting of a bright point of light surrounded by concentric rings of decreasing brightness, known as an Airy disc.

Dione, a satellite of Saturn, discovered by Giovanni Domenico *Cassini in 1684. It has a nearly circular orbit above Saturn's equator at 377,400 km from the planet's

Doppler effect

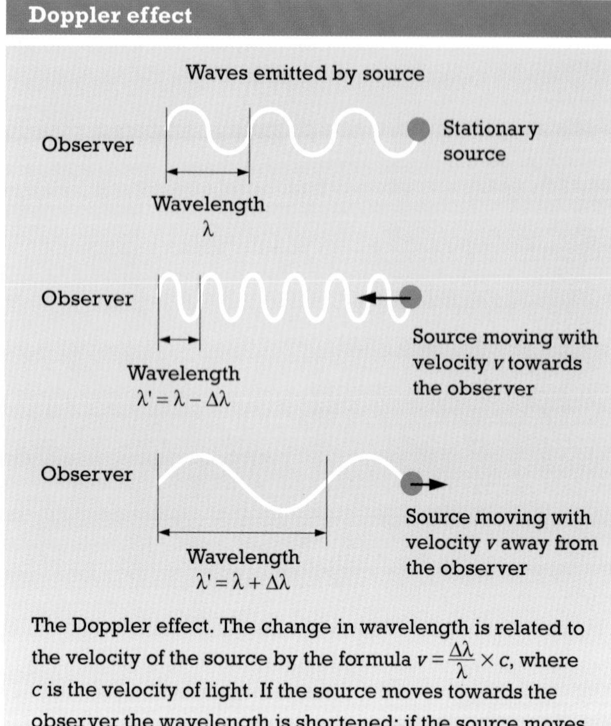

The Doppler effect. The change in wavelength is related to the velocity of the source by the formula $v = \frac{\Delta\lambda}{\lambda} \times c$, where c is the velocity of light. If the source moves towards the observer the wavelength is shortened; if the source moves away from the observer the wavelength is lengthened.

centre. Its diameter is 1,120 km and it has a density of 1,400 kg/m^3 implying a predominantly water ice composition. *Voyager images have shown that as Dione moves in its orbit the trailing hemisphere is covered with wispy bright markings which may consist of surface deposits of ice ejected through a global *tectonic system of fractures. Although many impact craters are visible on Dione, extensive plains of relatively recent origin, as well as troughs and valleys hundreds of kilometres in length, indicate a significant level of activity within the interior of the satellite.

dish, the radiation-collecting curved structure of a *radio telescope. It acts as a reflector of radio waves and is usually mounted so that it can be pointed in any desired direction. The dish focuses incoming radio waves on to an aerial (US antenna), often called a feed. A paraboloidal reflector receives radio waves from a small angle along the direction of its axis; the larger the telescope, the smaller is this receiving beam at a given wavelength and the higher the *resolving power. For example, the Lovell radio telescope at *Jodrell Bank with a dish of 76.2 m (250 feet) diameter has a beam-width of 12 arc minutes at 20 cm wavelength.

distance modulus, the difference between the apparent and absolute *magnitude of a celestial object such as a star or galaxy, used as a way of measuring its distance from the Earth. Distance modulus is particularly used for objects more remote than a few hundred parsecs whose *parallax is too small to measure. Often the absolute magnitude as well as the amount of absorption of light by the gas between the star and the Earth can be estimated; the apparent magnitude can be observed directly. For example, there is a well-defined relationship between pulsation period and absolute magnitude of *Cepheid variable stars. Hence by observing such stars in a nearby galaxy the distance of the galaxy can be estimated.

diurnal libration *libration.

diurnal motion, motion performed in the course of a day. Because the Earth rotates on its axis once in a *sidereal day in an eastward direction, objects such as the Sun, Moon, planets, and stars appear to revolve about the Earth from east to west. The diurnal motion is the basic observed movement of these bodies. However, because all natural celestial bodies have their own intrinsic movement through space, this apparent motion is more complicated, particularly for the Sun and Moon.

Dog Star *Sirius.

Dolphin *Delphinus.

Donati's Comet, a *comet discovered in 1858 by the Florentine astronomer Giambattista Donati (1828–73). The comet was bright and well placed for northern hemisphere observation; astronomers used large refracting telescopes to view it and many drawings were made. The comet displayed a fountain effect in its *coma. It was as if the cometary nucleus was spraying out material mainly towards the Sun but that the solar radiation was pushing that material back towards the tail. It seemed to be throwing off these fountain-like envelopes regularly for several weeks. Measurements of the speed of recession of the envelopes during a subsequent reappearance enabled the US astronomer Fred Whipple to calculate that the nucleus was active on only one side and was spinning round every 4 hours 37 minutes.

Doppler effect, in astronomy, the apparent change of *wavelength of *electromagnetic radiation with the relative velocity of the observer and the source. It is a familiar experience to stand by the side of a road and hear the sound of an oncoming car's horn lower its pitch when it passes. This is an example of the Doppler effect, in this case acting on sound waves. In the case of a moving object, the sound waves are either lengthened or shortened, depending upon whether the object is receding from or approaching the observer, thus changing the pitch of the sound. The same effect holds for light where *spectral lines are shifted towards the red or blue ends of the spectrum according to whether the source is moving away from or towards the observer, respectively. Since the shift in wavelength is dependent on the relative velocity of source and observer, its measurement enables the velocity to be found according to the formula $v = c\Delta\lambda/\lambda$, where v is the velocity of the source with respect to the observer, c is the velocity of light, λ is the wavelength, and $\Delta\lambda$ is the shift in the wavelength. The Doppler effect is a valuable means of measuring the velocity of a star or binary star, and also the velocity of a galaxy where the *red shift is large.

Dorado, the Goldfish, a *constellation of the south celestial hemisphere. It was introduced by the Dutch navigators Pieter Dirkszoon Keyser and Frederick de Houtman at the end of the 16th century; it is sometimes also known as the Swordfish. Dorado contains most of the Large *Magellanic Cloud, including the *Tarantula Nebula.

Double Cluster in Perseus, a double open *cluster, also known as h and Chi Persei, in the constellation of Perseus. Each cluster is about 0.5 degrees in extent and both are visible to the naked eye. They are designated NGC 869 and NGC 884 and are also known as the Sword Handle. The clusters are very young, the constituent stars being only about 10 million years old.

double star *binary star.

Draco, the Dragon, a large *constellation in the north polar region of the sky, one of the forty-eight constellations known to the ancient Greeks. In mythology it represents the dragon Ladon, who was guardian of the golden apples of the Hesperides, and who was shot by Hercules; in the sky Hercules is depicted with one foot on the dragon's head. The fourth-magnitude star Alpha Draconis is also known as *Thuban. The second-magnitude Gamma Draconis, the brightest star in the constellation, is the star that *Bradley observed when he discovered the *aberration of starlight. There are several interesting double stars in Draco for users of binoculars and small telescopes.

Draper classification, a system of classifying stars according to their *spectrum. Stellar spectra contain lines and bands depending upon the atoms and (for *late-type stars) molecules present in the stars' atmospheres. In addition, a star's *surface temperature dictates to a large extent how strong these lines and bands appear in the star's spectrum. The **Secchi classification** of *spectral classes was superseded by the **Harvard classification** introduced at the Harvard Observatory in the 1890s. It was used in the *Henry Draper Catalogue* of stars and is essentially an arrangement of stars in order of decreasing surface temperature. The system was refined as more information was collected about the different spectral classes and about the wide variety of chemical compositions and atmospheric densities and pressures existing in stars of all kinds including giant, supergiant, dwarf, and subdwarf stars. The Draper classification is often used in plotting the *Hertzsprung–Russell diagram.

Both h and Chi Persei in the **Double Cluster in Perseus** are open star clusters visible to the naked eye. They are separated from each other by some 15 parsecs and lie at the end of one of the Galaxy's spiral arms at about 2,000 parsecs from the Sun.

Dumb-bell Nebula, a *planetary nebula of irregular dumb-bell shape in the constellation of Vulpecula, also known as M27. It is the result of the ejection by an old star of its outer layers. Over 200 parsecs distant, its shape suggests its name. It is an excellent object for small telescopes and is the second-largest of the bright planetary nebulae. For photograph see planetary nebula.

dust, in astronomy, fine particles in the Earth's atmosphere or in space. Dust can be found throughout the universe, for example, in the *interplanetary medium, the *interstellar medium, and in dust shells surrounding stars. When dust rises in the Earth's atmosphere, sometimes from volcanic origins, its absorption of starlight affects the quality of stellar observations. Earth is not the only planet to have dust in its atmosphere; Mars is also subject to dust storms that blur out details otherwise visible by telescope. Dust pervades the space between planets and its scattering effect on sunlight can be seen as the *zodiacal light. *Interstellar absorption and *interstellar reddening is particularly noticeable in the plane of the galaxy. Furthermore, alignment of the grains by the interstellar magnetic field sometimes causes weak *polarization of the starlight. Dust sheets are seen in various galactic regions by the light they scatter from nearby stars or by the infra-red radiation emitted on account of the temperature of the grains of dust. Some stars are cocooned in dust shells, the particles affecting the colour of the perceived light from the star, with the shell re-radiating its captured energy in the infra-red region of the spectrum. Spiral galaxies, particularly when viewed edge-on, show dusty lanes embedded in the stars and gas. A substantial mass of the universe is bound up in small particles of cosmic dust.

dwarf star, a normal star on the main sequence in the *Hertzsprung–Russell diagram. More than 90 per cent of the stars are found in this stage of *stellar evolution, in which they obtain their energy by the *nuclear fusion of hydrogen to helium in their cores. The Sun is a fairly typical dwarf star. However, most dwarfs are smaller and cooler than the Sun, and of lower luminosity; these are the **red dwarfs** which form the most numerous stellar component of our own galaxy. The duration of the dwarf stage varies inversely as the third power of the star's mass. The masses of dwarf stars range from about 0.1 to sixty times that of the Sun, and the duration of the dwarf stage is thought to range from one million years or less for the most massive stars, up to many billions of years for the least massive. When the helium produced at the star's core exceeds the *Schönberg–Chandrasekhar limit of about 12 per cent of the star's total mass it begins to evolve into a *giant star. Dwarf stars are distinct from *white dwarf stars and the *cataclysmic variable stars known as dwarf novae. Both of these classes of stars have a very different structure from dwarf stars.

early-type star, a hot star of *spectral class O, B, or A. The name originates from an early theory of stellar evolution which proposed that such stars were in an early stage of their evolution. Although this theory has now been shown to be wrong, the term is still used.

Earth, the planet third in distance from the Sun and the only one in the solar system on which life is known to exist. The Earth is 12,756 km (7,926 miles) in diameter at the equator and has a mean density of 5,520 kg/m^3. It rotates in 23 hours 56 minutes about an axis tilted at 23 degrees 26 minutes to a line perpendicular to the Earth's orbit about the Sun. This pronounced axial tilt produces the *seasons, governed at each place by the annual variation in the amount of heat received from the Sun. The Earth moves in an almost circular path averaging 149.6 million kilometres (93.5 million miles) from the Sun, taking 365.25 days to complete one orbit. It has one natural satellite, the *Moon.

From space the Earth appears as a bluish globe with dense cloud formations and brilliant polar ice caps. The rapid rotation of the planet produces complex weather patterns in the atmosphere, seen from space as cloud bands and huge rotating cloud patterns. Seventy per cent of the surface is covered with vast oceans of water, the continental land masses making up the other 30 per cent. The continents are rafts of light rock 35 km (20 miles) thick floating on plastic underlying rock, that is, rock that can be deformed under pressure. The ocean basins are plates 6 km (4 miles) thick constantly forming at mid-ocean ridges and being consumed elsewhere under continental plates, a process described by *tectonics. This ocean floor spreading leads to the slow movement of the continents known as continental drift. The surface of the Earth is not covered with the ancient craters which all the *inner planets once had, attesting to the relative youth of the surface and the severe weathering effects acting on it. The Earth's atmosphere consists of 78 per cent nitrogen, 21 per cent oxygen, and with trace gases making up the other 1 per cent. This life-supporting atmosphere is quite a recent development. Extensive ancient volcanic activity produced vast amounts of steam, which condensed to form the oceans, and carbon dioxide, later incorporated into rocks. This allowed an atmosphere of nitrogen and oxygen to develop. The Earth's surface is slowly being blanketed by dust from the disintegration of interplanetary debris which collides with the Earth's atmosphere to produce *meteors. A few bodies reach the ground as *meteorites. The fluid metallic core of the Earth, carried round by the daily rotation, forms a magnetic field which interacts with the *solar wind to form the Earth's *magnetosphere. This envelope round the Earth forms a tail away from the Sun and shields the planet from much harmful cosmic radiation. Atomic particles entering the tail are accelerated and dumped into the Earth's atmosphere near the poles to produce the *aurora. In space around the Earth are the two *Van Allen radiation belts, two giant doughnut-shaped envelopes of trapped charged particles at heights of about 3,000 and 22,000 km (2,000 and 14,000 miles).

It is the vast concentration of surface water at just the right temperature which makes the Earth such a haven for life. It is unlikely that these conditions are unique to Earth. There should be thousands if not millions of similar life-sustaining planets in the universe.

Earth, structure of

The Earth's interior

The inner regions of the Earth consist of concentric layers, each different from the others in size, density, and chemistry.

Crust (between 6 and 40 km thick); consists of the continents and ocean basins

Combined thickness of 2,900 km: Upper mantle; Lower mantle

Liquid outer core; 2,200 km thick

Inner core; 2,500 km in diameter

earthshine, the faint glow illuminating the dark part of the lunar disc when the Moon is at its thin crescent *phase near new moon. It is sometimes referred to as the 'old moon in the new moon's arms', and is caused by sunlight reflected from the Earth to the Moon and back again. Its brightness is partly controlled by the *albedo of the Earth which depends on the amount of cloud cover, particularly over the Pacific Ocean.

eccentricity *ellipse.

eclipse, the total or partial disappearance of one celestial object behind another, or within the shadow cast by the other body. Eclipses of the Sun (**solar eclipses**) occur when the Sun is partially or totally hidden by the new moon coming between the Earth and the Sun; eclipses of the Moon occur when the full moon enters the shadow cast by the Earth. The shadow cast by the Moon on the Earth during an eclipse has a dark central area of total shadow, known as the **umbra**, surrounded by a partly shaded region, called the **penumbra**. Because the angular radius of the Earth's

When seen from space, as in this satellite image, the **Earth** could well be described as 'the blue planet' by members of an extraterrestrial civilization. They would be able to infer that the blue areas are water, the white polar caps are ice, and that the Earth has an atmosphere which supports large, white clouds of water vapour. By examining the spectrum of light reflected from the Earth, they may also be able to deduce that the atmosphere is mainly nitrogen and oxygen.

Eclipse

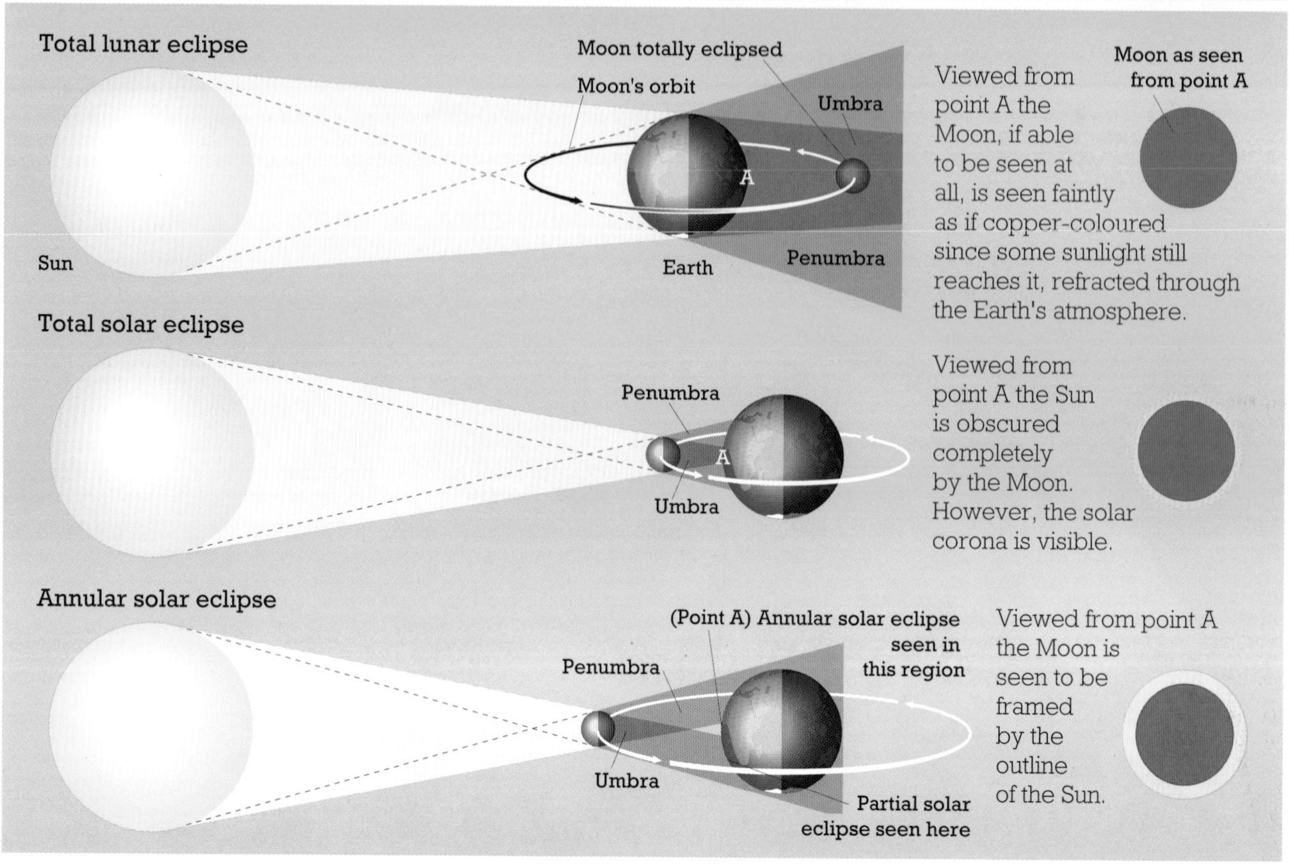

Total lunar eclipse

Moon totally eclipsed
Moon's orbit
Umbra
A
Sun
Earth
Penumbra
Moon as seen from point A
Viewed from point A the Moon, if able to be seen at all, is seen faintly as if copper-coloured since some sunlight still reaches it, refracted through the Earth's atmosphere.

Total solar eclipse

Penumbra
A
Umbra
Viewed from point A the Sun is obscured completely by the Moon. However, the solar corona is visible.

Annular solar eclipse

(Point A) Annular solar eclipse seen in this region
Penumbra
Umbra
Partial solar eclipse seen here
Viewed from point A the Moon is seen to be framed by the outline of the Sun.

conical shadow (the umbra) at the Moon's distance is almost two-and-a-half times the Moon's angular diameter, the Moon at favourable times can spend as long as an hour and forty minutes totally immersed in the shadow. This is the maximum duration of a total lunar eclipse. If the full moon does not enter the shadow completely, then a partial lunar eclipse is experienced by anyone on the night side of the Earth as illustrated in the figure.

The normally invisible delicate corona of the Sun is revealed during a total **eclipse** of the Sun. The ray-like structure is due to the alignment of gas along the Sun's magnetic field. Instruments such as the coronagraph enable the corona to be studied during the times between total eclipses.

It happens that, by chance, the Moon's and the Sun's angular diameter are almost the same so that the new moon cannot hide the Sun for more than a few minutes. Indeed if the Moon is at *apogee, the Moon's angular size is slightly smaller than that of the Sun. At such times an eclipse of the Sun may still be produced if the line joining the centres of the Sun and Moon intersects the Earth's surface. Within a few hundred kilometres radius of the point an observer may see the dark lunar disc surrounded by a brilliant ring of light. Such an event is termed an **annular** solar eclipse. If, however, the angular size of the Moon exceeds that of the Sun, and if the line joining their centres intersects the Earth's surface, an observer at or near that point of intersection will see a **total eclipse** of the Sun. Because the Earth is rotating, and the Moon is revolving round the Earth while the Earth revolves about the Sun, the point of intersection moves rapidly across the Earth's surface from the place where the line of centres first meets the Earth's surface to the place where it leaves. This point, and the region within which a total or annular eclipse of the Sun is visible, sweeps out a narrow band called the *path of totality or annularity. Surrounding it is a wider region within which an observer will see a partial solar eclipse with only part of the Sun's disc being hidden by the Moon. The time of occurrence, the duration, and the exact circumstances of a solar or lunar eclipse depend upon the changing geometry of the Earth–Moon–Sun system. The inclination of the Moon's orbital plane to the *ecliptic means that a solar or lunar eclipse will not occur at every new and full moon. Comparison of observation with prediction provides an opportunity to check up on the theory of the Moon's motion. Because the geometry almost exactly repeats itself after an interval of 18 years and 10 days, an eclipse is almost exactly repeated after this interval of time, called the *Saros. Ancient eclipse records also provide an opportunity to check upon theories of how *tides affect the Moon's orbit.

The four large Galilean satellites of *Jupiter also exhibit eclipses by Jupiter at regular intervals. In 1675 the Danish astronomer *Roemer deduced the finite velocity of light by comparing the observed times of eclipse with those predicted by theory. In a number of binary star systems, such as Algol, the line of sight from the observer to the system lies so close to the orbital plane of one component star about the other that each star suffers partial or total eclipse at regular intervals. The resulting periodic diminution in the light of the apparently single star reveals its double or binary nature.

eclipsing binary, a close *binary star whose orbital plane is nearly in the line of sight, so that one component passes in front of the other, causing the *magnitude to vary in a period equal to that of the orbital motion. Primary minima occur when the component of lower surface brightness passes in front of the one of higher surface brightness. Secondary minima may also be observed, when the brighter component passes in front of the fainter, but the change in magnitude may be too slight to detect. The *eclipses, which are more strictly *occultations, may be total, annular (transits), or partial. Over 5,000 eclipsing binaries are known. The three main types are *Algol stars, type EA; *Beta Lyrae stars, type EB; and W Ursae Majoris stars, type EW.

ecliptic, the great circle on the *celestial sphere giving the apparent annual path traced out by the Sun against the stellar background. The two intersections of the ecliptic with the celestial equator define the *First Point of Aries and the *First Point of Libra. When the Sun reaches the former, spring begins in the northern hemisphere and autumn in the southern hemisphere; when it reaches the latter, autumn begins in the northern hemisphere and spring in the southern. The planets and the Moon are always found within a few degrees of the ecliptic. Because of this the ancients attached great significance to it, dividing it into the houses of the *zodiac. The angle between the planes of the ecliptic and the equator is called the **obliquity of the ecliptic** and has a value of 23 degrees 26 minutes at the present era. Because of the gravitational influences of the other planets the value changes slowly with time. The ecliptic corresponds to the plane of the Earth's orbit about the Sun.

Eddington, Sir Arthur Stanley (1882–1944), British astronomer, physicist, and mathematician who applied atomic theory to problems of stellar constitution and evolution. In doing so he discovered the great importance of *radiation pressure in defining stellar structure. He studied stellar motion and the dynamics of pulsating stars, and suggested that spiral nebulae were in fact galaxies like the Milky Way. He became a leading exponent of *Einstein's general theory of relativity and in 1919 obtained experimental support for it by observing the bending of light rays by the Sun during an eclipse.

effective temperature, a measure of the total power radiated by a star per unit area of its surface. It is a uniform way of describing *stellar temperature and is defined as the temperature of the *black body which, by *Stefan's Law, would radiate the same power per unit surface area as the star itself. It is a useful concept because it enables observed stellar properties to be compared with the predictions of *stellar evolution theory (which refers primarily to the stellar interior) without reference to the details of the stellar atmosphere. Although the *spectrum of a star is always very different in detail from that of a black body, the effective temperature is nevertheless indicative of the gas temperatures in the outer parts of a stellar *photosphere, and *Wien's Displacement Law can be used to determine at what wavelengths the stellar emission is strongest. For example, the Sun has an effective temperature of 5,780 K and a maximum emission at a wavelength of 5×10^{-7} m.

Effelsberg, the site near Bonn in Germany of the world's largest fully steerable *radio telescope, completed in 1971. The dish is built in a valley where the surrounding hills offer natural shielding against artificial radio interference. The dish is 100 m (328 feet) across and yet can operate at wavelengths as short as 1 cm. This impressive accuracy is achieved by a special design technique called homology. The dish is flexible but is supported by a rigid double pyramid structure. As the dish is tipped it deforms from one parabolic shape to another and by moving the radio receivers to follow the changing position of the *focus the optimum performance is maintained.

Egyptian astronomy Egypt, for all the antiquity of its civilization and brilliant night skies, does not seem to have reached the level of ability shown by the Chinese and the

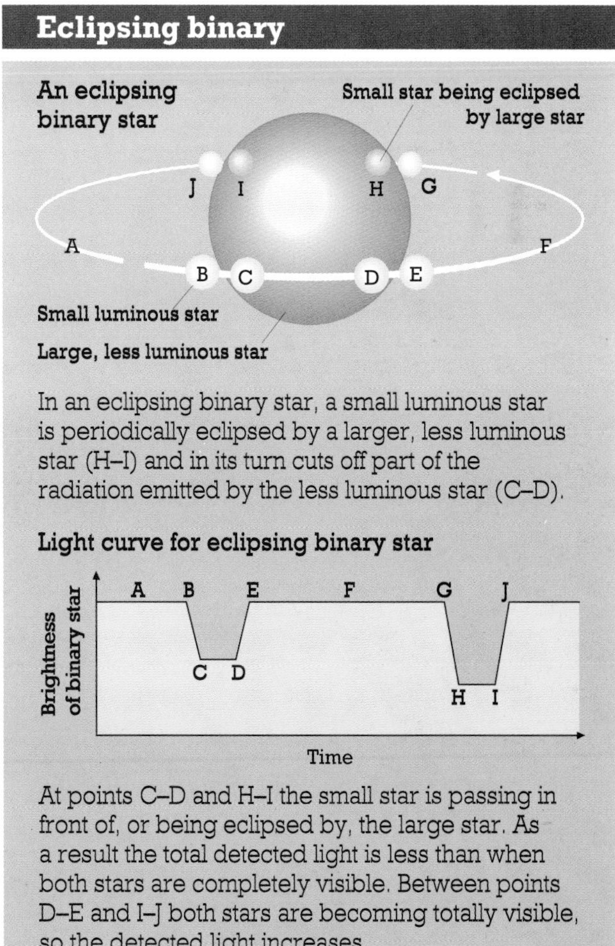

In an eclipsing binary star, a small luminous star is periodically eclipsed by a larger, less luminous star (H–I) and in its turn cuts off part of the radiation emitted by the less luminous star (C–D).

At points C–D and H–I the small star is passing in front of, or being eclipsed by, the large star. As a result the total detected light is less than when both stars are completely visible. Between points D–E and I–J both stars are becoming totally visible, so the detected light increases.

Sumero-Akkadian peoples in understanding, recording, and predicting the movements of the Sun, Moon, and planets. The Egyptians, living in the strip of land bordering the River Nile, led lives bound up in the river's yearly cycle of activity. The flooding of the fields, the sowing of seeds, and the growth and harvest of crops led their learned class to become practical surveyors, builders, geometers, and timekeepers. In their early astronomical work they used a lunar *calendar. They also divided the year into three parts: the inundation, from July to November; the season of sowing and growth from November to March; and the harvest season from March to July. It was discovered quite early on in their civilization that at their capital, Memphis, the Nile began to rise at the heliacal rising of Sirius, the brightest star in the sky, that is, at a time when Sirius first became visible in the morning twilight above the eastern horizon. The Egyptians therefore took Sothis (as it was called) to be a divine star responsible for or signalling the Nile's rising. The great Sothis period of 1,460 years, used by the Egyptian priests, is a consequence of the fact that, whereas the heliacal rising of Sirius occurs every 365.25 days, the calendar they used contained 365 days. Thus the heliacal rising of Sirius occurred one day earlier every four years with a return to the same calendar date in 1,460 years.

Various gods and goddesses were linked to particular stars with Isis, the goddess of agriculture and fertility, identified with Sothis. The sun-god Ra represented the Sun; Osiris, the god of the dead, was represented in the constellation of Orion and there was certainly a system of constellation figures, some of which are quite different from the ones with which we are familiar. For example, they had a hippopotamus, a crocodile, and a sparrow-hawk. Lists of stars painted on the lids of mummy cases give ten-day periods of the month, each under the protection of a deity. The use of such lists possibly enabled the hour of night to be found. It would seem then that as far as Egyptian astronomy is concerned, its cultural environment was not a suitable one in which a scientific flowering of the subject could take place.

The **ecliptic**, the apparent path of the Sun, Moon, and planets across the sky, passes through the zodiacal constellations, interpreted here as their mythical characters. Although the ecliptic appears to be a wavy line, it is a great circle when seen against the celestial sphere.

Einstein, Albert (1879–1955), German-born US physicist who developed the special and general theories of *relativity. After graduating from the Swiss Federal Polytechnic, Einstein worked in the Swiss Patent Office in Berne. In 1905 he published three remarkable papers: one on molecular

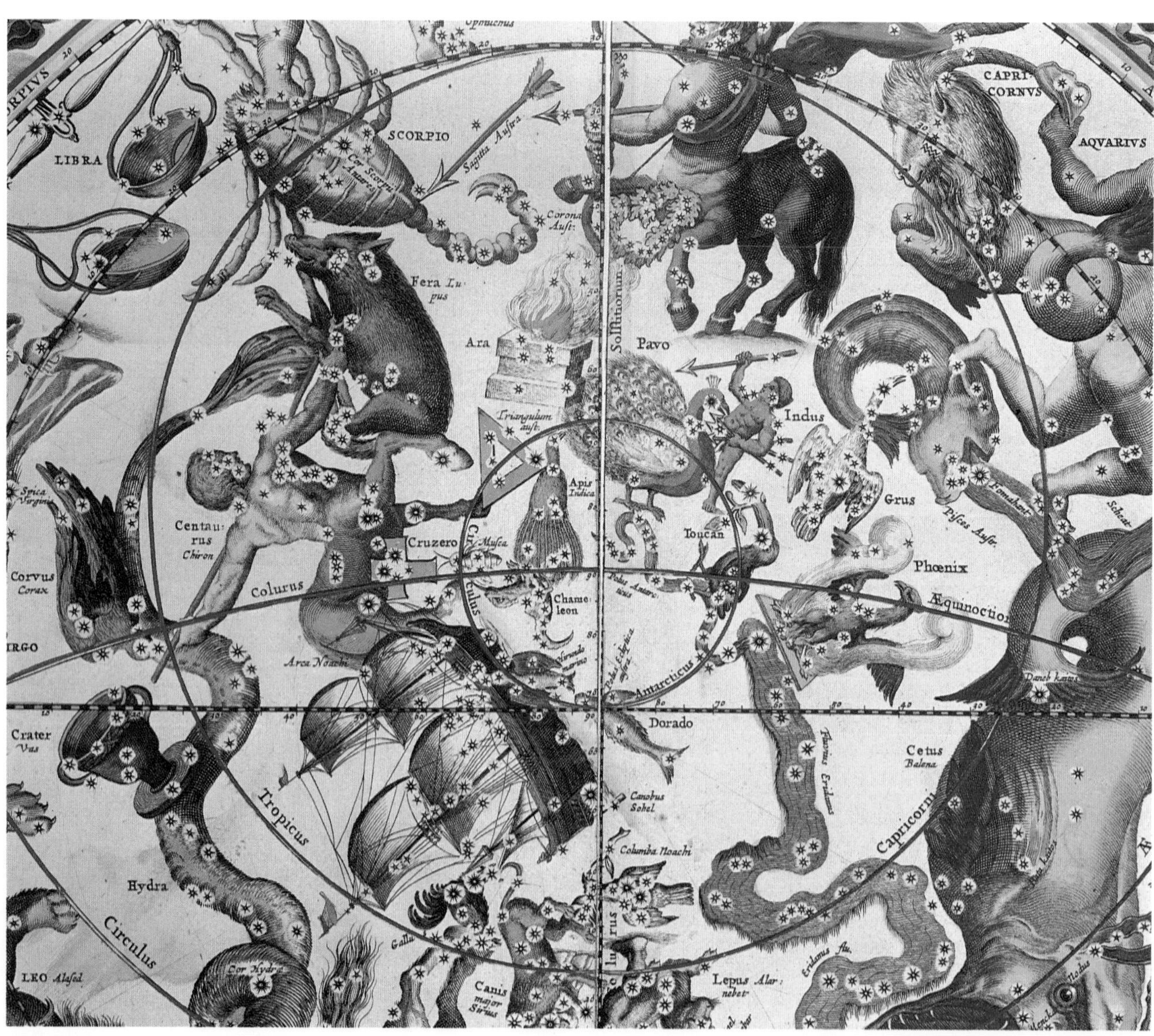

Electromagnetic radiation

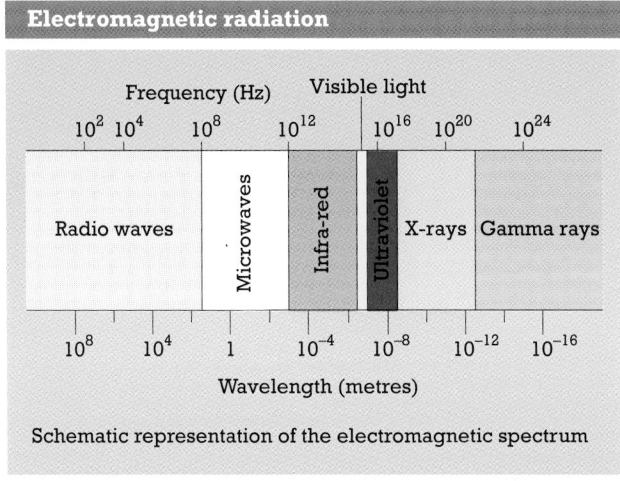

Schematic representation of the electromagnetic spectrum

motion, one on the *photon nature of light, and one which described the special theory of relativity. For the next ten years he worked continuously to extend his theory of relativity to include gravitation and in 1915 published his general theory of relativity. The first independent verification of general relativity was obtained in 1919 when the bending of light was observed during an eclipse. Einstein also made important contributions to quantum mechanics. However, he was unable to accept as final the probabilistic description of physics which quantum theory involved. His political and religious beliefs made it necessary for him to emigrate from Germany in 1933 to settle in the USA where he remained for the rest of his life. During his later years he attempted, without success, to develop a unified theory of electromagnetism and gravity. He was awarded the Nobel Prize for Physics in 1921.

Einstein–de Sitter universe *cosmology.

ejecta, the material thrown out from an impact *crater. When the incoming *meteoroid strikes the surface of the planet or satellite, the surface rock is momentarily converted into a fluid-like material that is pushed both downwards and outwards in a steady flow as the explosive excavation proceeds. The rock of the surface and the meteoroid are fragmented and lumps and particles of the material are ejected from the crater that is formed. The speed of ejection depends upon the angle of ejection. Low angles, that is, nearly tangential to the surface, have low velocities and the material falls close to the crater. Material ejected at higher angles has a higher velocity and travels further. Most of the ejecta is deposited as a blanket that extends out to between two and four times the diameter of the crater. Some falls back into the crater and if the gravity is weak enough some escapes into space.

Elara, a small satellite of Jupiter. It has a diameter of 76 km and revolves about the planet at a mean distance of 11,737,000 km in a highly eccentric *orbit in a period of 260 days. Elara was discovered in 1905 and is one of four Jovian moons orbiting the planet at approximately the same distance.

electromagnetic radiation When electrically charged bodies are accelerated they produce electromagnetic radiation which travels in a vacuum at the *velocity of light. The **electromagnetic spectrum** consists of the enormous range of *wavelengths and frequencies that can be present in electromagnetic radiation; the product of wavelength and frequency is equal to the velocity of light in a vacuum which is approximately 3×10^8 m/s. The radiation possesses both electric and magnetic properties. As the radiation travels, it creates electric and magnetic disturbances which will produce various effects on materials the radiation falls on. The electromagnetic spectrum is classified in broad regions known as the radio, microwave, infra-red, visible, ultraviolet, X-ray, and gamma-ray regions. Most of the electromagnetic spectrum, apart from two ranges of wavelengths, the visible and the short radio wavelength 'windows', are prevented by the Earth's atmosphere from reaching the Earth's surface. Until *radio telescopes were developed, almost all our information about the universe came from visible light, collected by telescopes and analysed by instruments such as *spectrometers. With the advent of artificial satellites and spacecraft the whole electromagnetic spectrum of radiation from the universe is now accessible for study.

element, one of the 105 or so different kinds of atoms that make up the matter of the universe. Hydrogen is the most abundant, comprising almost two-thirds of all the mass in the universe, with helium contributing almost one-third. The elements carbon, oxygen, nitrogen, and silicon account for almost all of the remaining 2 per cent with all other chemical elements making up the remainder.

The **element of an *orbit** is one of the six quantities that describe the size, shape, and orientation of an orbit and enable the position of the body performing that orbit to be found using suitable formulae.

elements, the abundance of The different elements that make up the composition of the nearby stars and the interstellar clouds of gas and dust are the same. This pattern, with certain easily explicable exceptions, is also seen in the planets and minor bodies of the solar system. All seem to have been made out of the same mixture of hydrogen and helium that was initially produced by the *Big Bang but to which have been added elements of higher atomic number generated by the nuclear transmutations that take place in stellar interiors and supernovae. Hydrogen is the most abundant element. The age of the universe and the violence of the majority of the processes that act in it have resulted in everything becoming well mixed. The relative proportions (abundance) of the elements can be determined by studying the chemical composition of the solar atmosphere and of certain primitive meteorites known as carbonaceous *chondrites, which appear to have been made out of very ancient, undifferentiated solar system material. Stellar and planetary material contains eighty-three stable chemical elements. For every silicon atom there are 3,180 hydrogen, 221 helium, 21.5 oxygen, 11.8 carbon, 3.7 nitrogen, 3.4 neon, 1.06 magnesium, 0.83 iron, and 0.5 sodium atoms.

ellipse, a closed curve, one of the three conic sections, the other two being the parabola and hyperbola. The *orbit of a planet about the Sun, a satellite about its planets, and the components of a binary star closely resemble ellipses. In an elliptic planetary orbit the Sun S is at one **focus** of the ellipse, the radius vector SP joining the Sun to a planet P moving according to *Kepler's Laws. The size of an ellipse is usually given in astronomy as the size of the ellipse's semi-major axis. The **major axis** is the longest diameter A_1A of the ellipse. The **minor axis** is the shortest, B_1B. The shape of an ellipse is defined by its **eccentricity** e, where

$e = CS/CA$. For ellipses, e takes values from 0 (when the ellipse is a circle) to very near unity (when it is very elongated). If $e = 1$, the ellipse has turned into a parabola.

ellipsoid, a three-dimensional surface obtained by uniformly stretching or compressing a sphere along three perpendicular axes (the principal axes), in general by different amounts. Cutting an ellipsoid with a plane yields an *ellipse. The surface shape of a rotating celestial body is ellipsoidal when the rigidity of the material is negligible compared to its gravity. In this case, if the rotation is slow enough and all other bodies are far away, the ellipsoid is a flattened spheroid with a bulge at the equator arising from centrifugal force. A spheroid approaches the shape of a sphere when the spin period becomes very long. Natural *satellites deformed by the tidal force of their planet (the same force that gives rise to *tides) are ellipsoidal, with the longest axis (the tidal bulge) pointing approximately towards the centre of the deforming body.

elliptical galaxy *galaxy.

elongation, the angle between the direction of a celestial object and the direction of the Sun measured with respect to the Earth. It is usually used to express the angular distance between the Sun and either the Moon or a planet. Inferior planets, that is, Mercury and Venus, stay relatively close to the Sun in the sky. Mercury has a maximum elongation of between 18 degrees and 28 degrees, the range being due to its eccentric orbit, and Venus has a maximum elongation of 47 degrees. Conjunction, quadrature, and opposition are *planetary configurations that occur when the elongation is 0 degrees, 90 degrees, and 180 degrees, respectively. Mercury and Venus are most conveniently observed at or near greatest elongation. When they are visible in the evening they are said to be at **eastern elongation**; when they are visible in the morning they are said to be at **western elongation**.

Elysium Planitia, a volcanic lava flood plain on Mars, north of the volcano Elysium Mons. The source vents have become hidden by the multitude of individual lava flows that have built up the region. Ash and water have also fashioned the plain. Small volcanoes and the rims of old impact craters just manage to break through the lava surface.

emission spectrum *spectrum.

Ellipse

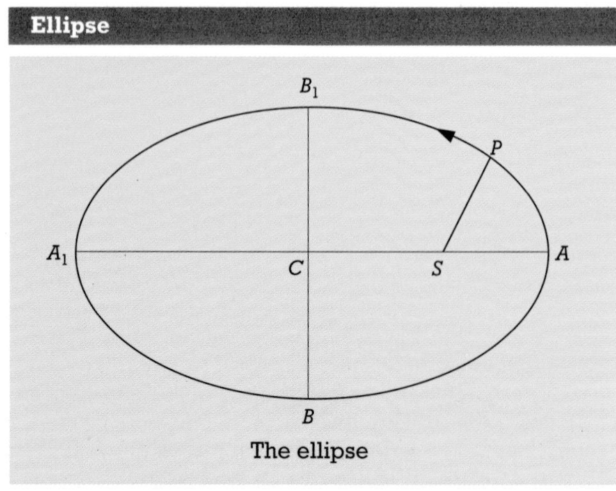

The ellipse

This mosaic of **Enceladus** was made from images taken by Voyager 2 in August 1981. Some regions of Enceladus are heavily cratered, whereas others are smooth, suggesting that in the relatively recent past Enceladus was geologically active and had a heated liquid interior.

Enceladus, a satellite of Saturn, discovered by William *Herschel in 1789. It has a nearly circular orbit above Saturn's equator at 238,000 km from the planet's centre. The orbit of Enceladus is near the tenuous E ring of Saturn. Enceladus is nearly spherical, with a diameter of about 500 km. The mean density of 1,000 kg/m^3 implies that the main constituent is water ice. The albedo, close to 1.0, is higher than that of freshly fallen snow so the surface of Enceladus is the brightest in the entire solar system. The surface shows clear evidence of large-scale activity in the satellite interior with a wide diversity of terrains. Heavily cratered zones are cut by a global system of curved ridges, up to 1 km in height, intermixed with lightly cratered smooth plains. Many craters are flattened indicating that the surface has slowly moved; the source of the heat in the satellite's interior is unknown.

Encke's Comet, a *comet with the shortest known orbital period of 3.31 years. Because of the unusual non-parabolic orbit, early attempts to calculate the elements of its *orbit met with great difficulty. By 1818 the German scientist Karl Gauss had developed a method for calculating the orbits of asteroids and German astronomer Johann Franz Encke (1791–1865) applied this method to observations that Jean-Louis Pons had made of a comet in November and December of that year. As the period was found to be so small Encke concluded that the comet must have been seen before. In fact it had been observed by, for example, Pierre Méchain in 1786 and the British astronomer Caroline Herschel in 1795. It has now been seen fifty-four times. It is one of the few comets named not after the discoverer, but after the astronomer who accurately predicted its return.

energy, the ability to perform work. Sources of energy used by mankind are usually directly or indirectly derived from sunlight, the main exception being nuclear energy. The burning of coal, oil, and wood, and the use of water falling from a higher to a lower place can be used directly to release energy or to drive various mechanical and electrical installations to produce electricity. Coal, oil, and wood were created from living organisms whose lives depended upon the energy in sunlight to fuel the processes building their structures. The weather within the Earth's atmosphere together with the heating of the world's oceans to evaporate vast quantities of water produce lakes above sea-level. Energy in the last case exemplifies the concept of *potential and *kinetic energy. While the water in the lake is above sea-level it has potential energy, energy it possesses by virtue of its position within the Earth's gravitational field. Once released from the lake and allowed to gather momentum by flowing downhill under gravity, its potential energy is transformed to kinetic energy, that is, energy possessed because of the water's momentum. The water then drives a water-wheel or turbine, losing kinetic energy to these man-made devices. Windmills also indirectly use solar energy in its effect upon weather. The use of the tides in estuaries, however, employs the movement of tidal waters caused by the Moon's attraction on the oceans of the Earth.

In a system of mutually gravitating bodies isolated in space, a close approximation being the solar system, the total energy is a constant, being the sum of the potential and kinetic energies. Each body has its own amount of energy, an amount that is not constant but varies as the body's velocity and position in the system change. Similarly the total momentum of the system remains constant but the momentum of each of the bodies within it changes because of their changing velocities as described by *Newton's Laws of Motion.

ephemeris (pl. ephemerides), a table giving the positions of a celestial object at various future times. Thus an ephemeris of the Moon would provide its future geocentric right ascension and declination in the equatorial system of coordinates or its geocentric *ecliptic latitude and longitude in the ecliptic system at future times separated by one half day. The *Nautical Almanac*, published by the *Royal Greenwich Observatory, is a set of ephemerides of Sun, Moon, planets, and other celestial objects published yearly some years in advance.

epicycle *Ptolemaic system.

Epimetheus, a small irregular satellite of Saturn with an average diameter of 60 km. The radius of the orbit is only 2.52 times larger than the radius of Saturn. It is a *co-orbital satellite with *Janus, their orbits being within 50 km of each other. Janus and Epimetheus are thought to be the remnants of a large satellite that was fragmented by an impact. This occurred early in the history of the solar system and the entire surface of the satellite is heavily cratered.

equation of time, the difference in time between the *transit across the observer's *meridian of the actual Sun and the *mean sun. The actual Sun moves along the *ecliptic at a varying rate, this change in speed being due to the fact that the Earth orbits the Sun in an elliptical path. When the Earth is at perihelion (about 4 January) it is moving at its fastest and when it is at aphelion (about 4 July) it is moving at its slowest. The mean sun is a fictitious object used to define mean time and moves along the celestial equator at a uniform rate. Actual time is indicated by the actual Sun and can be, for example, measured by a *sundial. During the course of the year the equation of time varies between −14.2 minutes and +16.3 minutes.

equator, the great circle on the surface of a planet, satellite, star, or *celestial sphere that is perpendicular to the axis of rotation. A great circle is the line on the surface of a sphere produced by an intersecting plane that passes through the centre of the sphere. The equator divides the sphere into northern and southern hemispheres. All points on the equator are equidistant from the *poles.

equatorial coordinate system *celestial sphere.

equilibrium, a state of balance of a system in which the forces operating do not produce any change in the system's state. Equilibrium can be **static** or **dynamic**; in the former case the parts of the system are at rest; in the latter they may be moving but cause no overall change in the system's state as, for example, the molecules in a gas. Equilibrium can also be neutral, stable, or unstable. A ball at rest on a plane horizontal surface is in **neutral** equilibrium since, when displaced to another position on the surface, it is equally in equilibrium in its new state. A ball at the bottom of a surface curved upwards (a bowl) is in **stable** equilibrium since, if displaced, it returns to its former position. If, however, the ball is balanced on the highest point of a surface curved downwards (a dome) it is in **unstable** equilibrium since, if slightly displaced, it moves further and faster from that position. Equilibrium, or lack of it, is an important property in many kinds of astronomical situation such as the structure of a star, the atmosphere and interior of a planet or satellite, nuclear fission and nuclear fusion, stellar motion, celestial mechanics and so on.

equinox, one of two instances of time when the centre of the Sun crosses the celestial equator on its annual journey around the *ecliptic on the *celestial sphere. These crossings occur around 21 March and 23 September each year and are related to the *seasons. For inhabitants of the northern hemisphere, the point where the Sun crosses from south to north is called the **vernal** (**spring**) **equinox** and heralds the official beginning of spring; the other crossing point is called the **autumnal equinox**. The names are reversed for the southern hemisphere. On the equinoctial days the hours of daylight and of darkness are very nearly equal. The vernal equinox is the zero point for both the equatorial and the ecliptic coordinate systems. This point is not fixed but is slowly moving westward at the rate of about 50 arc seconds per year. This movement is due to the *precession of the Earth's spin axis and it will take the equinox 25,800 years to make one revolution of the equator. In the days of *Hipparchus, during the 2nd century BC, the vernal equinox was in Aries; it is now in Pisces. For this reason the vernal equinox is still called the *First Point of Aries. Similarly the autumnal equinox is also called the *First Point of Libra.

Equuleus, the Little Horse, the second-smallest *constellation in the sky; it lies next to *Pegasus, just north of the celestial equator. Equuleus was among the forty-eight constellations listed by the Greek astronomer Ptolemy in his **Almagest*. Its brightest stars are of only fourth magnitude.

Eratosthenes (*c*.275–*c*.194 BC), Greek scholar of many disciplines, chiefly remembered for his estimate of the size of the Earth. From observations of the Sun's *meridian altitude at Syene (modern Aswan) and Alexandria he derived a value for the Earth's circumference accurate to within a few per cent. None of his writings survive, but they were used by Ptolemy (the author of the **Almagest*) and the geographer Strabo. A lunar crater, 58 km (36 miles) in diameter, is named after him.

Eridanus, the River, a large *constellation that meanders from the celestial equator deep into the southern hemisphere of the sky. At its southernmost end lies its brightest star, *Achernar. Eridanus was one of the forty-eight constellations known to the ancient Greeks, and was often identified with the River Po in Italy. Epsilon Eridani is a fourth-magnitude dwarf star similar to the Sun, 3.3 parsecs away. Omicron-2 Eridani, also known as 40 Eridani, is a remarkable triple star consisting of a Sun-like yellow dwarf, a red dwarf, and the most easily observable *white dwarf in the sky.

Eros, an *asteroid discovered by the German astronomer Georg Witt and one of the first of the *Amor group to be found. Eros orbits the Sun and spends most of the time inside the orbit of Mars, and actually passes close to the Earth. In 1975 Eros occulted the star Kappa Geminorum A and the three-second duration of the *occultation indicated that Eros has a diameter of 12–23 km. Studies of the colour and polarization of the light reflected from its surface indicate that the asteroid is too small to have a soil-like regolith covering. The rotation period of 5.27 hours has been measured by noting how the strength of the reflected radar signal varies as the asteroid rotates.

ESA *European Space Agency.

escape velocity, the minimum *velocity that an object, such as a spacecraft or atmospheric molecule, needs to escape from the gravitational field of a massive body. This velocity has a value of $\sqrt{(2GM/R)}$, where G is the constant of gravitation, M is the mass of the body, and R is the distance between the object and the centre of the body. The escape velocities from the surfaces of the Earth and Moon are 11.2 and 2.4 km/s, respectively. The low value for the Moon explains why it has lost all its atmosphere.

ether, a hypothetical substance formerly supposed to pervade the whole of space and be the medium in which light travelled. The **Michelson–Morley experiment**, which used interferometry to detect the presence of this luminiferous ether and to measure the movement of the Earth with respect to the ether, has led to the abandonment of this hypothesis and the acceptance that electromagnetic radiation is a wave motion without a medium.

Europa, a *Galilean satellite of Jupiter, discovered by Galileo in 1610. It has a nearly circular orbit in the plane of Jupiter's equator at 670,900 km from the planet's centre. Its diameter is 3,138 km and its density of 2,970 kg/m^3 implies that the interior is composed of mainly rocky materials. *Voyager images of Europa show a very bright and smooth icy surface with few impact craters, implying that the surface is being, or has recently been, renewed. The most characteristic feature of Europa's surface is an intricate system of dark lines, possibly resulting from *tectonic stresses. Pits, mounds, domes, and low ridges have also been identified. The thickness of Europa's ice crust is unknown but it is possible that a liquid layer lies beneath.

European Southern Observatory (ESO), a large *observatory in the Atacama Desert in Chile financed by Belgium, Denmark, France, Germany, Italy, The Netherlands, Sweden, and Switzerland. Located under clear skies, the ESO has more telescopes and a greater variety of instruments than any other observatory. The ESO's dozen or so telescopes include a Schmidt *reflecting telescope which has surveyed the southern sky, 2.2 m (87-inch) and 3.6 m (142-inch) reflecting telescopes, and jointly run instruments such as the Swedish–ESO submillimetre telescope for radio astronomy. A second 3.6 m (142-inch) 'New Technology telescope' was commissioned in 1989. It incorporates a thin primary mirror mounted on computer-controlled pistons which actively correct for optical distortions resulting from mirror flexure and misalignment. A 'Very Large telescope' to comprise four 8 m (315-inch) reflectors is under construction. Technical and other support is co-ordinated at the European headquarters near Munich. ESO also sponsors scientific meetings and publishes a major research journal.

European Space Agency (ESA), an organization run initially by a group of European nations to provide for and promote collaboration in space research and technology. ESA has enabled Europe to launch its own spacecraft. It has its headquarters in Paris, its technical centre in The Netherlands, its satellite operations centre in Germany, and its data centre in Italy. Though ESA officially began to operate in May 1976, it was formed out of a partial merger of the European Space Research Organization (ESRO), founded in 1964, and the European Launcher Development Organization (ELDO) which was concerned with the construction and launching of large rockets capable of putting European artificial satellites into orbit. The latter has been subsumed into Arianespace, an organization created in 1980 to take responsibility for the Ariane Launcher, the European rocket used to put satellites into orbit. Astronomically, ESA has been responsible for the launching of satellites such as the International Ultraviolet Explorer (IUE), an ultraviolet telescope and spectrometer; Cos-B, a gamma-ray detector; TD-1, an ultraviolet, X-ray, and gamma-ray detector; *Giotto, the planetary probe that intercepted Halley's Comet; and *Hipparcos, a satellite that measures stellar distances.

evection, a perturbation of the Moon's motion arising from the variation in its orbital eccentricity caused by the Sun's gravitational attraction. The effect, which has a period of 31.8 days, is one of many perturbations that must be taken into account when developing a *lunar theory which seeks to describe the Moon's motion precisely.

evening star *Venus.

event horizon, the boundary surrounding a *black hole beyond which the black hole is invisible to an external observer. At the event horizon, which is at the *Schwarzschild radius, the *escape velocity from the black hole equals the velocity of light with the consequence that all electromagnetic radiation coming from it is trapped. The presence of the black hole can still be detected by its powerful gravitational influence.

Evershed effect, the flow of gas outwards from the edge of the darker central area (the umbra) of a *sunspot to the lighter outer area (the penumbra). The gas flow is observed spectroscopically, the *Doppler effect on the spectral lines indicating speeds of about 2 km/s.

evolution, in astronomy, the change in an object with time. For example, *stellar evolution describes the successive forms taken by the structure of a star from its formation out of the interstellar material to its final state. In *cosmology, some models of the universe are static and non-evolving as in the *Steady State theory, or are evolving, as in the currently accepted *Big Bang theory where the universe began with an explosion.

exobiology *extraterrestrial life.

expanding universe *cosmology.

extinction, the reduction of radiation from an observed object after the radiation has passed through an intervening absorbing body. For example, the Earth's atmosphere gives extinction to all stellar brightness measurements. Corrections have to be made for this according to the elevation of the star above the horizon. Starlight may also suffer extinction by interstellar dust.

extragalactic nebula, an obsolete term for a *galaxy. The term was used when such objects were suspected of being outside our own galaxy, but were not definitely known to be so remote. In fact galaxies are not *nebulae at all. The most famous example is M31, which was known as the Great Nebula in Andromeda but is now called the *Andromeda Galaxy.

extraterrestrial life, life-forms that may have evolved on other planets. There is no hard evidence at present that life exists other than on the Earth; most *UFOs have been satisfactorily explained as being natural or man-made, and the *Viking missions to Mars were inconclusive in testing for the existence of life on that planet. Nevertheless, searches have been and are being made for signs that life has arisen in other parts of the universe. Certain knowledge either that life is confined to planet Earth or has been found elsewhere would have the profoundest philosophical implications for mankind. Factors contributing to any assessment of the probability that life exists elsewhere must include the size, age, and structure of the universe, and the conditions under which life as we know it can originate and evolve. Other factors of relevance in the search for extraterrestrial life include an assessment of the probability that intelligence leading to scientific and technological civilizations similar to our own may arise. In the known universe there are approximately 10^{10} galaxies. Most contain about 10^{11} stars. Only one star, the Sun, is definitely known to have planets, though there is some evidence that bodies of planetary size orbit some of the nearer stars. The universe is about 15 billion years old, the Earth is about 4.5 billion years old, and life has been on our planet for 3 billion years, though creatures resembling modern man have only existed for some 4 million years. Our present civilization of science and technology is only 300 years old. We also know that in general there is a sameness about the universe, in that everywhere the same kinds of bodies, made up of the same elements and in much the same proportions, are acted upon by the same forces, gravitational, electromagnetic, nuclear, and so on. The probability that the events leading to life on Earth have never been duplicated elsewhere is accepted by many as being very small indeed.

Scientists have attempted to duplicate, in laboratories, the conditions they believe existed on Earth just before life appeared. Passing ultraviolet radiation and electric discharges through sterilized flasks containing mixtures of carbon dioxide, water vapour, ammonia, and methane, they have synthesized the amino acids and purine and pyrimidine compounds that comprise the nucleic acids, the building-blocks of living cells. The implication is that somehow the first living cells evolved on the primitive Earth from such conditions. Even if the amino acids did not evolve on Earth, there is evidence from a study of meteorites such as the Murchison meteorite that amino acids can arrive from outside the Earth. Again the discovery that complicated molecules such as water, ammonia, and formaldehyde exist in the interstellar gas strengthens the probability that life is not confined to our planet. For these reasons several lines of research to detect the existence of extraterrestrial life are being pursued and names such as **exobiology** or **astrobiology** are used to cover such researches. The name of *SETI (search for extraterrestrial intelligence) is used for searches confined to finding intelligent species elsewhere in the universe.

F

faculae, bright regions on the *Sun's surface which usually form before sunspots appear. After the spots have disappeared, faculae can persist in the same region for some weeks. Faculae are hotter than their surroundings. In addition, there is a slight strengthening of the *magnetic field in those areas.

false colour, the use of different colours in the image of an object to indicate different properties of that object. For example, the variation of temperature over the surface of a planet can be represented by assigning different colours to each range of temperatures. Similarly, the types of substances distributed on it, or the wavelengths of the electromagnetic radiation coming from it, can also be represented in this way. It is therefore a false image but can contain a great deal of information in a readily understandable form.

Faye's Comet, a long-period *comet, discovered in 1843, that was captured by Jupiter into a typical short-period orbit. It was discovered shortly after it had a close approach to Jupiter, and has now been seen eighteen times. It has an orbital period of 7.34 years. The solar heating of Faye's spinning dirty snowball nucleus is such that the gas and dust are not emitted from the subsolar point, that is, where it is noon on the comet, but from a point in the afternoon section. The jet effect caused by this emission changes the comet's orbit slightly. Furthermore, the direction of the jet is thought to be changing because of the *precession of the comet's spin axis causing further changes in its orbit.

The Sun's surface is in a state of continual change, as this image of **faculae** taken with a spectroheliograph shows. Solar activity is also manifested in flares, sunspots, and prominences.

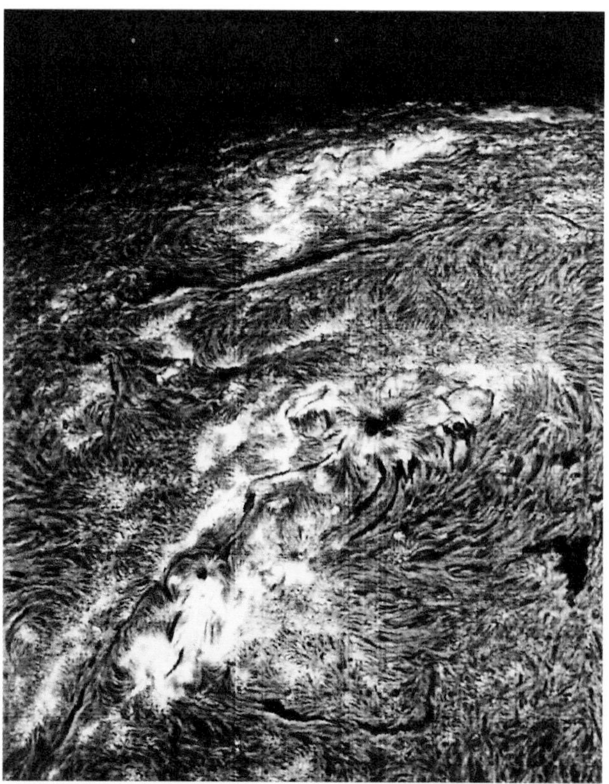

FG Sagittae, a rapidly evolving star in the constellation of Sagitta. The stages of *stellar evolution are frequently measured in millions of years but FG Sagittae is in a very rapid phase of its evolution, visibly changing over a few decades. It is a supergiant star nearing the end of its life and is surrounded by a shell of gas which it ejected perhaps 6,000 years ago. This shell now forms a *planetary nebula about a third of a parsec (1 light-year) across and is expanding at 34 km/s. Over the last forty years FG Sagittae has changed colour from blue to yellow, indicating that it has cooled from 12,000 K to 5,000 K, and in that time has increased in size by a factor of six; it is now some 60 times the diameter of the Sun. These dramatic changes probably began about a hundred years ago, and it is expected that the star will eventually shrink and heat up again, and that it will end its life as a white dwarf. Astronomers are therefore watching it carefully in order to increase their understanding of this phase of stellar evolution.

field equations *relativity, general theory of.

field of view, the range of angles of incoming electromagnetic radiation that an instrument can detect simultaneously. It is a general term used for all kinds of astronomical telescopes and subsidiary analysing equipment. For an optical telescope it represents the angular spread in the sky that can be viewed. For example, this is of the order of 5 degrees for binoculars.

filament, a slender thread-like feature on the surface of the Sun. Filaments take different forms depending on their origin. For example, the penumbra of *sunspots shows filaments that radiate away from the central umbra. These are due to the *Evershed effect by which gas flows from the umbra to the penumbra. Filaments may also be *prominences that are seen from above and so their loop-like structure is not evident.

filar micrometer, an instrument for measuring the angular separation of double stars or the angular diameter of a planet, satellite, or other celestial object. It consists of a set of fine cross-wires attached to a telescope. These cross-wires can be moved by a micrometer screw gauge. The instrument may previously have been calibrated on objects of known angular diameter.

fireball, a very bright meteor. Some tens of thousands of these enter the Earth's atmosphere every year. Although they are often brighter than magnitude −10 they are usually small pebble-sized objects.

The **primordial fireball** is the explosive beginning of our universe as described by the *Big Bang theory.

First Point of Aries, the point of intersection of the *celestial equator and the *ecliptic where the Sun crosses the equator from south to north. The Sun reaches the First Point of Aries about 21 March, the time of the vernal (spring) *equinox. Over the past 2,000 years the First Point of Aries has moved westwards along the ecliptic from Aries to its present location in Pisces by *precession at a rate of some 50 arc seconds per year.

First Point of Libra, the point of intersection of the *celestial equator and the *ecliptic where the Sun crosses the equator from north to south. The Sun reaches the First Point of Libra about 21 September, the time of the autumnal *equinox. Because of the *precession of the equinoxes the First Point of Libra slips westwards along the ecliptic at a rate of some 50 arc seconds per year.

first quarter *phase.

Fitzgerald contraction *Lorentz–Fitzgerald contraction

Flamsteed, John (1646–1719), the first director of the *Royal Greenwich Observatory and the first *Astronomer Royal. The observatory was founded as a result of a report from him to the Royal Society. Flamsteed had a passion for astronomical precision. His early work concentrated on compiling lunar and solar tables and in developing theories to explain the apparent motion of the Moon and Sun. He was the first to set the *equation of time on a sound footing. Between 1676 and 1689 he made 20,000 observations at Greenwich with his great sextant, which was accurate to better than 10 arc seconds. He devised a meridian *transit timing method to give the accurate right ascension of forty reference stars. His observations were published posthumously as *Historia Coelestis Britannica* (1725) and *Atlas Coelestis* (1729). His *Historia Coelestis Britannica* star catalogue gave the positions of 3,000 stars much more accurately than previous works and assigned numbers to them. Some stars, such as 61 Cygni, are still known by their Flamsteed numbers.

Flamsteed numbers, a sequence of numbers used to identify stars in a *constellation. Examples are 61 Cygni and 70 Ophiuchi. They are allocated in order of right ascension to stars that are listed in a catalogue called *Historia Coelestis Britannica*, compiled at Greenwich by *Flamsteed, the first Astronomer Royal. However, the stars were not actually numbered by Flamsteed himself; the numbers were added after Flamsteed's death by Joseph Jerome le Français de Lalande (1732–1807) in a French edition of the catalogue published in 1783.

flare, a sudden disturbance in the Sun's upper *chromosphere in which a considerable release of energy heats the surrounding material, resulting in a brilliant flame-like cloud of gas. The increase to maximum brightness may be accompanied by the ejection of electrified particles such as protons, ions, and electrons. These particles travel at speeds sufficient to impinge upon the Earth's atmosphere. This heats the atmosphere and in addition causes geomagnetic storms and *aurorae.

flare star, a star that exhibits sudden transient increases in its total brightness. Many stars, and particularly cool *main-sequence stars of which the Sun is one, are flare stars. The irregularity and short duration of these stellar flares, and the exceedingly strong magnetic fields exhibited by some stars, suggests the phenomenon is akin to a solar *flare. However, while the Sun's total brightness is enhanced by only 0.01 per cent during a solar flare, some flare stars more than double in brightness. These energetic outbursts last only a few minutes. Flare stars may have wholly convective interiors and this characteristic, together with the complex and strong magnetic fields, is believed to be the cause of the eruptions in these stars.

flash spectrum, the spectrum of the Sun's atmosphere taken during a solar *eclipse at the moment when only a thin crescent of the Sun, consisting of the solar atmosphere, is still visible. The pattern of bright crescent images, spread throughout the visible range of radiation in the spectrum, reveals the elements present in the solar atmosphere. Observable only for a few seconds as the Moon moves in front of the Sun's atmosphere, the flash spectrum was first observed by the US astronomer Charles Young during the eclipse of 1870.

flocculi, bright patches in the *chromosphere of the Sun. In early solar studies, flocculi were thought to be regions of incandescent calcium vapour. However, the term has subsequently been applied to a number of different features of the Sun's surface including *plages and *filaments.

focal length *focus.

focal ratio *focus.

focus (pl. foci), in geometry, one of usually two points such that the sum or difference of their distances from any point of a given curve or solid surface is constant. For example, the *ellipse and the parabola, as well as their three-dimensional counterparts, the *ellipsoid and the paraboloid, all have foci. The shape of a planet's orbit around the Sun is an ellipse with the Sun at one of the foci.

In a *telescope the focus is the point at which rays of electromagnetic radiation meet after reflection or refraction. In an optical telescope light from a celestial body is collected by the objective lens or primary mirror and an image is formed at the focus where it can be magnified by the eyepiece for viewing or for analysis by other instruments. The distance between the lens or mirror and its focus is called its **focal length**. The **focal ratio** of a lens or mirror is found by dividing its focal length by its diameter (aperture). Some radio telescopes use a paraboloidal *dish which reflects incident radio waves on to an aerial (US antenna) at the geometrical focus of the paraboloid.

Fomalhaut, a first-magnitude star in the constellation of Piscis Austrinus, and also known as Alpha Piscis Austrini. It is a hot *dwarf star at a distance of 7 parsecs, with about thirteen times the luminosity of the Sun. Infra-red radiation detected by the IRAS satellite in 1983 is due to a disc of cool material that may be forming planets.

forbidden line, a *spectral line found in some astronomical spectra, but which is not present under ordinary terrestrial conditions where an excited atom would collide with other atoms before it had a chance to radiate energy away. In some nebulae, however, the density of the dust and gas is so low that collisions are infrequent and the atom has time to make a transition between its excited state and a lower state. In doing so it produces a 'forbidden' line in the material's spectrum.

force, the physical agency that changes the *velocity of an object. Quite often an object is acted on by several forces simultaneously and a change in velocity occurs only if the net force is not equal to zero. The second of *Newton's Laws of Motion defines the net force acting on an object of constant *mass as being the product of that mass and the acceleration that is induced in it. There are two major classes of force. Contact forces act between objects that are touching;

The **Fornax** cluster of galaxies, some 18 million parsecs away, is one of the nearest clusters of galaxies and is therefore regularly studied. This photograph clearly shows spiral galaxies, which contain many young blue stars. It also shows elliptical galaxies, which are composed mainly of old, yellowish stars.

for example, the force exerted by a gas on the walls of its container is the result of the collisions between the gas molecules and the walls, and the *momentum they lose during that collision. (**Pressure** is defined as being the force per unit area.) Action-at-a-distance forces, such as *gravity, act through empty space; so does the electromagnetic force between electrically charged particles. The forces between atomic particles also fall into this category but are usually effective only over very short atomic distances.

Fornax, the Furnace, a *constellation of the southern sky introduced by the 18th-century French astronomer Nicolas Louis de Lacaille. Its brightest star is only of fourth magnitude and its main claim to fame is that it contains a dwarf galaxy (the *Fornax System) of our *Local Group of galaxies, and the Fornax cluster, an important galaxy cluster.

Fornax System, a dwarf galaxy in the constellation of Fornax. A member of the *Local Group of galaxies it is an elliptically shaped system lying at a distance of some 200,000 parsecs and contains a number of globular *clusters. The Fornax galaxy itself is of ninth magnitude while the brightest globular cluster in it is of thirteenth magnitude.

Foucault's knife-edge test, a means of detecting errors in the shape of an optical surface such as a concave mirror. By illuminating the surface under test, an image is formed of a small bright pinhole object. To the eye positioned just behind this image, the whole optical surface appears bright because rays are converging from every part of the surface towards the image (and therefore the eye) position. For a correctly shaped surface, these rays all focus to a point-like image. If a knife edge is passed through this perfect image, the entire optical surface will dim uniformly as all rays are cut simultaneously. If the image is imperfect, some rays will be cut before others, and the effect is to reveal in exaggerated relief those regions where the optical surface deviates from the desired shape. These zones can then be corrected by additional polishing. The test was invented by the French physicist Léon Foucault in the 1850s, and enabled him to construct excellent *reflecting telescopes.

Foucault's pendulum, an experiment, devised in 1851 by the French physicist Léon Foucault (1819–68), that demonstrates the rotation of the Earth. After watching a rod vibrate as it was turned in a lathe, Foucault realized that a pendulum would swing in a fixed plane in space unless an additional force caused it to deviate. The pendulum in a grandfather clock is constrained to swing parallel to the clock case. However, to an observer on Earth, a freely suspended pendulum will appear to swing in a plane that rotates as the Earth turns under it. For a pendulum at the north or south pole this rotation is at the same rate as the Earth's but in the opposite direction. At lower latitudes, the rotation is slower because the pendulum is also being carried around by the turning Earth. Foucault first demonstrated the Earth's rotation with pendulums in his cellar. He

Foucault's pendulum

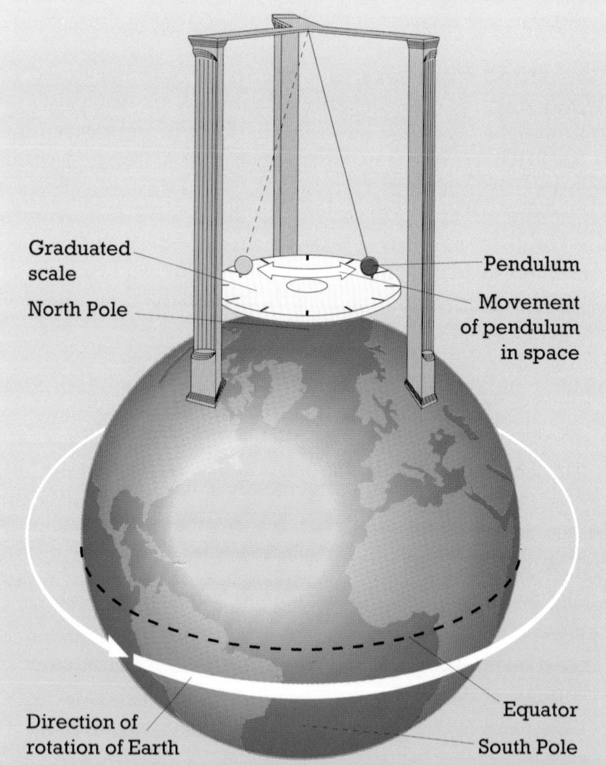

If a pendulum is set swinging, over time the vertical plane in which it swings appears to rotate with reference to a graduated scale directly underneath it. During a twenty-four hour period, the plane in which the pendulum swings rotates once above the scale. In actual fact this plane is fixed in space and it is the Earth that is rotating below the pendulum.

The Apollo 14 lunar mission landed at the site marked by the arrow, in the **Fra Mauro** region of the Moon. The Fra Mauro formation, the fine lumpy material extending across the picture from the lower edge, is the ejecta from the Imbrium Basin just over the horizon.

repeated his demonstration in the Paris Observatory, and, with a giant 67-m (220-foot) long pendulum, in the Panthéon in Paris. Foucault's pendulum caused a world-wide sensation: no one in 1851 had doubted that the Earth turned (the daily motion of the stars was proof of that), but here was a demonstration of the Earth's rotation which could be made without referring to the external universe, and by which inhabitants of a cloudy world could determine if their planet turned.

Fra Mauro formation, a mountainous equatorial region in the Moon's *Oceanus Procellarum. Fra Mauro is actually part of the *ejecta blanket that was scattered over the lunar surface by the impact that produced the *Imbrium Basin. The region is covered with small craters and contains ridges and furrows that radiate from the centre of the distant Imbrium Basin. It is named after the adjacent crater and was visited by Apollo 14, the first manned mission to a region other than a *mare.

Fraunhofer line, a dark *spectral line that can be seen in the *spectrum of the light from the Sun. Fraunhofer lines were first noted by William Wollaston in 1802, but were independently recorded and more fully investigated by Joseph Fraunhofer (1787–1826), a famous Bavarian glass-maker and optical instrument manufacturer. He used them, in the main, as fixed-wavelength points at which he could measure the refractive indices of different sorts of glass. He denoted the most striking lines by the letters of the alphabet, from A, which approximately marked the red end of the spectrum, to I which marked the violet end. By 1834 the British physicist David Brewster had correctly suggested that the lines were produced by atomic absorption in the upper (cooler) parts of the solar atmosphere. The most prominent lines are due to neutral hydrogen, sodium, and magnesium, and to singly ionized calcium. Many of the weaker lines are produced by iron.

Friedmann, Aleksandr Aleksandrovich (1888–1925), Russian mathematician who offered an alternative development of Einstein's general theory of *relativity. Friedmann spent most of his working life at the University of St Petersburg where his work included various fields of mathematics and physics. Einstein had originally calculated that the universe was unchanging, but Friedmann produced different solutions to his field equation which showed that the density and radius of the universe could change with time; in other words, that the universe was continuously expanding or contracting.

full moon *phase.

galactic equator, the intersection of the plane containing the disc of the *Galaxy with the *celestial sphere. It is inclined at about 63 degrees to the celestial equator and is used as a reference plane in a *coordinate system to locate objects within the Galaxy. Such objects are referred to by their galactic longitude and galactic latitude.

galactic halo, the large spherical volume of the *Galaxy enclosing the more luminous disc. It contains old, faint *Population II stars and globular *clusters, as well as the outer regions of the galactic magnetic field, electromagnetic radiation, cosmic rays, and tenuous, highly ionized gas at a temperature of about 10^6 K. This gas, possibly heated by *supernova explosions in the disc, rises into the halo, cools, and then falls back to the disc in a so-called **galactic fountain**. The stars and clusters in the galactic halo do not have any general systematic motion, but travel in randomly orientated orbits, and from the vantage point of the Earth often have large *radial velocities. Many other galaxies are thought to have a halo and analysis of the rotation of the discs of *spiral galaxies suggests that our own and other galaxies may contain as much as ten times more material than is accounted for by known constituents, this **dark matter** being presumed to reside in galactic haloes; it may be that most of the mass of a galaxy is of an as yet unknown nature.

galactic latitude, the angle of an object on the *celestial sphere measured along an arc perpendicular to the *galactic equator. Galactic latitude is measured from 0 degrees corresponding to the galactic equator to 90 degrees north or south of the galactic equator corresponding to the north and south *galactic poles.

Abell 1060, a large cluster of **galaxies** in the constellation of Hydra, is a strong X-ray source. The X-rays do not originate from a single galaxy but are probably emitted by hot gas filling the whole of the central region of the cluster. One of the members of the cluster may be ploughing through this hot gas and being stripped of its relatively cool gas.

galactic longitude, the angle of an object on the *celestial sphere, measured eastwards along the *galactic equator from the galactic centre to the point where an arc through the celestial object drawn perpendicular to the galactic equator meets it. The galactic centre has equatorial coordinates: right ascension 17 hours 45.6 minutes and declination 28 degrees 56.3 minutes south.

galactic pole, one of two points at *galactic latitude 90 degrees. The north galactic pole lies in the constellation of Coma Berenices with equatorial coordinates: right ascension 12 hours 51.4 minutes, declination 27 degrees 7.7 minutes north. The south galactic pole lies in the constellation of Sculptor with coordinates: right ascension 0 hours 51.4 minutes, declination 27 degrees 7.7 minutes south.

galactic rotation, the rotation of a *galaxy about an axis within it. The *proper motions of stars in our own galaxy and *stellar dynamics and kinematics, show that stars, clusters, and molecular clouds all orbit about the galactic centre under *Newton's Law of Gravitation. Spectroscopic studies of nearby galaxies like the Andromeda Galaxy reveal that they too are in rotation. Study of the orbits of stars in the Sun's neighbourhood shows that they can be circular, slightly elliptical, or highly elliptical. All of the stars within the central bulge of a galaxy like our own have almost the same period of revolution about the galactic centre, a consequence of the fact that they orbit within an almost spherical mass of stars. Only those stars outside the bulge have periods of revolution described roughly by *Kepler's Laws.

galaxy, an association of stars, dust, and gas, with a total mass ranging from 10^6 to 10^{13} times the mass of the Sun. The Milky Way is our own galaxy, and the Sun is only one star of the 100 billion stars in it. The true character of galaxies was not discovered until the 1920s when the very intense debate concerning their nature was finally resolved. Telescopes prior to this period showed them as diffuse areas of light, resembling *nebulae, but the 100-inch (2.5 m) reflector at the Mount Wilson Observatory, first used in the 1920s, gave images of some individual stars in the Andromeda Galaxy showing it to be a galaxy rather than a nebula. This telescope was also used by *Hubble to measure the periods of *Cepheid variable stars in the same galaxy. These variable stars are sufficiently well understood to enable their distances to be accurately measured. Hence the Andromeda Galaxy was shown to be at a distance of about 700,000 parsecs (2.25 million light-years), well outside our own galaxy. However, individual stars can only be seen in the nearest galaxies. The distances of remote galaxies can sometimes be measured by their *red shift using the *Hubble constant.

There are many types of galaxy, the principal ones being spiral and elliptical galaxies, and there have been several attempts to classify them using strict lettering and numbering schemes such as the *Hubble classification. However, many types fall outside the scope of these schemes and are named after the astronomer who first listed them as a distinct type (such as Seyfert and Markarian galaxies) or letters from classification schemes (for example, N- and cD-type galaxies). Galaxies with no obvious form are classified as irregular. All galaxies contain stars, gas, and dust, but in very

These four photographs show various aspects of the structure of a **galaxy**. (1) NGC 3031 (M81) is a spiral galaxy and is part of a nearby group of galaxies in the constellation of Ursa Major. (2) NGC 4565, in the Virgo Cluster of galaxies, is a spiral galaxy seen edge-on; the disc and the central bulge are both clearly visible. (3) The elliptical galaxy NGC 4472 (M49) is the largest galaxy in the Virgo Cluster. (4) NGC 175 is a good example of a barred spiral galaxy.

different proportions, and even within one galaxy the distribution of these constituents may vary enormously. Many have a distinct nucleus or central concentration of material which may be active, that is, emitting large amounts of energy, and even exploding. Some emit jets of material at speeds close to the velocity of light. The more modern branches of astronomy such as *radio astronomy and *XUV astronomy have caused attention to be focused on these active galaxies in recent years and it is only by combining information from all parts of the spectrum that they are being understood. Although galaxies contain enormous amounts of matter by terrestrial standards, they usually occur in even larger groupings or **clusters**, with the smaller clusters having only a few members but larger clusters may contain several thousand galaxies. However, the origin and evolution of galaxies is still not fully understood.

*Spiral galaxies are probably the best-known type of galaxy. The term is used to describe galaxies which have an obvious spiral structure, such as the Andromeda Galaxy. Our own galaxy is believed to be a spiral. Their appearance is similar to that of a Catherine wheel with the majority of the stars and luminous material forming the spiral arms. The spiral arms also contain an interstellar medium of dust and neutral hydrogen gas; the latter is now extensively studied using its *twenty-one centimetre line which is only accessible to radio astronomers. The masses of almost all spiral galaxies fall in the range 10^9 to 3×10^{11} times the mass of the Sun. **Elliptical galaxies** are also a common type of galaxy. They appear as ellipses on astronomical photographs, with no evidence of spiral structure. The classification covers a wide range of sizes, from small dwarf ellipticals with masses of only a few million times that of the Sun to giant ellipticals and **cD-type galaxies** with masses of 10^{13} times that of the Sun. Most of the matter in an elliptical galaxy is in the form of stars and hot gas. The largest known galaxies are cD-type galaxies. These are massive elliptical galaxies, and occur at the centre of a few of the largest clusters of galaxies. They are distinguished by having a large nucleus, or they may have several nuclei in rapid motion with respect to each other, but all within an extensive envelope. Often they are fairly strong radio sources. *Markarian galaxies were first classified by the Soviet astronomer Beniamin Markarian in 1967. They have very strong emission in the ultraviolet part of the spectrum. Their spectrum is similar to that of *HII regions within our own galaxy. An **N-type galaxy** has a star-like nucleus and a very faint haze around it. These galaxies are usually strong radio sources, and it has been suggested that they evolve into *quasars. On an astronomical photograph, a *Seyfert galaxy looks like a spiral galaxy except for its nucleus, which is very much brighter than in a normal spiral. Seyfert galaxies are also distinguished by their spectra which show broad bright emission lines, implying the presence of a large amount of a rapidly moving hot gas in the nucleus. This type was discovered by the US astronomer Carl Seyfert in 1943.

Galaxies that can be observed optically and which are also strong radio sources are called *radio galaxies. They include Seyfert, cD- and N-type galaxies, as well as quasars.

Galaxy, the *galaxy to which the Sun, and therefore the solar system, belongs. A star system containing at least 100 billion stars, it has three main regions: the central bulge, the disc, and the *galactic halo. The bulge consists of old *Population II stars and may even have a *black hole at its centre (the nucleus), responsible for the energetic processes seen there. With interstellar matter, the bulge is some 6,000 parsecs (20,000 light-years) in diameter in the plane of the disc and 3,000 parsecs (10,000 light-years) in diameter at right angles to the disc. The disc, containing young *Population I stars, interstellar material, open (or galactic) *clusters, and spiral arms, is of diameter 30,000 parsecs (100,000 light-years) but only 1,000 parsecs (3,000 light-years) in thickness. The halo, roughly spherical and concentric with the disc and the bulge, consists of stars, most of which are members of globular clusters. These stars are old Population II stars.

The Sun is about two-thirds of the way from the disc centre to the edge; its position in the plane of the disc (the galactic equator) ensures that the disc stars seen from the Earth are concentrated into a narrow band, the **Milky Way**, seen right round the celestial sphere and inclined at about 63 degrees to the equator. The Milky Way is uneven in brightness, being brightest in the constellations of Cygnus and Aquila in the northern hemisphere and in Scorpius and Sagittarius in the southern hemisphere. The galactic centre lies in Sagittarius but is impossible to observe in visible light because of the dark *nebulae obscuring the light from the many stars beyond them. Since most of these dark nebulae lie near to or in the Galaxy's equatorial plane, they form conspicuous gaps in the Milky Way. Radio and infra-red radiation can penetrate this galactic 'smog' enabling astronomers to map out not only the galactic centre but also the structure of the disc. A strong radio source, *Sagittarius A, lies in the centre. The hydrogen scattered through the disc emits radiation in the radio region of the electromagnetic spectrum at the *twenty-one centimetre line and studies of its distribution have enabled the spiral nature of the disc to be mapped. It is in the spiral arms that new stars are still being formed.

Galilean satellites, the four satellites of Jupiter named *Io, *Europa, *Ganymede, and *Callisto, first observed by Galileo in 1610. They have nearly circular orbits almost directly in the plane of Jupiter's equator. The orbital periods of three of the satellites, Io, Europa, and Ganymede, have a precise relationship to each other. As a consequence, the three satellites can never line up on the same side of Jupiter. Moreover, during each orbit, Io is deformed by tides created by Jupiter and this heats and melts Io's interior. The inner satellites are denser and have less ice than the outer ones. The outer satellites have more impact craters and are more geologically stable.

Galileo Galilei (1564–1642), Italian astronomer and physicist who was the first person to observe astronomical objects through a *refracting telescope. This telescope,

The Milky Way, the popular term for the **Galaxy** which we inhabit, appears as a faint cloudy band of light encircling the night sky. This photograph shows a bright region of the Galaxy in the constellation of Aquila.

invented by himself and now known as a Galilean telescope, had a planoconvex objective lens and a planoconcave eyepiece. By the end of 1609 he had developed it to its maximum practical magnifying power of 30. During that year Galileo discovered that the Moon was mountainous and estimated the height of the mountains by the length of the shadows they cast. He also discovered that the Milky Way was actually a band of stars crossing the sky and his discovery that there was a miniature planetary system of four satellites in orbit around Jupiter strengthened his belief in the *Copernican system as it indicated that there were at least four bodies that certainly did not orbit the Earth. His results were published in March 1610 in *Sidereus Nuncius* ('The Starry Messenger'). By late 1610 he had discovered the phases of Venus which provided more evidence for the Copernican system and he observed that Saturn was oval in shape. This observation was due to the rings of Saturn but these could not be resolved in his telescope. He later did work on sunspots and the rotation of the Sun. His *Discorsi e Dimostrazioni Matematiche intorno a due Nuove Scienze* ('Dialogue Concerning Two New Sciences') (1638), a critical examination of the geocentric cosmology as expounded by the Greeks, led to a direct confrontation with the Roman Catholic Church, his inquisitorial trial in 1633, and his subsequent house arrest for the last eight years of his life.

Galileo is considered to be the founder of the experimental method in astronomy. When he faced the Inquisition over his belief in the Copernican system in which the Earth moved about the Sun, it is said that when he was forced to recant, he muttered under his breath 'And yet it does move'.

gamma-ray astronomy, the observation of gamma rays produced by a variety of extraterrestrial events such as matter–antimatter collisions, radioactive decay, *synchrotron radiation, *bremsstrahlung, collisions of high-energy particles, and *plasmas at high temperatures. Gamma rays have wavelengths shorter than X-rays and it is customary to place the boundary between gamma rays and X-rays at *photon energies of one million electron volts. Celestial sources producing gamma rays of low energy have to be observed by scintillation counters or spark chambers in satellites, but high-energy gamma rays penetrate to the Earth's surface and can be measured by ground-based detectors of *Cerenkov radiation. Positron and electron annihilation by collision has been detected in the direction of the galactic centre by the gamma rays produced. It is suggested that the short-lived positrons are created by gamma rays from their interactions with the interstellar gas or that material falling into a massive *black hole results in the production of positron–electron pairs.

Gamov, George (1904–68), Russian-born US physicist noted for his theories on how heat and radiation are generated by stars. After his defection to the West in 1933, Gamov worked on the development of the atom bomb during World War II. In 1948, together with Ralph Alpher and Hans Bethe he published his most famous work which proposed that the early universe consisted entirely of very hot neutrons and their decay products, which then cooled to form nuclei, thus explaining the abundance of helium in the universe. This work became one of the foundations of the *Big Bang theory. Gamov also suggested that all the matter in the universe is in constant rotation.

Ganymede, a *Galilean satellite of Jupiter, discovered by Galileo in 1610. It has a nearly circular orbit in the plane of Jupiter's equator at 1,070 million km from the planet's centre. With a diameter of 5,262 km it is the largest satellite in the whole solar system and is in fact larger than Mercury. Its density of 1,940 kg/m^3 implies that the interior contains both water ice and rocky materials, probably in different regions. *Voyager images show a geologically complex surface displaying heavily cratered zones with bright craters in different stages of preservation; and brighter areas of recent origin with sets of linear, parallel grooves which intersect and overlap. Evidence of *tectonic movement and faulting is common on these younger geological fractures.

Gargantuan Basin, a huge ancient lunar crater. It contains the *Oceanus Procellarum and its diameter of 2,400 km (69 per cent of the Moon's diameter) makes it the largest crater in the solar system. The innermost ring of this basin shows up as a series of *wrinkle ridges in the overlying lava. The basin was initially covered by low-density aluminium (US aluminum) basalts that are poor in iron. These have been coated at a later time by a thin layer of iron-rich lavas. The borders of the Gargantuan Basin are defined to the south by the edges of the Oceanus Procellarum and to the north by the fringes of the Mare Frigoris. It formed around 4,200 million years ago. The *Imbrium Basin is contained within it and was formed there some 500 million years later. The production of the basin could have thinned the lunar crust over a very large region and thus resulted in the disparity between the large amount of lava that flowed over the near side of the Moon compared with that on the far side.

gegenschein, that part of the *zodiacal light seen in the region of the sky opposite to the Sun. The gegenschein (German for **counterglow**) is a faint oval patch of light about 6 degrees by 10 degrees in extent, the long axis being in the *ecliptic. Although it is seen in the opposite direction to the

Sun, that is, looking out of the solar system, it is not (as was once suggested) due to the Earth having a comet-like dust tail.

Gemini, the Twins, a *constellation of the zodiac. In mythology it represents the twins Castor and Pollux, sons of the god Zeus and Queen Leda of Sparta. The two brightest stars in the constellation are named *Castor and *Pollux. Gemini lies on the edge of the Milky Way and contains the large open *cluster M35 (NGC 2168), well seen through binoculars, plus NGC 2392, an eighth-magnitude *planetary nebula popularly known as the Eskimo Nebula. The Sun passes through the constellation from late June to late July each year. The *Geminid meteors appear to radiate from the constellation every December.

Geminids, an active *meteor shower that reaches its peak on 14 December each year, when about 60 meteors per hour can be seen. Meteors from this *radiant (right ascension 7 hours 28 minutes, declination +32 degrees) appear from 7–16 December. The meteoroids responsible for the Geminids are relatively close together in their *meteor stream and have an orbit similar to the asteroid 1983TB. Geminids were first seen in about 1862.

geocentric coordinates, the *coordinate system in which all the positions and velocities of objects are computed with respect to the Earth's centre of mass. It is particularly relevant for the calculation of the orbits of artificial Earth satellites and for Earth-based measurements within the solar system. Early models of the solar system, such as the *Ptolemaic system, were *geocentric systems but the *Copernican system gave a more satisfactory explanation of the movement of the planets.

geocentric system, a theoretical working model of the solar system which has the Earth in a near central position and the planets and Sun in orbit around it. It is an ancient system and was first proposed by Apollonius of Perga in the 3rd century BC. It was subsequently developed by *Hipparchus and reached its most refined form with the analysis of Ptolemy around AD 200. The seven major moving celestial bodies were ordered according to the speed with which they cross the sky. These were, in order from the central Earth, the Moon, Mercury, Venus, the Sun, Mars, Jupiter, and Saturn. This *Ptolemaic system enabled planetary positions to be predicted to an accuracy of about 1 degree of arc. There were, however, considerable complications. Each planet was thought to move around a small epicycle circle, the centre of which travelled along a larger deferent circle; and the Earth was slightly away from the centre of the deferent. The geocentric system remained in favour until the times of Copernicus and the Renaissance when it was gradually replaced by the physically correct *Copernican system.

geodesic *curvature.

geology, planetary *planetology.

geophysics, the study of the physical processes that take place in the interior of the Earth. Using *seismology, data can be obtained by studying the velocity of the P-waves (where the vibration of sound waves is parallel to the direction of propagation) and the S-waves (where the vibration is transverse), and the way in which these waves are affected by crossing the boundaries between the interior regions of the Earth. Geophysics has recently been extended to cover the interiors of the other planets and satellites in the solar system. The systematic exploration of a planet's interior concentrates on mapping the variations of composition, temperature, mass motion, and pressure as a function of depth and age. Seismology can again be used and the density distribution in a planet's interior can be obtained by plotting a satellite's movement in the planet's gravitational field. Geophysics also investigates the ways in which planets generate magnetic fields, surface plate *tectonics, continents, and gravitational anomalies. Mountain building, the occurrence of seismic activity and glaciation, as well as the influence of tidal forces, are also considered. Modern space technology has enabled our detailed understanding of the Earth's geophysics to be compared with our growing knowledge of the conditions inside Mercury, Mars, Venus, and the Moon, as well as the solid satellites of the outer planets.

geostationary orbit, the orbit of an artificial satellite about the Earth such that the point on the Earth's surface directly below the satellite (the subsatellite point) is almost stationary. This greatly simplifies spacecraft-to-Earth communications, a requirement for most telecommunication and many meteorological satellites. To be geostationary, the orbit must be geosynchronous, with low eccentricity and inclination so these orbits are confined to a narrow belt, and traffic problems arise there. Over time the subsatellite point gradually drifts away in the east–west direction because of the perturbations produced by the non-spherical shape of the Earth, and manœuvres with rockets are required to keep the satellite on station.

geosynchronous orbit, the orbit of an artificial satellite about the Earth such that the orbital period is almost exactly one day, so that the satellite remains approximately fixed over a point on the Earth's surface. By the third of *Kepler's Laws, this occurs when the length of the semi-major axis is about 42,160 km. The point on the Earth's surface directly beneath the satellite describes a daily oscillation (called a Lissajous figure). This oscillation is small for the particular case of a *geostationary orbit. Superimposed on this oscillation there is a slow drift from day to day, mainly due to *libration. This drift can be used to determine, for example, the shape of the Earth. Geosynchronous orbits are often used for astronomical and military satellites.

giant planet, one of the *planets Jupiter, Saturn, Uranus, or Neptune. They are described as giants because they have radii that are 11.18, 9.42, 3.84, and 3.93 times that of the Earth. They are also 317.8, 95.1, 14.5, and 17.2 times more massive; in fact Jupiter and Saturn together account for around 92 per cent of the total mass of the planetary system. They orbit the Sun in the outer regions of the solar system, Jupiter and Neptune having mean orbital radii of 5.20 and 30.06 *astronomical units. The Jovian and Neptunian years are 11.86 and 164.79 Earth years, respectively. The giant planets have densities in the range 700–1,700 kg/m^3, these being much lower than those of the inner planets. In the main the giant planets consist of hydrogen and helium, with a relatively low percentage of their mass forming a rock–iron core.

giant star, a star located to the upper right of the main sequence in the *Hertzsprung–Russell diagram. When a *dwarf star has exhausted the hydrogen available in its core for nuclear fusion to helium, it continues to consume the

Giotto

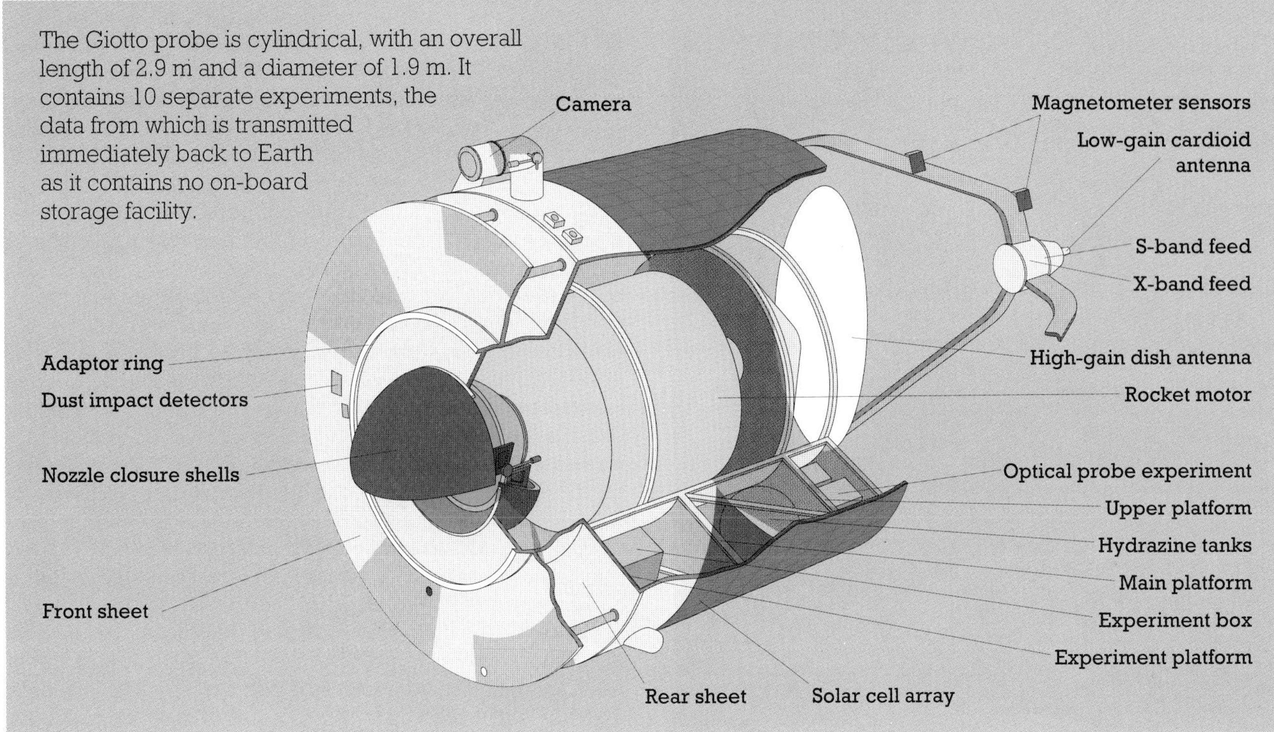

The Giotto probe is cylindrical, with an overall length of 2.9 m and a diameter of 1.9 m. It contains 10 separate experiments, the data from which is transmitted immediately back to Earth as it contains no on-board storage facility.

hydrogen in a shell around the core, and begins to evolve to the giant stage of *stellar evolution, expanding to conserve energy. The masses of typical giant stars in our own galaxy range from two to ten times that of the Sun. Stars with a mass much less than that of the Sun evolve much more slowly, so few of them have yet had time to evolve to the giant stage. The diameters of giant stars are mostly in the range from five to a hundred or more times that of the Sun, and their luminosities range from about twenty-five to a thousand or more times that of the Sun. Many of the bright naked-eye stars are giants, including Aldebaran, Arcturus, Becrux, Capella, and Pollux. *RR Lyrae stars and stars of the *Mira Ceti class are giant variable stars. Stars with luminosity intermediate between the giants and dwarfs are called **subgiants**; those with greater luminosity than giants are *supergiants.

gibbous moon *phase.

Giotto, a spacecraft launched in 1985 by the *European Space Agency to make a close fly-past of *Halley's Comet. The minimum distance of 605 km between the spacecraft and the comet's nucleus was attained on 14 March 1986 when the spacecraft was travelling at 68 km/s relative to the comet. The on-board camera produced our first view of a cometary nucleus and showed it to be a smooth pear shape, 16 km long and 8 km wide. The nucleus is solid, made up of dust and ice, with a dark grey surface which reflects only 6 per cent of the light that falls on it, making it one of the darkest surfaces in the solar system. This darkness is due to the surface being coated with black tar-like organic material and to the crumbly nature of the dust. During the Giotto fly-past only 10 per cent of the surface was actively emitting gas and dust. Dust particles in the vicinity of the comet were found to contain snow. The gas emitted by the comet was estimated to be (by mass) 84 per cent water, 3 per cent formaldehyde, and 3 per cent carbon dioxide, the remainder being made up of small organic molecules. The spacecraft was named after the Italian artist who depicted the appearance of Halley's Comet in AD 1301 as the Star of Bethlehem in the Padua *Adoration of the Magi*.

globular cluster *cluster.

Goodricke, John (1764–86), British amateur astronomer who founded the study of *variable stars. A severe illness in infancy left him a deaf-mute. His astronomical work, in which he collaborated with his neighbour Edward Piggott in York, was confined to the last five years of his short life, and was largely devoted to the study of variable stars visible to the naked eye. He investigated *Algol, and in 1782 discovered its period of 2.87 days, correctly speculating that its variation in magnitude is due to eclipses by a darker object orbiting the star. In 1784 he discovered the variation of *Beta Lyrae and of Delta Cephei (the first of the class of *Cepheid variable stars), and measured their periods. He was awarded the Copley Medal of the Royal Society in 1784 and became a Fellow in 1786 but died two weeks later, aged 21, apparently from pneumonia contracted while observing Delta Cephei.

Gould's Belt, an expanding, disc-like concentration of young *Population I stars which is inclined at an angle of 20 degrees to the plane of the Milky Way. The southern part of Gould's Belt is easily visible to the naked eye as a band of bright, blue stars that stretches from the constellation of Ophiuchus, across the Milky Way, and into Orion, but the Belt is also delineated by molecular clouds, neutral hydrogen, gaseous *nebulae, *stellar associations, and dust. The Belt, also known as the **Local System**, has a diameter of approximately 1,000 parsecs or 3,000 light-years (almost 10 per cent of the radius of the Galaxy). The rate of expansion

The encounter of the spacecraft **Giotto** with Halley's Comet in 1986 was a spectacular success and a wealth of scientific information was transmitted back to Earth for analysis. However, the encounter caused considerable damage to the satellite's superstructure and left some instruments inoperable.

and the ages of the oldest stars in the Belt suggest that the Belt originated 40–80 million years ago. There is as yet no convincing indication of what caused the formation of so large and energetic a system. First clearly recognized by John *Herschel, the Belt is named after the US astronomer Benjamin Apthorp Gould, who investigated the system in the late 19th century.

grand unified theory (GUT), an attempt by cosmologists and physicists to produce a theoretical model of the universe which unifies all four fundamental forces of nature: gravitation, electromagnetism, and the weak and strong nuclear forces, describing their relationship at all ages of the universe. Such a theory would enormously increase our understanding of the universe and its evolution.

granulation, fine markings on the surface of the Sun as if the surface is made up of grains or granules 700 to 1,000 km across. Granulation results from the turbulent and violent convective activity in the Sun's surface material. Each granule is short-lived; after a few minutes the granule has cooled, darkened, and sunk back towards the solar interior.

gravitation, the attraction exercised by every particle of matter on every other. More than 300 years ago *Newton proved three things: that this gravitational acceleration is directly proportional to the masses of the interacting bodies, it decreases as the inverse square of their mutual distance, and is directed along the line joining the centres of mass. The fact that unsupported objects fall to the ground, as well as that we can stand without floating away from the surface of our planet, are the most evident examples of the effects of the gravitational attraction of the Earth on every object on its surface.

Gravity has a peculiar property known as **non-saturation** by which it becomes bigger and bigger as the masses of the interacting objects increase. Thus, although gravity is a relatively weak force (for example, the gravitational attraction between the proton and the electron inside the hydrogen atom is exceedingly small compared to their electrical attraction) the combined effects of the gravitational pull exerted by all the bodies in the universe is the dominating force in *cosmology. The gravitational attraction exerted by a mass is **spherically symmetric**; that is, it depends only on the distance from its centre of mass but not on direction. This is why celestial bodies like the Moon, which are dominated by gravity, tend to be round, whereas smaller less massive bodies which are dominated by, for example, electrical forces can retain their non-spherical shape. In the solar system, minor planets orbiting around the Sun between Mars and Jupiter form a useful laboratory for understanding gravity: the small ones are not much different from large stones, while the big ones are mostly round. For very large masses, small deviations from roundness can be explained by taking into account the proper rotation and the particular constitution of the body.

The motion of the planets around the Sun as well as the satellites around them, can be predicted to an extremely high degree of accuracy on the basis of gravity alone. The centripetal acceleration, which is a consequence of a body's orbital velocity, arises through this gravitational attraction and prevents the bodies from falling towards each other, as non-orbiting unsupported masses would do.

*Newton's Law of Gravitation begins to break down for very strong gravitational forces, and the general theory of *relativity provides a more accurate description.

gravitational waves, tiny wave-like disturbances in the curvature of *space–time. Einstein's general theory of *relativity predicts the existence of gravitational waves travelling from a source at the speed of light. They interact with matter in a manner somewhat similar to that of travelling tidal forces oscillating at right angles to the direction of propagation. Because *gravitation is a much weaker force than the electromagnetic force, the task of building gravitational wave detectors sensitive enough to measure such waves is a very difficult one. The achievement of such sensitivity is the goal of a number of laboratories at the present time using laser *interferometry between freely suspended masses separated by at least 1 km (0.6 miles) and much progress has been made since the pioneering work of the US physicist Joseph Weber in the 1970s. Only extreme astrophysical situations such as sudden stellar collapses, coalescing compact *binary star systems, or rotating *neutron stars are able to produce gravitational radiation strong enough to be detected. Evidence that gravitational radiation exists comes from studies of the binary *pulsar PSR 1913+16 whose orbit is decreasing exactly as predicted by Einstein's theory, the cause being the emission of gravitational radiation. Gravitational wave astronomy is potentially important in its promise of providing a new window on the universe.

gravity, the *force of attraction between two masses of material. The gravitational force was first expressed mathematically by *Newton in his *Principia* (published in 1687). He found that the force F between two bodies of masses m_1 and m_2, these bodies being a distance d apart, was given by $F = Gm_1m_2/d^2$. The constant of proportionality, G, is known as the **gravitational constant.** Newton also proved that a uniformly dense spherical body behaves gravitationally as if its mass were concentrated at a point, known as the **centre of mass** of the body. The gravitational force between the bodies acts along a line joining the two centres of mass.

The force of gravity is the weakest of the known forces but is the only force that acts between neutral bodies, that is, bodies that are not electrically charged. It is thus of extreme importance in astronomy because planets, stars, and galaxies are neutral and their orbits are governed by the gravitational law given above. *Kepler's Laws are a simple consequence of the inverse square relationship.

Newton's Law of Gravitation breaks down in regions where the gravitational force is very strong. In the solar system this was exhibited by an unexplained advancement of the *perihelion of the orbit of Mercury by 43 arc seconds per century. Under these circumstances the general theory of *relativity has to be introduced. This was first expounded by *Einstein in 1915. Here the force of gravity changes the geometry of space from being linear into being curved. Ripples in the surface of the space–time continuum are known as gravitational waves.

Great Bear *Ursa Major.

Great Looped Nebula *Cygnus Loop.

Great Nebula in Orion, an emission *nebula also known as M42, in which new stars are being formed. It lies near the middle star in *Orion's Sword and can, under good seeing conditions, be seen by the naked eye as a fuzzy glow. Even moderate telescopic power reveals the famous *Trapezium stars in its interior and the exquisite veining in white and pink that have made the fan-shaped nebula one of the most observed, drawn, and photographed celestial objects for over a century. It is about 6 parsecs (20 light-years) across and about 400 parsecs (1,200 light-years) distant. Extensive observations have shown that the bright nebula is located on the front of a much larger gas cloud which is composed mainly of dust and neutral hydrogen with traces of many other more complex molecules. Virtually all the known interstellar molecules have been detected in this cloud.

Great Red Spot, a huge oval hurricane system in the outer atmosphere of Jupiter's southern hemisphere at a latitude of 20 degrees south. 14,000 km wide and 24,000–40,000 km long, it has persisted for at least three centuries, being first recorded by the British scientist Robert Hooke in 1664. It is not a fixed feature but drifts westwards at the rate of about 0.5 degrees per day and oscillates relative to the north–south axis by about 1,800 km every 90 days. Infra-red observations indicate that the spot is an anticyclonic high-pressure region that is much colder than its surroundings and elevated about 8 km above the adjacent clouds. The gas at the centre of the spot is moving upwards from the surface of Jupiter at a few millimetres per second. It then flows out to the edge of the spot before descending. The spot is slowly shrinking and is now half the size that it was a century ago. The colour fluctuates slowly from pale pink to bright red; this colour is possibly due to condensed phosphorus.

Greek astronomy The Greeks were, so far as we know, the first people to make planned observations in order to determine the size and structure of the universe. The earliest Greek philosophers, who lived in the Greek colonies of Asia Minor and Italy in the 6th century BC, made various speculations about the organization of the universe, but their theories were not yet based on careful observation. Thales of Miletus is said to have improved the art of navigation by the stars, borrowing ideas from the Phoenicians, and to have used Babylonian tables to forecast an eclipse of the Sun in 585 BC. He argued that the Sun, Moon, and stars were incandescent vapours floating on a watery firmament above the flat Earth. His pupil Anaximander thought the heavenly bodies were holes in tubes of mist containing wheels of fire circling the Earth; eclipses and the Moon's phases were due to temporary closing of the holes. Pythagoras of Samos and his followers thought that the Earth and other heavenly bodies (including the Sun) were spherical, moving in perfect circles about an invisible central fire. A second phase in Greek astronomy was the more systematic teaching at Plato's Academy and Aristotle's Lyceum in Athens. Plato (*c.*429–347 BC) adopted the Pythagorean ideal of circular motion but preferred a *geocentric system; he encouraged astronomical speculation as an exercise to improve the mind, but discouraged observation. Aristotle (384–322 BC), originally a follower of Plato, gradually moved away from this position, but retained a geocentric model. The flowering of Greek observational astronomy came in the third phase, the Hellenistic Age (3rd century BC to 2nd

The red colour of the huge cloud of luminous gas in the **Great Nebula in Orion** is caused by glowing hydrogen. The dark areas are large patches of dense cool dust which can be seen only in silhouette against the background of the glowing gas. The gas and dust will probably condense to form stars, as has already occurred in the brightest regions of the nebula.

century AD), when the main centre of learning was Alexandria. Outstanding astronomers of this period were *Aristarchus, the first known proponent of a *heliocentric system; *Eratosthenes, who determined the size of the Earth; *Hipparchus, probably the greatest astronomer of antiquity; and Ptolemy, whose *Ptolemaic system, described in the *Almagest*, was accepted as essentially correct for well over a thousand years. This was the last achievement of Greek astronomy, as the intellectual decline of Greece under Roman rule led to an interruption in original astronomical research.

greenhouse effect, the heating of a planetary atmosphere by the trapping of infra-red radiation. The greenhouse effect is responsible for the exceedingly high surface temperature of Venus and is also in operation on the Earth and in the lower atmospheres of the *outer planets. Most of the energy radiated by the Sun is in the visual and near infra-red region of the spectrum. This passes easily through the transparent atmosphere and heats the planetary surface. The surface then radiates in the far (long-wavelength) infra-red. Gases like carbon dioxide and *ozone are called greenhouse gases because they are relatively opaque to this radiation and absorb it. Half of the subsequent re-radiation is reflected back towards the surface of the planet, leading to an increase in the surface temperature. The extensive reduction of the amount of carbon dioxide that occurred during the Carboniferous period of the Earth's evolution cut off the greenhouse effect and led to the Permian ice ages. An increase in carbon dioxide would increase the surface temperature which may be sufficient to melt some of the polar ice caps and so raise the sea level.

Greenwich Mean Time (GMT), also known as *Universal Time, the mean solar time in the time zone that is centred on the former *Royal Greenwich Observatory, London. Noon (12.00 hours GMT) occurs when the mean sun crosses the *Greenwich Meridian. Today, with very few exceptions, all the countries of the world keep time within an even hour or half hour of Greenwich. The surface of the Earth is divided approximately along lines of equal longitude into twenty-five time zones, each of these zones being 15 degrees wide except zones +12 and −12 which are 7.5 degrees wide. Therefore local time is 4 minutes behind GMT for every degree of longitude travelled west. The Greenwich Meridian was adopted internationally as the zero of longitude at a conference in Washington in 1884. Its acceptance was facilitated by the overwhelming use of the Greenwich Meridian in navigation and the adoption, in the US and Canada, of time zones based on Greenwich. Originally, different towns in Great Britain kept their own local time, varying according to longitude. In the mid-19th century Greenwich time was adopted by railways throughout Britain for the sake of uniformity. However, it was only in 1880 that Greenwich Mean Time became the legal time throughout Great Britain. The international reference time-scale for civil use is now based on atomic clocks, but is subject to step adjustments (**leap seconds**) to keep it close to mean solar time on the Greenwich Meridian. The formal name of this time-scale is UTC (a language-independent abbreviation of co-ordinated universal time) but it is still widely known as Greenwich Mean Time.

Greenwich Meridian, the *prime meridian on the Earth's surface. It passes through *Airy's transit instrument at the old *Royal Greenwich Observatory, London. This meridian is the circle of zero longitude and it divides east from west. It was accepted as the prime meridian of the world at the Washington Conference of 1884. The international date-line is at 180 degrees longitude from the Greenwich Meridian.

Gregorian calendar, the modified *calendar, also known as the 'New Style', introduced by Pope Gregory XIII in 1582. It is a modification of the *Julian calendar and is now in use throughout most of the Christian world. The Gregorian calendar adapted the Julian calendar to bring it into closer conformity with astronomical data and to correct errors which had accumulated because the Julian year of 365.25 days was 11 minutes 10 seconds too long. Ten days were suppressed in 1582 and, to prevent further displacement, Gregory provided that of the centenary years (1600, 1700, etc.) only those exactly divisible by 400 should be counted as leap years. The Gregorian calendar was adopted in Great Britain in 1752 by which time eleven days needed to be suppressed.

Grus, the Crane, a *constellation of the southern half of the sky. It was introduced by the Dutch navigators Pieter Dirkszoon Keyser and Frederick de Houtman at the end of the 16th century, and represents the long-necked water bird, the crane. Its brightest star is Alpha Gruis, known as Alnair, of magnitude 1.7. Delta and Mu Gruis are both naked-eye double stars.

Gum Nebula, the largest *nebula yet discovered. It is named after its discoverer, the Australian astronomer Colin Gum, and extends over 30 degrees in the constellations Vela and Puppis. Its light is unlikely to be due to the stars within it and it has been suggested that the hydrogen in the nebula is still emitting light as a result of the Vela *supernova explosion thousands of years ago.

Hadley Rille, the only *rille investigated during the *Project Apollo programme. The rille is in the Palus Putredinis region of the *Imbrium Basin, one of the maria of the Moon, and at the base of the lunar Apennine mountains. The Apollo 15 crew visited the area in 1971 and the lunar roving vehicle was used for the first time there. Observations showed that the mare surface had subsided by around 100 m and that the rille provided a suitable lava channel. The wall of Hadley Rille showed that the soil-like *regolith of the mare was about 5 m deep. The remainder of the wall was made up of units that were 10–20 m thick, indicating that the filling of the mare basin has occurred many times, the lava solidifying and shrinking between each flow.

Hale, George Ellery (1868–1938), US astronomer. Hale was a man of enormous energy and powers of persuasion. Three times in his career he obtained funds to build a telescope which was at the time the largest in the world. The first was the 40-inch (1.02 m) refracting telescope at the *Yerkes Observatory which is still the largest of its type. In 1904 he planned the *Mount Wilson Observatory near Pasadena, California, and was its director until 1923. Under his directorship powerful solar instruments were built, as well as the 60-inch (1.5 m) and 100-inch (2.5 m) reflecting telescopes. In 1928 he began work on a 200-inch (5.1 m) reflecting telescope, now known as the Hale telescope at the *Mount Palomar Observatory; it took twenty years to complete. Many flourishing scientific institutions such as the *IAU (International Astronomical Union) received the benefit of his talent for organization and administration. Hale was also an accomplished scientist. He discovered the magnetic fields of sunspots and their twenty-two-year cycle of polarity reversal, and he invented the *spectroheliograph which enabled photographs of the Sun, or solar features, to be obtained in light of a narrow band of wavelengths.

Hale telescope, the 200-inch (5.1 m) *reflecting telescope sited at the *Mount Palomar Observatory, USA. The US astronomer *Hale conceived the idea of a telescope with a 200-inch mirror more than fifteen years before it was built. It was first used in 1947, but regular observing only began in 1949. The mirror, cast in 1934 from low-expansion Pyrex glass, is aluminized and has a *focal ratio of 3.3. The open-girder tube is so big that it contains an observing cage at the focus in which the observer can sit. Telescopes, unlike many other pieces of equipment and mechanisms, actually improve with age as new auxiliary equipment is added to them and the Hale telescope, almost half a century after it was completed, works better than ever thanks to modern electronic equipment attached to it.

Halley, Edmond (1656–1742), British astronomer and mathematician who was the first to discover a periodic comet, which was subsequently named *Halley's Comet. He applied *Newton's method of calculating cometary orbits to twenty-four comets and found that the comets of 1531, 1607, and 1682 had very similar orbits. He concluded that this was the same object returning to the inner solar system every seventy-six years or so. He published his results in 1705 as *A Synopsis of the Astronomy of Planets*. Cometary studies, however, occupied only a small part of his scientific life. Apart from being *Astronomer Royal and Savilian Professor of Geometry at Oxford, Halley produced a chart of the southern sky in 1678, showed in 1716 how the distance between the Earth and the Sun could be calculated from the *transits of Mercury and Venus, and discovered the proper motion of stars in 1718. He discovered the relationship between barometric pressure and height above sea-level, mapped the Earth's magnetic field at the Earth's surface, accurately predicted the paths of solar eclipses, and offered the first reasoned account of the aurora borealis. Halley also spent some of his time in monetary, naval, and diplomatic pursuits and was instrumental in the production of Newton's *Principia*.

Halley's Comet, a bright intermediate-period *comet that returns to the inner solar system about every 76 years. It orbits the Sun in the opposite direction to the planets and has a perihelion distance of 0.59 astronomical units; at aphelion it is beyond the orbit of Neptune. It was the first comet to be recognized as being periodic, a discovery made by *Halley in 1696. The comet was first recorded in 240 BC and has been visible to the naked eye on all of its thirty recorded appearances. It was brighter than the brightest star in the northern celestial hemisphere in AD 374, 607, 837, and 1066. The 1066 appearance was embroidered on the Bayeux Tapestry. The brightness of the comet around perihelion has been interpreted as indicating that it loses about 3×10^{11} kg of gas and dust at each appearance, this being about 0.1 per cent of its total mass. The larger dust particles form a *meteor stream which is intersected by the Earth twice a year. This results in the Eta Aquarid *meteor shower in late April and the Orionids in late October. The mass of this stream indicates that the comet is middle-aged. It was captured by Jupiter into its present orbit some 200,000 years ago, at a time when its nucleus was around 19 km in diameter. This nucleus is now about 11 km across and in 300,000 years' time will have decayed away completely. Halley's successful prediction of the 1759 return of his comet was regarded as a sensational proof of *Newton's Law of Gravitation. The 1986 reappearance of the comet was twenty-nine years into the space age and a fleet of spacecraft, including *Giotto, flew past it.

The 200-inch **Hale telescope** at the Mount Palomar Observatory was the largest optical telescope in the world when it was completed in 1947. The observer rides in a cage at the top of the telescope lattice and enters the cage by the stairs at the left as the telescope is swung to meet him or her.

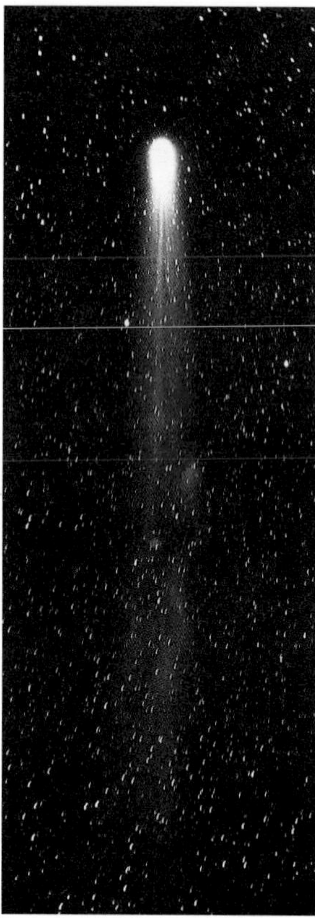

The scientific measurements of the structure and composition of **Halley's Comet** transmitted to Earth by spacecraft such as Giotto in 1986 have confirmed many of the deductions about comets made from Earth-based observations. These photographs show Halley's Comet as it appeared in 1910; the 1986 reappearance was disappointing as seen from Earth.

Hawking, Stephen William (1942–), British mathematical physicist famous for his work on *black holes and *cosmology. His theories are based on a combination of quantum mechanics and *relativity. Although suffering from a progressive neurological illness and confined to a motorized wheelchair, he travels widely and produces outstanding theoretical research on these topics. He was one of the youngest persons ever to be elected to the Royal Society at the age of 32. His book *A Brief History of Time* was an immediate best-seller when it was published in 1988.

Hayashi line, a nearly vertical track in the *Hertzsprung–Russell diagram thought to be followed by a young star as it evolves on to the main sequence. After its initial gravitational collapse from gaseous material in a *nebula, the star is thought to follow this line as it contracts from the point where its material becomes opaque to radiation, up to its becoming a *dwarf star on the main sequence. During this stage the star has a uniform temperature throughout, owing to the mixing of its material by convection. The luminosity of the star decreases because of the contraction, but the temperature remains nearly constant. The concept was introduced by the Japanese astronomer Chushito Hayashi. Owing partly to its rapidity, this stage of *stellar evolution is difficult to observe in practice, and the theoretical calculations of astronomers differ widely.

Hektor, the largest of the *Trojan group of asteroids. It oscillates about the *Lagrangian point 60 degrees ahead of Jupiter. Its orbit has a sizeable inclination, 18.26 degrees to the ecliptic. Hektor has a very dark surface with an albedo of 0.03, a spin period of 6.92 hours, and a mean diameter of about 240 km. Its brightness varies by more than one magnitude during a rotational cycle which has led astronomers to suspect that Hektor may be a binary system made of two nearly equal-sized bodies almost in contact with one another.

heliacal rising, setting, the rising or setting of a star or planet just before (for rising) or at the time of (for setting) sunrise. This is an annual event for most celestial objects. For the rising, the object has just moved above the eastern horizon into the pre-dawn morning sky. The Sun is close behind it and very soon the sky becomes too bright for the object to be seen. A planet at this time would have an *elongation close to zero. Heliacal risings and settings have a strong astrological significance. The heliacal rising of Jupiter, in a Piscean triple conjunction with Saturn, is thought by most astronomers to be the Star of Bethlehem. The heliacal rising of *Sirius (Alpha Canis Majoris) was thought by the ancient Egyptians to be the herald (and cause) of the flooding of the Nile.

Less strictly, the term is also applied to the earliest date when an astronomical body is capable of being seen in the dawn sky.

heliocentric, the *coordinate system in which all the positions and velocities of objects in the solar system are computed with respect to the Sun's centre of mass. It is the most physically meaningful reference system, but *Newton's Laws

Stephen **Hawking** is one of the most brilliant theoretical physicists of the 20th century. His crippling physical handicap means that he must rely on clear powerful thinking rather than pen and paper and this clarity of mind is reflected not only in his many technical publications but also in his *Brief History of Time* which presents the difficult subject of relativity for the general reader.

of Motion do not directly apply. Therefore computations of perturbations in the orbits of the planets are somewhat more difficult in this system.

The **heliocentric system** is a model of the solar system in which the Sun is at the centre and all the planets orbit around it. Early models of the solar system were *geocentric systems but the *Copernican system gave a more satisfactory explanation of the movement of the planets.

heliometer, a *refracting telescope with a split objective lens. It was used for measuring the angular separation of two stars. The two halves of the lens each gave images of the two stars and, by moving the half-lenses until the images coincided, the stars' angular separation could be measured. The instrument is now obsolete.

heliopause, the boundary of the *heliosphere. If the Sun were the only star and there were no interstellar wind, the heliosphere would extend indefinitely. Because of the presence of the interstellar wind the heliosphere is bounded by it at a distance of between 50 and 100 astronomical units where the solar and interstellar winds merge in strength.

heliosphere, the region surrounding the Sun, occupied by the *solar wind, a flow of charged particles, mainly protons and electrons, that are ejected from the Sun into the interplanetary medium. The region is bounded by the *heliopause. Studies of this region using instruments in the Pioneer and Voyager spacecraft confirm that the heliosphere extends well beyond the solar system, possibly to a distance from the Sun of 100 astronomical units.

heliostat, an optical device that keeps an image of the Sun in a fixed position as the Sun crosses the sky. Like the *coelostat, it continuously directs the image into the aperture of a telescope. Though the image produced by the heliostat rotates slowly, this is not normally a disadvantage. The McMath solar telescope at the *Kitt Peak National Observatory in Arizona employs a heliostat whose mirror is 1.5 m (59 inches) in diameter and provides an image of the Sun 0.8 m (31 inches) across.

helium, an element making up about one-third of the mass of the universe. Although it is thought that most helium in the universe was created in the first few minutes of the life of the universe after the *Big Bang, it is also the end-product of the *carbon–nitrogen cycle and the *proton–proton reaction within main sequence stars whereby hydrogen is converted into helium with the production of energy which the stars radiate.

Helix Nebula, a *planetary nebula in the constellation of Aquarius. It is the largest planetary nebula, being about one-fifth of a degree in diameter. The Helix Nebula is only about 140 parsecs away and its filamentary ring is rapidly expanding at about 30 km/s. It is one of the most beautiful objects in the sky. For photograph see planetary nebula.

Hellas Planitia, a large impact basin on the surface of Mars. Over 1,500 km in diameter and 3 km deep, it can be seen through telescopes, under good seeing conditions, from the Earth. Occasionally fog, condensed from water vapour in the thin Martian atmosphere, forms in the region.

Helmholtz–Kelvin contraction *Kelvin–Helmholtz contraction.

Hayashi line

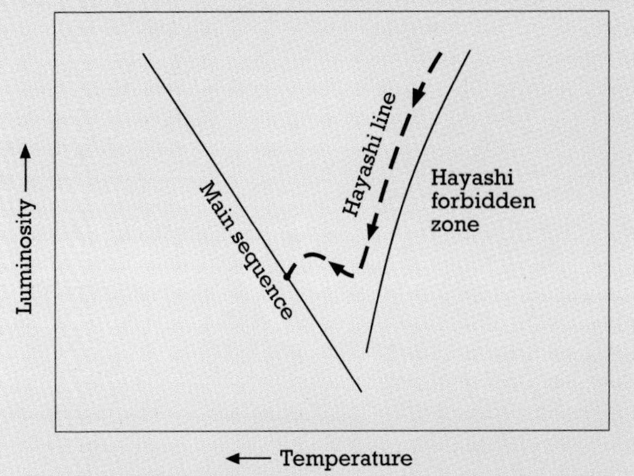

Portion of a Hertzsprung–Russell diagram, showing the contraction of a young star along a Hayashi line to its position on the main sequence. The Hayashi forbidden zone is a region in which there are no reasonable solutions to the equations of stellar structure.

Herbig–Haro object, a small bright concentration of dust and gas often associated with *T Tauri stars. The high-energy radiation from the young T Tauri stars illuminates the dust and gas of the *nebula in which they reside. In some cases the stars emit a beam of radiation in a particular direction in space, producing a bright string of Herbig–Haro objects.

Hercules, a large *constellation of the northern hemisphere of the sky, one of the forty-eight figures known to the ancient Greeks. It represents the mythical character who carried out twelve labours, and it is the fifth-largest constellation. In the sky, Hercules is seen kneeling with one foot on the head of Draco, the dragon. The head of Hercules is marked by Alpha Herculis, called Rasalgethi from the Arabic meaning 'kneeler's head'. It is a red supergiant star that varies irregularly between third and fourth magnitudes. It is also double, with a fifth-magnitude companion visible in small telescopes. The body of Hercules is marked by a distinctive keystone shape of four stars, Epsilon, Zeta, Eta, and Pi Herculis. Between Zeta and Eta Herculis lies M13 (NGC 6205), the brightest globular *cluster in northern skies, visible to the naked eye and in binoculars. M13 is estimated to contain 300,000 stars and lies 7,000 parsecs (22,500 light-years) away. Another globular cluster, M92 (NGC 6341), is visible in binoculars.

Hermes, an *Apollo–Amor object discovered by the German astronomer Karl Reinmuth in October 1938. That year it passed within a million kilometres of the Earth. The orbit of this small asteroid of 0.8 km diameter has a perihelion distance of 0.62 astronomical units, a period of 2.1 years, and an inclination of 6 degrees to the *ecliptic. The figures lack precision because the asteroid was only seen between 25 and 29 October 1938 and has since been lost.

Herschel, Sir John (Frederick William) (1792–1871), British astronomer and only son of William *Herschel. John Herschel was awarded a Fellowship of the Royal Society by the age of 21 and was a founding member of the Royal

This portrait of William **Herschel** was painted by William Artaud in 1818–19. During his early life Herschel was an accomplished musician but took up astronomy after reading some books on the subject. He borrowed a small telescope but, finding it inadequate, went on to build his own.

Astronomical Society. His early work included a calculation of the orbits of the binary star Gamma Virginis and a measurement of the difference in longitude between the Greenwich and Paris Observatories. In 1833 he published a monumental catalogue of 2,307 nebulae and stellar clusters, 525 arising from his own observations. In 1836 he published, in six volumes, the parameters of 3,346 double stars. Later on he went to Cape Town, South Africa, and his telescopic observations led to catalogues of 1,707 nebulae and 2,102 binary stars. He also studied the Orion region, the Eta Carinae Nebula, and the Magellanic Clouds. He founded the subject of stellar *photometry by inventing an astrometer, an instrument that enabled star brightnesses to be measured by comparing stellar images with an image of the full Moon. He also made many advances in the practical development of photography and introduced the terms 'positive' and 'negative'.

Herschel, Sir William (Frederick) (1738–1822), German musician who moved to Bath, England, and turned to astronomy. His desire to study the nature and distribution of distant stars led him into the field of telescope design and production. Herschel concentrated on *reflecting telescopes with mirrors made of *speculum and he built for himself the most powerful telescope of the day. On 13 March 1781 Herschel thought he had discovered a comet but its orbit quickly indicated that it was a new distant planet. He named it Georgium Sidus after his patron King George III but it was later renamed Uranus. Herschel completed his 48-inch (1.2 m) mirror in 1789 and discovered two new satellites of Saturn (Mimas and Enceladus) soon afterwards. It was not until the age of 43, after his discovery of Uranus, that he became a professional astronomer. Herschel's main achievement was in the field of stellar distribution. He discovered that most *nebulae were star systems and developed a theory of stellar evolution; he also concluded that the Milky Way was a flat and finite system of stars. Herschel discovered that *binary stars were not just chance alignments but actually two stars in orbit around their common centre of mass. He also found that the Sun was moving through space in the direction of the constellation Hercules. Herschel's sister, **Caroline Lucretia Herschel** (1750–1848), independently discovered eight comets and several nebulae and star clusters. In 1798 she published a star catalogue.

Hertzsprung, Ejnar (1873–1967), Danish observational astronomer who initiated several important developments in astronomical photography. In 1905 he investigated the relationship between the surface temperature (or colour) and absolute brightness of stars. A few years later *Russell illustrated this relationship graphically in what is now known as the *Hertzsprung–Russell diagram, which has become fundamental to the study of stellar evolution. Hertzsprung also applied photography to *photometry and the study of binary stars and variable stars.

Hertzsprung gap *Hertzsprung–Russell diagram.

Hertzsprung–Russell diagram (H–R diagram), a chart on which stars are plotted according to their temperature (or *colour index or *spectral class) and brightness (*luminosity or absolute *magnitude). Also known as a **colour–luminosity diagram**, it is named after the Danish astronomer *Hertzsprung and the US astronomer *Russell, who published the first such diagram in 1914. The hottest stars are towards the left of the H–R diagram, and the most luminous stars are towards the top. When a representative sample of stars is plotted, most are found to form a

Hertzsprung–Russell diagram

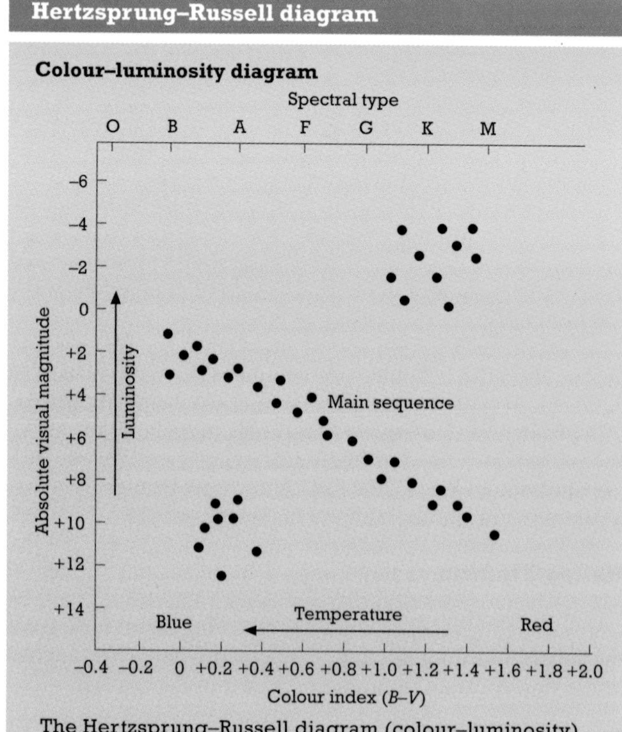

The Hertzsprung–Russell diagram (colour–luminosity) for the local stars.

Hertzsprung–Russell diagram

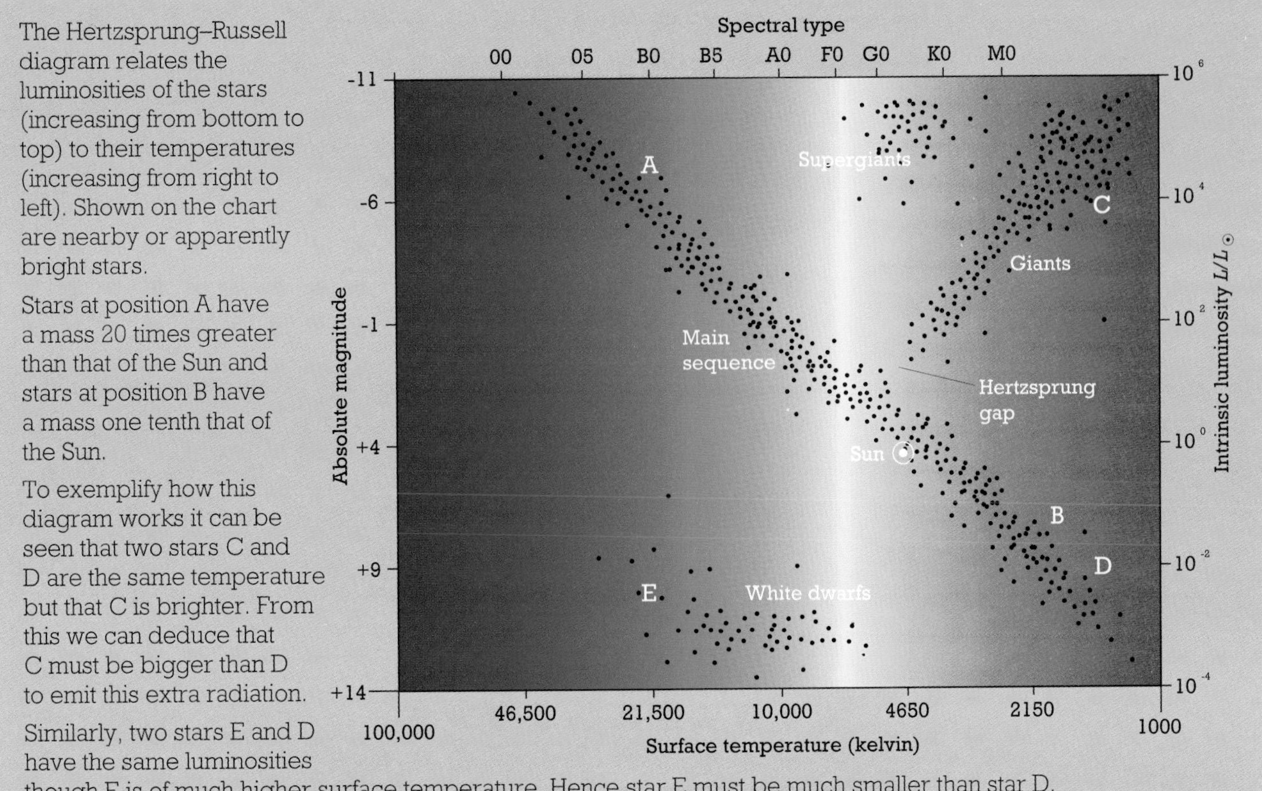

The Hertzsprung–Russell diagram relates the luminosities of the stars (increasing from bottom to top) to their temperatures (increasing from right to left). Shown on the chart are nearby or apparently bright stars.

Stars at position A have a mass 20 times greater than that of the Sun and stars at position B have a mass one tenth that of the Sun.

To exemplify how this diagram works it can be seen that two stars C and D are the same temperature but that C is brighter. From this we can deduce that C must be bigger than D to emit this extra radiation.

Similarly, two stars E and D have the same luminosities though E is of much higher surface temperature. Hence star E must be much smaller than star D.

band running from upper left to lower right, called the **main sequence**, or *dwarf stars. The Sun is a main sequence star. The most luminous main sequence stars also have the largest *stellar mass. Above the main sequence are found, successively, the subgiants, *supergiants, and *giant stars; below it are the *subdwarfs and *white dwarfs. These groups are called **luminosity classes**. The H–R diagram is a useful aid to the understanding of *stellar evolution. For example, in young *clusters of stars, formed together at about the same time, nearly all members are on the main sequence. In older clusters there are no hot stars on the main sequence, but the H–R diagram shows a **horizontal branch** of cool giant stars, as the more massive stars have evolved to become giants. The older the cluster, the lower the *turn-off point where stars have begun to evolve off the main sequence. Some H–R diagrams have a gap, called the **Hertzsprung gap**, between the main sequence and the giant branch.

Hesperus, the Greek name for the planet Venus when it appears as the evening star seen near eastern *elongation. Both Venus and Mercury are only visible to the naked eye as morning or evening objects. Venus was therefore called by the Greeks Hesperus as an evening object and *Phosphorus as a morning object.

Hevelius formation, a relatively smooth part of the radial *ejecta blanket of the *Orientale Basin on the Moon. The crater Hevelius is partly buried by this ejecta, which is about 100 m thick in this region. As this ejecta settled back on to the lunar surface it acted rather like a fluid, filling in small craters, banking up in front of obstacles, and flowing around them, just like a landslide on the Earth. Strong flow patterns can be seen on the Hevelius formation and it is the youngest major ejecta blanket on the Moon. The crater is named after the Polish astronomer Johannes Hevelius who produced, in 1647, a lunar map that attempted to include the effects of *libration.

Hidalgo, an asteroidal body moving in a very eccentric orbit. It has a semi-major axis of 5.861 astronomical units, an eccentricity of 0.656, and an inclination of 42.4 degrees to the ecliptic. While the perihelion lies near the inner edge of the asteroid belt, the aphelion distance is close to the semi-major axis of Saturn. Its orbit is governed by *chaos theory and it undergoes relatively frequent close encounters with the giant planets. Hidalgo has a very dark surface and a diameter of about 50 km. It has been suggested that this object is actually an extinct cometary nucleus rather than an asteroid.

Himalia, a small satellite of Jupiter. It has a diameter of 186 km and revolves about the planet at a mean distance of 11,480,000 km in a highly eccentric orbit in a period of 251 days. It is one of a group of four Jovian satellites revolving about the planet at almost the same distance.

Hipparchus (Hipparchos), Greek astronomer of the 2nd century BC, probably the greatest astronomer of antiquity. He attempted to measure the distances of the Sun and Moon by geometrical methods; although his estimate of the Sun's distance, 15 million km (9 million miles), was much too small, it at least demonstrated that the Sun was far larger than the Earth. The appearance of a *nova in 134 BC prompted him to compile the earliest known comprehensive star catalogue, covering 850 stars. On comparing his measurements with those made by earlier astronomers, he discovered the *precession of the equinoxes. He also made

The dark shape of the **Horsehead Nebula** is a dense cloud of dust that obscures the light from the glowing gas behind. The horse's head is really only a small part of a huge dark cloud covering the whole of the lower area of the photograph, hiding the stars behind it. The two bluish nebulae NGC 2023 (nearest to the horse's head) and IC 435 (below it) are regions of dust illuminated by two bright stars embedded in the dark dust cloud. The faint fine structure of the bright nebula NGC 2024 just below the brilliant star Zeta Orionis can also be seen.

careful measurements of the lengths of the seasons and of the year, and made proposals for the improvement of the calendar. Few of his writings survive, but he was probably the most important source used by Ptolemy in writing his **Almagest*. A prominent lunar crater, 150 km (94 miles) in diameter is named after him. It is near the apparent centre of the Moon's hemisphere that faces the Earth. The *Hipparcos satellite (High Precision Parallax Collecting Satellite), launched in 1989, also recalls his name.

Hipparcos, a satellite, which contains a specially designed telescope, launched by the *European Space Agency on 8 August 1989 to provide highly precise measurements of the position, *parallax, and proper motion of 120,000 selected stars in a mission scheduled to last thirty months. The accuracy hoped for, about ±0.002 arc seconds, is only attainable above the Earth's atmosphere. A fault in the apogee motor prevented the package being transferred from a highly elliptic transfer orbit into the planned-for almost circular orbit but, in a skilful rescue operation, the scientists and engineers in charge managed to obtain data for 70 per cent of the time (about seventeen hours per day) enabling a catalogue to be created of the positions of the selected stars to an accuracy of 0.005 arc seconds.

HI region *interstellar medium.

HII region *interstellar medium.

Hohmann transfer orbit, an interplanetary orbit used by spacecraft to reach the neighbourhood of other planets. Normally the transfer orbit is nearly elliptical and is chosen to minimize the energy and fuel consumption required. To reach an outer planet, the Earth needs to be near perihelion and the planet near aphelion. To reach Mercury or Venus, the Earth needs to be near aphelion and the planet near perihelion.

horizon, the line that an observer sees which divides the surface of the Earth from the sky. The observer's horizontal plane is tangential to the Earth's surface at the observer's position, the Earth's surface being taken as spherical. The horizon is the great circle produced by the intersection of the observer's horizontal plane and the *celestial sphere. The upper and lower *poles of the horizon are the *zenith and the *nadir. The zenith is that point on the celestial sphere vertically above the observer. The nadir is opposite to the zenith on the celestial sphere.

Horologium, the Pendulum Clock, a faint *constellation of the southern sky, introduced by the 18th-century French astronomer Nicolas Louis de Lacaille. It is a barren area of sky, consisting of no stars brighter than fourth magnitude. R and TW Horologii are red giant variable stars that reach fifth magnitude at their brightest.

Horsehead Nebula, a dark *nebula in the constellation of Orion. It resembles a horse's head and neck in shape, and is conspicuous because of its much brighter background. The nebula is near the bright star Zeta Orionis, the easternmost of the three stars forming *Orion's Belt, and can be seen in moderate-sized telescopes of about 20 cm (8 inches) aperture.

hour angle *celestial sphere.

Hoyle, Sir Fred (1915–), British mathematician and astronomer and the foremost supporter of the *Steady State theory of cosmology. In 1948 he developed a mathematical basis for this within the context of Einstein's theory of general *relativity. Since then, observational evidence has tended to confirm the predictions of the alternative *Big Bang theory of the universe. Because the Steady State theory postulates a continuous creation of matter in the form of pure hydrogen, Hoyle was led to investigate the nuclear processes of heavy element formation in stars. In this way, his often controversial theories have stimulated much fruitful research both by himself and others. Hoyle's research has been exceptionally wide ranging, covering solar physics, the origins of the solar system, stellar structure, galactic structure and interstellar material, galaxy formation, and highly condensed objects. Most of his work has been carried out at Cambridge, UK, where he was Plumian Professor of Astronomy from 1958 to 1973. Hoyle has also become involved in the debate about the origins of life by suggesting that life may have entered the Earth's atmosphere from meteors.

Hubble, Edwin Powell (1889–1953), US astronomer whose researches laid the foundations of modern observational *cosmology. After embarking briefly on a legal career, Hubble undertook photographic research of faint nebulae at the *Yerkes Observatory in 1913. Following an interruption for military service in World War I, he joined the *Mount Wilson Observatory in 1919 just as the 100-inch (2.5 m) telescope had become fully operational. With this instrument, in 1923 he discovered *Cepheid variable stars in the outer parts of the *Andromeda Galaxy (M31) and using the *period–luminosity law, he determined the distance of these stars and therefore of the nebula itself. He had now established that M31 and other spiral nebulae were galaxies similar to our own; the implications of this discovery for cosmology were immense. Hubble devised a classification scheme for external galaxies, now known as the *Hubble classification. He also discovered that the universe is expanding by determining the *recession of the galaxies, and he deduced a fundamental relationship between their distance and velocity of recession which has become known as Hubble's Law.

Hubble space telescope

The delicate telescope assembly is protected by the outer shell known as the support systems module (SSM). The SSM also provides electrical power through the solar panels and transmission of data back to Earth via telecommunications satellites.

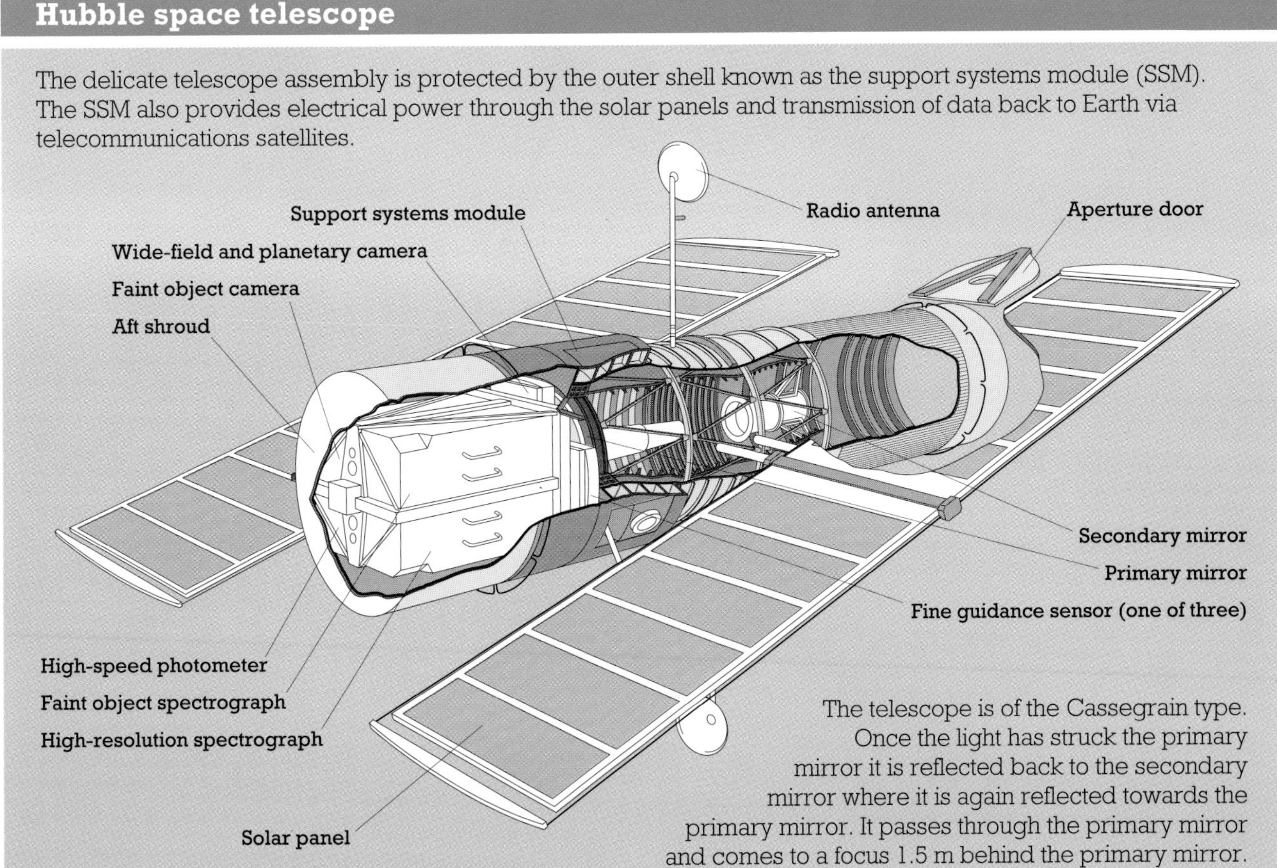

The telescope is of the Cassegrain type. Once the light has struck the primary mirror it is reflected back to the secondary mirror where it is again reflected towards the primary mirror. It passes through the primary mirror and comes to a focus 1.5 m behind the primary mirror.

Hubble classification, a scheme suggested by *Hubble for the classification of galaxies. It was originally thought that it gave some insight into the evolution of galaxies with time. Hubble thought that almost spherical elliptical galaxies (designated E0) gradually flattened until they resembled thin lenses (E7), then took one of two paths, becoming normal *spiral galaxies (S0 to SC) or *barred spiral galaxies (SBa to SBc). The S0 galaxies, known as *lenticular galaxies, have a central bulge and disc but no *spiral arms. A normal spiral galaxy was thought to develop spiral arms at the expense of the massive central nucleus, the arms becoming more loosely coiled from Sa to Sc. In the case of barred spiral galaxies it was thought that a brightly luminous bar developed across the nucleus, with a spiral arm springing from each end of the bar. More recent study of galaxies and the stars within them has cast doubt on this evolutionary scheme but as a handy classification of the different kinds of galaxy it is still useful.

Hubble constant, a mathematical constant which relates the speed of *recession of the galaxies with their distance from our own galaxy. It is one of the fundamental constants in *cosmology as it is directly related to the age of the universe. It appears in **Hubble's Law**, which relates the *red shift of a galaxy to its distance from us, giving a **velocity–distance relation**. Thus $V = H \times d$, where d is the distance of the galaxy, V is the galaxy's velocity of recession, measured from the *Doppler effect producing the red shift in the galaxy's spectrum, and H is the Hubble constant. It has a value of about 55 km/s per megaparsec.

Hubble–Sandage variable, an irregularly variable hot *supergiant star of high luminosity, among the brightest objects in its parent galaxy. Such stars are unstable and probably radiate close to the **Eddington limit** beyond which the radiation would blow apart the star's outer layers. Bright examples are Eta Carinae in our own Galaxy, which reached apparent magnitude −0.8 in 1843 but is now of seventh magnitude, and S Doradus in the Large Magellanic Cloud, one of the most luminous stars known with an absolute magnitude of −9 and which lends its name to the class of S Doradus variable stars. They are usually associated with diffuse *nebulae and surrounded by expanding envelopes of gas. They are massive young objects undergoing rapid *stellar evolution and are potential *supernovae. These rare stars are named after the US astronomers *Hubble and Alan R. Sandage who investigated them.

Hubble space telescope (HST), a 2.4 m (94-inch) *reflecting telescope put into Earth orbit on 25 April 1990 by the space shuttle *Discovery*. The telescope, which is controlled from Earth, is designed to make observations in the ultraviolet, optical, and infra-red regions of the spectrum and has a wide variety of auxiliary equipment including two spectrographs, a *photometer and two cameras (see the figure on page 69). The data are transmitted to Earth stations for analysis. From its position above the Earth's atmosphere, the HST should enable a volume of the universe some 350 times as large as the present observable universe to be examined, and will detect light from objects which are seven times further from the Earth than has been possible so far. Because of the finite velocity of light, therefore, it should be possible to observe objects as they were seven times further into the past. Unfortunately, although the HST is now in operation and doing useful work, a built-in error in shaping the secondary mirror has prevented the telescope from operating to its full potential. Plans have been made to overcome this defect, either by computer technology or by visiting the HST in a future space shuttle mission. Nevertheless, the images it is producing are in many cases very much better than those obtained through Earth-based telescopes.

The **Hubble space telescope**, shown in orbit in this artist's impression, was intended to overcome the blurring and blocking effects of the Earth's atmosphere to incoming electromagnetic radiation. However, it was quickly found that the secondary mirror was incorrectly shaped and so its usefulness has been reduced. Nevertheless, much useful work is being done and there are plans to correct the fault at some time in the future.

Hyades, an open *cluster of about 100 stars occupying an area about 20 degrees across. Probably about 400 million years old, it is so near to the Earth that the individual *proper motions of the cluster stars can be observed. From these the distance of the cluster from the Sun can be calculated to be about 40 parsecs (130 light-years).

Hydra, the Water Snake, the largest *constellation, winding more than one-quarter of the way around the sky. It straddles the celestial equator, and in Greek mythology represents the multi-headed serpent slain by Hercules as one of his labours. Its brightest star is Alpha Hydrae of second magnitude and known as Alphard, an Arabic name meaning 'the solitary one'. R Hydrae is a red giant variable similar to *Mira Ceti, fluctuating between fourth and tenth magnitude every 390 days or so. NGC 3242 is a ninth-magnitude *planetary nebula known as the Ghost of Jupiter. M83 (NGC 5236) is a face-on spiral galaxy of eighth magnitude.

hydrogen, the simplest of the chemical elements, its atom consisting of one proton and one electron. It accounts for two-thirds of the mass of the universe. It is thought that it was created in the earliest moments of the life of the universe after the *Big Bang when the temperature fell to such a level that protons and electrons could combine to form hydrogen atoms. Hydrogen is present in stellar interiors where some of it is converted to helium by nuclear fusion reactions; it also exists in interstellar space where it radiates at the *twenty-one centimetre line. There are three types (isotopes) of hydrogen: normal, deuterium, and tritium.

hydroxyl radical (OH), the first interstellar molecule to be detected by *radio astronomy, in 1963. Although unstable on Earth it is widespread in the *interstellar medium with a typical abundance of a few parts per 100 million. OH has also been detected in many other galaxies. The OH *spectral lines at 18 cm wavelength are sometimes boosted by powerful *maser action. The strongest OH masers produce 10^{30} watts of radio power.

Hydrus, the Little Water Snake, a *constellation of the southern sky introduced by Pieter Dirkszoon Keyser and Frederick de Houtman at the end of the 16th century, snaking between the Large and Small *Magellanic Clouds. Its brightest star, Beta Hydri, is of third magnitude but there are no objects of particular interest.

Hyperion, a satellite of Saturn discovered independently by the British amateur astronomer William Lassell and the US astronomer George Bond in 1848. Its orbit has a semi-major axis of 1.48 million km and a sizeable eccentricity of 0.104. Hyperion and *Titan are at similar distances from Saturn but their orbital motions are locked together and Hyperion is prevented from approaching Titan too closely because for every three orbits of Hyperion, Titan completes four and the conjunction always occurs when Hyperion is farthest from Saturn. Hyperion is probably the remnant of a catastrophic collision in the early solar system. Because of their locked motion, fragments ejected from Hyperion enter a region of *chaos where they can approach close to Titan and be swept away from Hyperion. Hyperion has a very irregular oval shape and its surface is marked by prominent steep slopes tens of kilometres high. It has an abnormally low albedo, about 0.2, compared with most other Saturnian moons, and fewer large craters. The eccentric orbit and irregular shape of Hyperion enables Saturn to exert a strong torque causing Hyperion's rotation period to vary.

Iapetus, a satellite of Saturn discovered by Giovanni Domenico *Cassini in 1671. Its orbit has a semi-major axis of 3.56 million km and is inclined to Saturn's orbital plane at 7.5 degrees though the inclination is variable due to perturbations from the Sun. The diameter of Iapetus is 1,460 km and its density is 1,200 kg/m^3. A peculiar feature of Iapetus, already known from Earth-based observations and confirmed by the *Voyager images, is the striking difference in the albedo of its leading and trailing hemispheres as it travels in its orbit. While the trailing side is bright, heavily cratered, and most probably covered by water ice, the leading side is very dark, with a mean albedo of about 0.05. The corresponding surface material is probably carbon-based but it is unclear whether it was emitted from the interior of Iapetus or swept up from space due to its orbital motion.

IAU (International Astronomical Union), an international association of the world's professional astronomers based in Paris, France. It has a large number of commissions covering all branches of astronomy and convenes its general assemblies at three-yearly intervals in a succession of different countries. It also organizes international colloquia and symposia.

Icarus, an *asteroid discovered by *Baade in 1949. Its orbital perihelion distance of 0.1867 astronomical units ensures that it crosses the orbit of Mercury. At one of its passages close to Earth, in 1968, it was detected by radar. For such a small object, of radius 0.3–0.7 km, it is unusually spherical indicating that it might have lost its extremities.

Imbrium Basin, a very large old circular *mare in the northern hemisphere of the Earth-facing side of the Moon. It is surrounded by the Apennine and Jura mountains and has a floor of layers of solidified basaltic lava, this lava having been produced over at least four widely differing periods of volcanic activity and having widely differing numbers of impact craters. The basin contains *wrinkle ridges and the *Hadley Rille region that was visited by Apollo 15. The basin is a *mascon, a region of higher than average surface gravity. This is produced by having layer upon layer of mare basalt lying on top of the lower-density lunar crust. The molten basalt lava rose up from about 150 km below the lunar surface through cracks in the crust rather like mercury rises in a barometer. On reaching the surface it spread out, cooled, and shrank, causing the surface to subside. This subsidence allowed more lava to rise and spread over the original surface, because the equilibrium fluid level was above the cold surface of the old lava layer.

Indus, the Indian, a *constellation of the southern sky, introduced by Pieter Dirkszoon Keyser and Frederick de Houtman at the end of the 16th century. It represents a North American Indian. Its brightest stars are of third magnitude. Epsilon Indi is a fifth-magnitude yellow dwarf similar to the Sun, 3.5 parsecs (11.2 light-years) away.

inertia, the property of a body that makes it resist any change in its state of rest or uniform motion in a straight line.

It requires *force to overcome inertia and the distinction between a body's *mass and *weight is bound up with the existence of inertia. Thus, in space, a moving body may have many tonnes mass and yet be weightless, that is, no force acts upon it. Such a weightless body will be able to crush an astronaut to death, caught between the body and another massive though weightless body as they collide. By *Mach's Principle, inertia may be due to the presence of the rest of the universe.

inferior planet, a planet revolving about the Sun in an orbit closer to the Sun than that of the Earth. Mercury and Venus are inferior planets. Because their orbits are smaller than the Earth's these two planets are always seen in the evening or morning and are never seen in directions far from that of the Sun.

infra-red astronomy, the observation of the extraterrestrial universe by detecting *electromagnetic radiation at infra-red wavelengths. The Earth's atmosphere is transparent to several wavebands of infra-red radiation particularly in the shorter wavelengths, and observations in selected wavebands can be made by ground-based telescopes at specially chosen sites. The United Kingdom infra-red telescope (UKIRT) stationed at *Mauna Kea, Hawaii, at a height of 4,200 m (13,800 feet) is well above the atmospheric water vapour which is the prime cause of opacity to the cosmic infra-red radiation. As infra-red energy is liberated thermally by materials at normal environmental temperatures, special care is given to the design of instruments such as telescopes and photometers so that the radiation emitted by the material of which the instrument is made does not interfere with the astronomical measurements. Critical parts of the instrumentation need to be cooled, particularly the detectors; liquid helium, at a temperature of 4 K, is used in many systems. Other infra-red observations are made from rocket flights and balloon platforms. However, the most successful enterprise has been the Dutch–UK–USA satellite carrying a telescope–photometer system. This Infra-Red Astronomical Satellite (IRAS) was launched in January 1983 and surveyed more than 96 per cent of the total sky area in four wavebands at 12, 25, 60, and 100 μm. Over its ten months of operation it catalogued some 250,000 point sources and some 20,000 small extended sources.

The study of cosmic infra-red radiation allows thermal objects to be investigated, that is, those sources that have temperatures from a few tens of degrees Kelvin to a few thousand. Of interest here are solar system objects, cool stars, and interstellar dust. Other infra-red radiation originates by non-thermal processes, particularly *synchrotron radiation emitted from fast-moving electrons within the magnetic fields of galaxies. Molecular and solid state spectra have many features in the infra-red domain and these are used to investigate stellar atmospheres and the galactic concentrations of gas and dust.

inner planet, a planet that is closer to the Sun than the asteroid belt, which has a mean radius of 2.8 times that of the orbit of the Earth. The term thus applies to Mercury, Venus, the Earth, and Mars (and some would argue to the Moon as well because of its size). These planets, also known as **terrestrial planets**, have much smaller masses than the *outer planets but have much larger densities, being in the main composed of rock and iron with very little hydrogen and helium, and very little of the volatile molecules like carbon dioxide, methane, and ammonia. The inner planets are dominated by Venus and the Earth, which have reasonably similar masses. Mercury and Mars have masses of around 10 per cent of the mass of the Earth and Venus. Mercury and Venus do not have satellites. The two satellites of Mars, Phobos and Deimos, are most likely to be captured asteroids. Mercury is the only inner planet without an atmosphere: the surface is too hot to retain one.

insolation, the amount of solar radiation received by a planetary, satellite, or cometary surface per square metre per second. The degree of insolation, coupled with the absorption, determines how hot the surface will become. At the distance of the Earth from the Sun, the insolation is called the *solar constant. This varies by around 1 per cent throughout the Sun's eleven-year *sunspot cycle. It also varies by about ±3.5 per cent as the Earth moves around its elliptical orbit. The insolation decreases as the inverse square of the distance between the Sun and the celestial object, and this means that the temperature of that object depends roughly on the inverse square root of its distance from the Sun.

interferometry, the analysis of interference patterns formed by *electromagnetic radiation. An instrument which combines two or more waves to create an interference pattern is called an **interferometer**. Interferometry is widely used in astronomy to give data on the size and structure of astronomical objects. For example, the *Michelson interferometer is an optical instrument used to measure the angular diameters of stars. Though interferometry can be applied to most parts of the electromagnetic spectrum, it is most commonly used in *radio astronomy. A radio interferometer gives an interference pattern obtained by adding together two signals from a star or other source of radiation with a clearly defined phase difference between them. This phase difference depends on the separation (baseline) between two aerials (US antennas). Interferometers can obtain the angular resolution (image clarity) of one large *dish telescope. The longer the baseline, the higher the *resolving power that can be achieved. Radio interferometry has progressed rapidly. Among the most important developments have been the introduction by the British astronomer Martin Ryle of the phase-switch interferometer, and his development of *aperture synthesis. In the latter technique the Earth's daily rotation effectively moves one aerial around the other; by altering the aerial separation each day and combining the signals in a computer, it is possible to generate an image that could otherwise only have been made with a telescope of the same diameter as the aerial separation. The main examples of synthesis telescopes are at Cambridge (UK), Westerbork (The Netherlands), and the *Very Large Array in New Mexico (USA). The maximum aerial separations of these are 5 km (3 miles), 1.4 km (0.9 miles), and 22 km (14 miles), respectively. The other significant development has been very long baseline interferometry (*VLBI), the basing of telescopes in different continents to give extremely high-resolution (0.001 arc seconds) pictures of radio sources. This resolution is much better than anything which can be achieved by a ground-based optical telescope. For example, the Very Large Baseline Array currently under construction will eventually cover 8,000 km (5,000 miles) from the Virgin Islands to Hawaii.

intergalactic medium, the material between the galaxies. In principle this may contain atoms, molecules, dust, gas, rubble, and stars though the overall density of the material must be very small. The intergalactic medium may account for some of the dark matter in the *missing mass problem.

interplanetary medium, the *dust and *plasma that occupies the space between the planets. The plasma is known as the *solar wind and consists of protons and electrons that are being blown away from the solar corona. Normally the plasma at the Earth's orbit contains around five protons and electrons per cubic centimetre and these are moving past at a velocity of 400 km/s. The medium also contains dust particles and these are found in a symmetrical flattened cloud surrounding the Sun. Sunlight scattered from this cloud creates the *zodiacal light. Dust is also found in a series of *meteor streams containing meteoroids which can have masses ranging from thousands of kilograms to less than a million millionth of a gram. The small particles are being blown away from the Sun by *radiation pressure whereas the larger bodies are slowly spiralling into the Sun as they lose momentum. Around one tonne of dust needs to be fed into the cloud every second by decaying comets to keep the cloud in equilibrium.

interstellar absorption, the absorption of starlight as it passes through clouds of dust between the stars and the Earth. As the clouds tend to concentrate in the plane of the Galaxy, light from stars in these zones tends to be more affected. The absorption makes stars appear less bright than they otherwise would be and their colours are changed by *interstellar reddening, much in the same way that the Sun is dimmed and reddened when it is close to the horizon.

interstellar medium, the dust and gas that make up about 10 per cent of the mass of the Galaxy. In some regions the material is relatively cool (100 K) and is detectable by infra-red telescopes. Such clouds contain neutral hydrogen (**HI regions**), molecular hydrogen, and other molecular radicals that can be identified by radio telescopes. In other regions, closer to luminous stars, the gas may be hot (1,000–10,000 K) with the hydrogen being ionized (**HII regions**). From measurements of the effects of the dust on the light of the background stars, the particles are thought to be about 0.1 μm in size, comprising carbon and silicates, and having icy surfaces. They originate from the atmospheric winds of cool stars and from *supernova explosions. This dust goes into the mixture which will eventually provide the material for the birth of new stars. The interstellar medium is also permeated by a weak magnetic field which aligns the grains of dust, causing weak *polarization of the stellar light transmitted through it. Flows of *cosmic rays of unknown origin are also found in the interstellar medium.

interstellar reddening, the absorption of starlight at short wavelengths by clouds of dust between the stars and the Earth. *Interstellar absorption of light is stronger at short wavelengths corresponding to blue light, and the effect is to produce a reddening of the starlight. Radio and infra-red radiation, because of their longer wavelengths, can pass easily through dust without absorption.

invariable plane, a plane passing through the centre of mass of the solar system and perpendicular to the *angular momentum vector of the solar system. Its position is not precisely known since the masses, positions, and velocities of the solar system's bodies used in calculating the angular momentum are not exactly known.

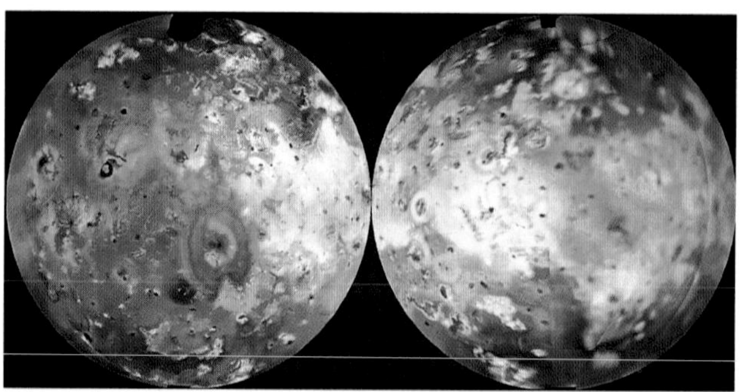

This enhanced natural colour image of Jupiter's moon **Io**, showing its two opposing hemispheres, was taken by Voyager 1 during its fly-past in 1979. The left-hand image shows the face that is always turned towards Jupiter. The volcanoes Babbar Patera (black, *below centre*), Loki (black lava ring), and Ra Patera (*left*, spider-like) are clearly visible. The right-hand image shows the volcano Prometheus (*left*, on the equator). The red areas are sulphur-rich lavas ejected by the larger volcanoes, whilst the white deposits are sulphur dioxide snow ejected by smaller volcanoes like Prometheus.

Io, a *Galilean satellite of Jupiter discovered by Galileo in 1610. Its almost circular orbit is nearly in Jupiter's equatorial plane with a semi-major axis of 421,600 km. The orbital motions of Io, *Europa, and *Ganymede are locked together which causes strong tidal effects resulting in the heating and melting of Io's interior. Io's diameter of 3,630 km and its density of 3,570 kg/m^3 are close to the values for the Earth's Moon. Io's surface, photographed by the *Voyager probes, is covered by vividly coloured sulphur compounds overlying the rocky interior of the satellite. The geology is dominated by volcanic processes, yielding three kinds of surface features: vent regions, with many volcanic craters and lava flows; plains; and mountains up to 9 km high. Nine active plumes of volcanic debris reach heights of 300 km and their deposits continually form new surfaces on the satellite. Pools of molten sulphur are probably common. Io has a very thin atmosphere of sulphur dioxide and its orbit is surrounded by a torus of plasma, mostly ionized sulphur and oxygen, which rotates in step with Jupiter's magnetic field and is supplied by material escaping from the satellite. A cloud of neutral sodium atoms precedes Io in its orbit.

ion, an atom or molecule which has lost or gained one or more electrons. For example, a singly charged (ionized) atom has lost or gained one electron, and a doubly charged atom has lost or gained two electrons. Such atoms have a net positive electric charge if electrons are lost and a net negative charge if they are gained. Normally the term ionization refers to electron loss. At the centre of a star an atom may have lost all or most of its electrons leading to a state where the stellar material is essentially a 'gas' consisting of nuclei and electrons.

ionosphere, a region in the Earth's upper atmosphere extending from a height of about 75 km (47 miles) to over 1,000 km (625 miles) where the air has an appreciable electrical conductivity due to the splitting of the constituent molecules into *ions and free electrons by solar ultraviolet radiation. The existence of such a region was first postulated by the British physicist Balfour Stewart in 1882. Because of its high conductivity, the ionosphere is an effective reflector for radio waves in the short-wave to long-wave band regions so making it possible for communications to take place over long distances around the Earth's surface. The ionosphere is however transparent to very high-frequency radio waves (including television) so these have to be transmitted around the Earth via artificial satellites.

Isaac Newton telescope, a reflecting telescope originally commissioned at the *Royal Greenwich Observatory at Herstmonceux, Sussex, in 1967. During its operation there, it used a 98-inch (2.5 m) primary mirror. In 1979 it was dismantled and, after a lengthy renovation which included the substitution of a new 100-inch mirror for the 98-inch one, was rebuilt and put into operation again in 1984 at the *La Palma Observatory in the Canary Islands to take advantage of the better observing conditions there. The new mirror, made of Zerodur which has a very low coefficient of expansion, retains its accurately figured shape at all temperatures experienced at the telescope site. The equatorial *mounting of the telescope had to be modified to allow for the change in latitude from Herstmonceux to La Palma (51 degrees north to 28 degrees north). In the quarter century since the original Isaac Newton telescope came into operation, enormous progress has been made in astronomical technology; in its armoury of detectors the refurbished telescope has a *CCD camera and an image photon counting system. It is also capable of being remotely controlled from the United Kingdom. Like many large telescopes nowadays its observing time is shared between groups from several countries. These include Britain, The Netherlands, and Spain.

Islamic astronomy The Arabian conquests, from the middle of the 7th century, produced an empire within which a rich flowering of science, astronomy, medicine, mathematics, chemistry, and philosophy resulted. Astronomy was used to determine the *calendar and also for astrological purposes. Observatories were set up; astronomical tables were published in the 9th century by Muhammad ibn-Mūsā al-Khwārizmī. The Caliph Hārūn al-Rashīd collected Greek texts as did his son al-Ma'mūn (813–33). Among the texts al-Ma'mūn obtained was Ptolemy's **Almagest* which is Arabic for 'the greatest'. The Islamic astronomers of this period included al-Farghānī who wrote a treatise on elementary astronomy, Ja'far Abū Ma'shar, who wrote on astrology, and Thābit ibn Qurrah (826–901), a theoretician of merit. In order to determine the exact diameter of the Earth, the Caliph al-Ma'mūn had the size of one degree of latitude measured. For many of their astronomical tasks the Arab astronomers used well-constructed *armillary spheres, *astrolabes, and *quadrants.

The greatest Arabian astronomer is reckoned to be al-Battānī, known in Europe as Albategnius, whose main works were carried out in the last quarter of the 9th century and the first quarter of the 10th century. A skilled observer and mathematician, al-Battānī carried the study of astronomy to great heights, making a number of important discoveries such as the sidereal movement of the apogee of the Sun's apparent orbit.

Many of the Arabian astronomers such as al-Battānī left astronomical tables. Among them are the Hakemite Tables, the Toledo Tables published by a Spanish astronomer Ibn al-Zarqālah (1029–87), and the Alfonsine Tables, in use until the middle of the 16th century and a product of the influence of Arabic astronomy in Spain. At Marāgheh, near Tabriz in Iran, Arabian astronomers such as Nāsir al-Dīn al-Tūsī, who died in 1274, and al-'Urdī built an observatory which included a library of 400,000 manuscripts. In the 15th century in Samarkand a further flowering of Islamic astronomy took place. A grandson of the Mongol ruler Tamerlane built a well-equipped observatory and, among other works, produced an original star catalogue. The number of learned Arabic astronomers, who often combined astronomy with mathematics and medicine or chemistry, is enormous. Their practical skills were great but in the end they did not bequeath to their successors a body of theory and cosmology worthy of their efforts. This may have been due to their inability to grasp fully the powerful tool of the scientific method. Nevertheless the contribution of Islamic astronomy to science is vast though indirect. During the Dark Ages in Europe, when astronomy was practically non-existent, Islamic astronomers preserved Greek science by keeping it alive in translation, extensions, and verifications by accurate observation.

isotropy, the property of an astronomical quantity where its value does not depend on the direction of measurement. Thus the *microwave background radiation is said to be isotropic (at least approximately) in that it has the same temperature no matter in which direction the detector is pointing. Again, in the interior of a spherical, non-rotating star, temperature, pressure, and density are isotropic in that their values are the same at the same distance from the star's centre and do not depend upon the direction from the centre.

J

Jansky, Karl Guthe (1905–50), US engineer who discovered radio waves from the centre of the Galaxy. This discovery eventually led to the development of *radio astronomy. Employed by the Bell Telephone Company in the US, Jansky was set the task of locating sources of radio interference which were disrupting radio communications. He built a steerable aerial (US antenna) which rotated on four wheels from a Model T Ford, and which came to be known as the 'merry-go-round'. With this instrument Jansky, in 1931, discovered an unusual source of radio 'noise' which appeared only when the centre of the Milky Way in Sagittarius passed through the beam of his aerial. Jansky published his results in 1932 but did not pursue his discovery. His name is given to an *astronomical unit which measures the amount of power received at the Earth's surface from a radio source.

Janus, a satellite of Saturn discovered in 1966 by the French astronomer Audouin Dollfus. Subsequently lost, it is now accepted to be satellite 1980S1 orbiting Saturn in an almost circular orbit of radius 151,472 km in a period of 0.69433 days. It is a small satellite of diameter no more than 220 km. Janus is a *co-orbital satellite with Epimetheus.

Jet Propulsion Laboratory (JPL), a laboratory directed by *NASA at Pasadena and Goldstone in California, USA. It specializes in space research, particularly in the guidance and tracking of interplanetary spacecraft. It operates a deep-space tracking system and was responsible for the Ranger, *Mariner, *Pioneer, and *Voyager spacecraft.

The **Jewel Box** cluster is one of the youngest known star clusters in the Galaxy, being only a few million years old. The central mass of approximately 50 stars contains some of the most luminous stars in the Galaxy; the brightest of these is about 80,000 times the luminosity of the Sun.

Jewel Box, a magnificent open *cluster situated close to *Crux in the far southern hemisphere. The red giant Kappa Crucis and other bright stars in the cluster were likened by John *Herschel to 'a casket of variously coloured precious stones', and the name derives from this description. The cluster lies at a distance of 2,300 parsecs (7,500 light-years).

Jodrell Bank, a radio *observatory near Manchester, UK, operated by the University of Manchester. It is the site of the world's first large steerable *radio telescope. The dish is 76.2 m (250 feet) in diameter and 92 m (300 feet) high. Founded by *Lovell, the telescope came into operation in 1957 and immediately made historic radar measurements on the carrier rocket of Sputnik 1, the first artificial satellite. In 1987, after thirty years of service, the telescope was renamed the Lovell telescope. It is still at the forefront of research, being a key element in *interferometry arrays, including the University of Manchester's *multi-element radio linked interferometer network (MERLIN) and in *VLBI experiments. Jodrell Bank is now formally known as the **Nuffield Radio Astronomy Laboratories**.

Jovian outer satellites, two groups of small satellites orbiting at the outskirts of Jupiter's system. The orbits of four satellites, Leda, *Himalia, *Lysithea, and *Elara, have semi-major axes of between 11 and 12 million km and inclinations of about 28 degrees to Jupiter's equator. Four other satellites, Ananke, Carme, *Pasiphae, and *Sinope, are about twice as distant from Jupiter and have retrograde orbits, that is, moving in the opposite direction to Jupiter's orbital motion. Diameters are in the range from 10 to 180 km and the composition is rocky. These objects are probably fragments of early satellites captured from the *asteroid belt just after the formation of Jupiter and subsequently broken up by impact events.

Julian calendar, the *calendar introduced by Julius Caesar in 46 BC and slightly modified under Augustus, in which the ordinary year has 365 days, and every fourth year is a leap year of 366 days. It has largely been superseded by the *Gregorian calendar, but it is still used in some of the republics of the former Soviet Union. Unlike more primitive calendars the Julian calendar contains months that are whole numbers of days in length.

Julian date, a time and calendar system introduced to provide unambiguous dates of celestial events. Julian day (JD) numbers are measured from mean noon on 1 January of the year 4713 BC, and give the number of days and fractions of a day that have elapsed since then. For example, the Julian date for 24 June 1962 is JD 2,437,839.5 when 24 June begins, while the time of an observation made on 24 June 1962 at 18.00 hours GMT is JD 2,437,840.25.

To create this composite photograph of **Jupiter** and its principal moons, Jupiter (*upper right* with its belts and zones clearly visible) and the four Galilean satellites were individually photographed by Voyager 1 during its fly-past in 1979, and assembled into this collage. Not to scale, but in their relative positions, are Io (*upper left*), Europa (*centre*), Ganymede (*lower left*), and Callisto (shown partially *lower right*). Twelve other much smaller satellites orbit Jupiter, four inside Io's orbit, and the outermost millions of kilometres from the planet.

The Lovell radio telescope at **Jodrell Bank** has a dish of 76.2 m in diameter. The small telescope in the foreground has a dish of 3 m in diameter which is permanently trained on the Crab Nebula supernova remnant in order to gather data on the Crab pulsar.

Juno, an *asteroid, discovered by the German astronomer Carl Harding in 1804. Its orbital semi-major axis, eccentricity, and inclination to the ecliptic are 2.671 astronomical units, 0.255, and 13 degrees, respectively. Juno's diameter is 244 km, and its albedo is 0.22. Light-curve *photometry has yielded a rotational period of 7.21 hours, and maximum brightness variations of 0.22 magnitudes.

Jupiter, the fifth *planet from the Sun and the largest planet of the solar system, easily visible with the unaided eye. Its mass is about twice as large as the sum of the masses of all the other planets, and is about one thousandth of that of the Sun. Like all the other giant outer planets, it emits more energy than it receives from the Sun, thus implying the existence of an internal energy source. Although Jupiter is essentially a gaseous body, it has an inner core composed of rocky material, with a mass of perhaps ten to twenty times that of the Earth, surrounded by a liquid envelope of lighter elements, and a massive atmosphere. The satellite system comprises four large satellites, discovered by Galileo at the beginning of the 17th century; they orbit about the planet in nearly circular, equatorial orbits. There are also a number of smaller satellites, divided into three groups. The first is composed of small satellites in orbits very close to the planet; the second and the third are on larger and more eccentric and inclined orbits; those in the case of the outermost group are retrograde, that is, move in the opposite direction to Jupiter's motion in its orbit. The abundance of light elements such as hydrogen and helium in Jupiter, compared with their relative deficiency in the inner planets, suggests that when the planet was formed the core was able to attract and trap the gas that fell on the forming planet and now constitutes most of its mass. The atmosphere is characterized by parallel bands that move with respect to each other, driven by Jupiter's internal energy source; darker bands are called **belts**, and lighter ones **zones**. At the boundaries between belts and zones, vortices are present. One of these is the long-known *Great Red Spot, a gigantic storm system larger than the Earth, in which clouds rise several kilometres above the clouds of surrounding regions. Jupiter is a strong emitter of radio signals. The magnetic field is much stronger than that of the Earth, with a correspondingly much larger *magnetosphere. The line joining the magnetic poles is tilted at about 10 degrees to the rotation axis, and the equivalent of the terrestrial *Van Allen radiation belts, in which electrically charged particles are trapped, extend out to include some of the major satellites. Observations from the *Voyager spacecraft led to the discovery of a tenuous *ring system about the planet, much less spectacular and massive than that of Saturn. It is mostly composed of particles about 10^{-6} m in size. Of limited lifetime because they escape into the surrounding space from the ring, they are probably replenished by the debris resulting from *meteoroids impacting on larger bodies. Of the two small innermost satellites of Jupiter, one happens to be within the ring and the other at its edge and it is possible that they are *shepherding satellites which play a role in maintaining the ring system.

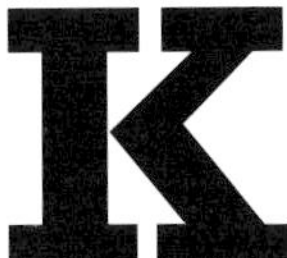

Kapteyn, Jacobus Cornelius (1851–1922), Dutch astronomer who founded and directed the Groningen Astronomical Laboratory. At this laboratory he undertook the measurement and analysis of photographic plates of the southern sky and produced the *Cape Photographic Durchmusterung*, a catalogue of 450,000 stars. By using statistical sampling techniques he was able to describe the structure of the Galaxy by measuring the *parallax, *magnitude, and proper motion of stars in small selected areas of the sky. Although Kapteyn's methods were sound, he made two erroneous assumptions. He neglected interstellar absorption and therefore underestimated the size of the Galaxy; also, as *Shapley later showed, Kapteyn was mistaken in assuming that the Sun was near the centre of the Galaxy.

Kelvin–Helmholtz contraction, the contraction of a star under its own gravitation, radiating energy into space as light and heat without deriving any energy from *nuclear synthesis. The concept was introduced by the British physicist Lord Kelvin and the German physicist Hermann von Helmholtz, who sought to explain the generation of stellar energy in this way. It fails to account for the very long lifetimes of the stars, but is thought to be the means by which stars contract in the earliest stages of *stellar evolution before nuclear synthesis begins. The **Kelvin time-scale** is the time a star would take to contract from infinite radius down to its present size. For the Sun this is about 25 million years, much less than the **nuclear time-scale** of 10^{10} years that it would take to convert its available hydrogen into helium.

The **Kitt Peak National Observatory**, located in the Arizona desert, is one of the most important US observatories. The aridity of the desert air and the high altitude of the observatory gives it very clear observing conditions, so that it can detect millimetre radio waves which would otherwise be absorbed by atmospheric water vapour. The millimetre radio telescope has detected most of the known types of interstellar molecules.

Kelvin scale of temperature, a temperature scale which has a degree size that is the same as the Celsius scale, but in which water melts at 273.15 K (this K representing the Kelvin scale) and boils at 373.15 K. At 0 K, a temperature known as absolute zero, the atoms and molecules of the body have zero *kinetic energy and have stopped moving. Absolute zero can be approached but not attained. Most celestial objects lose energy by radiation. Under these conditions the amount of energy that is lost obeys *Stefan's Law and is proportional to the fourth power of the Kelvin temperature of the object's radiating surface. Also the product of the Kelvin temperature and the wavelength of the radiation at which the maximum energy is lost is a constant, known as Wien's constant, given by *Wien's Displacement Law. This explains why cool stars have a red colour and the colour changes to orange, cream, white, and blue-white as one considers hotter and hotter stars.

Kepler, Johannes (1571–1630), German astronomer who was the first to describe accurately the elliptical orbits of the Earth and the planets around the Sun. Kepler worked with

Kepler's Laws

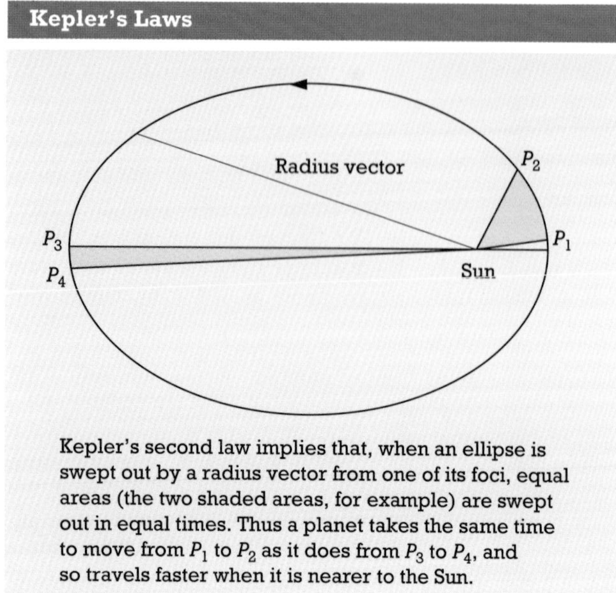

Kepler's second law implies that, when an ellipse is swept out by a radius vector from one of its foci, equal areas (the two shaded areas, for example) are swept out in equal times. Thus a planet takes the same time to move from P_1 to P_2 as it does from P_3 to P_4, and so travels faster when it is nearer to the Sun.

*Tycho Brahe at Tycho's observatory outside Prague and took over the observatory when Tycho died in 1601. Tycho left Kepler his tables of stellar and planetary positions. From these, after a great deal of analysis trying one mathematical model after another, Kepler deduced what are now known as *Kepler's Laws, but the physical explanation of these laws had to await *Newton's Law of Gravitation. Kepler also made discoveries in optics, general physics, and geometry.

Kepler's equation, an equation to find the position of a planet in an orbit described by *Kepler's Laws. It gives a relationship between the mean *anomaly M and the eccentric anomaly E. The equation is $M = E - e \sin E$, where e is the *eccentricity of the ellipse.

Kepler's Laws, three laws of planetary motion formulated by *Kepler from a study of the long series of naked-eye observations by the Danish astronomer *Tycho Brahe. They are:

The orbit of a planet about the Sun is an *ellipse with the Sun at one focus of the ellipse.

The **radius vector** (an imaginary line) joining a planet to the Sun sweeps out equal areas in equal times.

The squares of the *periods of revolution of the planets in their orbits are proportional to the cubes of their orbital semi-major axes.

Although subsequent more accurate observations have shown that celestial objects, such as planets orbiting the Sun or satellites orbiting planets, do not obey Kepler's Laws exactly, they are still a close enough approximation to the truth to remain useful.

Keyhole Nebula, a dark *nebula of dust embedded within a bright nebula in the constellation of Carina. The bright nebula surrounds the star Eta Carinae and is called the Eta Carinae Nebula. The Keyhole is named because of its shape and is part of the remnant of a large dark molecular cloud possibly two million years old containing many different kinds of molecules.

kinetic energy, the energy possessed by a body by virtue of its motion. It is the work the body could perform in coming to rest; for example, the ability of a moving hammer to drive in a nail. If the body is moving linearly with a velocity v and it has a mass m then its kinetic energy is $\frac{1}{2}mv^2$. If the body is rotating with angular velocity ω and has a moment of inertia I then its kinetic energy is $\frac{1}{2}I\omega^2$.

Kirkwood gap, any of several narrow regions in the belt of *asteroids where very few bodies are found. These gaps, named after the US astronomer Daniel Kirkwood who pointed them out in 1867, correspond to orbits characterized by resonances with Jupiter, that is, having *periods nearly equal to simple fractions ($\frac{3}{1}, \frac{5}{2}, \frac{7}{3}, \frac{2}{1}$) of Jupiter's orbital period. This causes the orbits of any bodies in the gaps to become unstable and, governed by *chaos theory, they may eventually make close encounters with planets. It is probably this mechanism that is responsible for putting *meteoroids and *Apollo–Amor objects into orbits crossing those of the inner planets.

Kitt Peak National Observatory, an *observatory in Arizona, USA, with a wide variety of telescopes. These include the *McMath solar telescope of 1.5 m (59 inches) aperture, the Mayall *reflecting telescope with an aperture of 4 m (157 inches), and two radio *dish telescopes of 11 m (36 feet) and 12 m (39 feet) diameter. The observatory's high altitude of 2,060 m (6,750 feet) and its wide range of facilities make it one of the most active and important observing stations in the world. Organizations operating its facilities include the Association of Universities for Research in Astronomy (a group of seventeen US university astronomy departments), the *National Radio Astronomy Observatory, and the *Steward Observatory. The Mayall reflector has a twin at the *Cerro Tololo Inter-American Observatory in Chile.

Kohoutek's Comet, a long-period *comet discovered in 1973. It was found by the Czech astronomer Lubos

These three photographs, spanning a period of twelve years, of the binary star **Krüger 60** clearly show the secondary star orbiting the primary star. The plane of the orbit is inclined to the line of sight so that the orbit appears to be more oval than it really is. However, the true shape of the orbit can be reconstructed from the apparent shape.

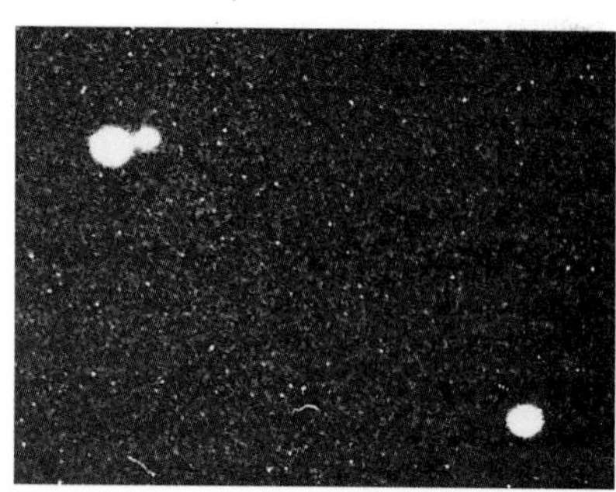

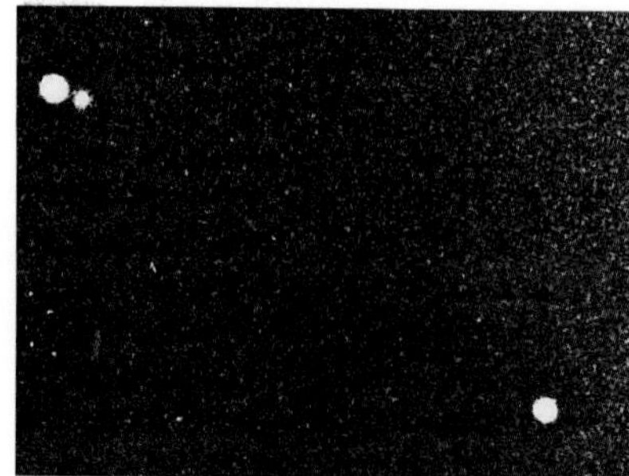

Kuiper Airborne Observatory

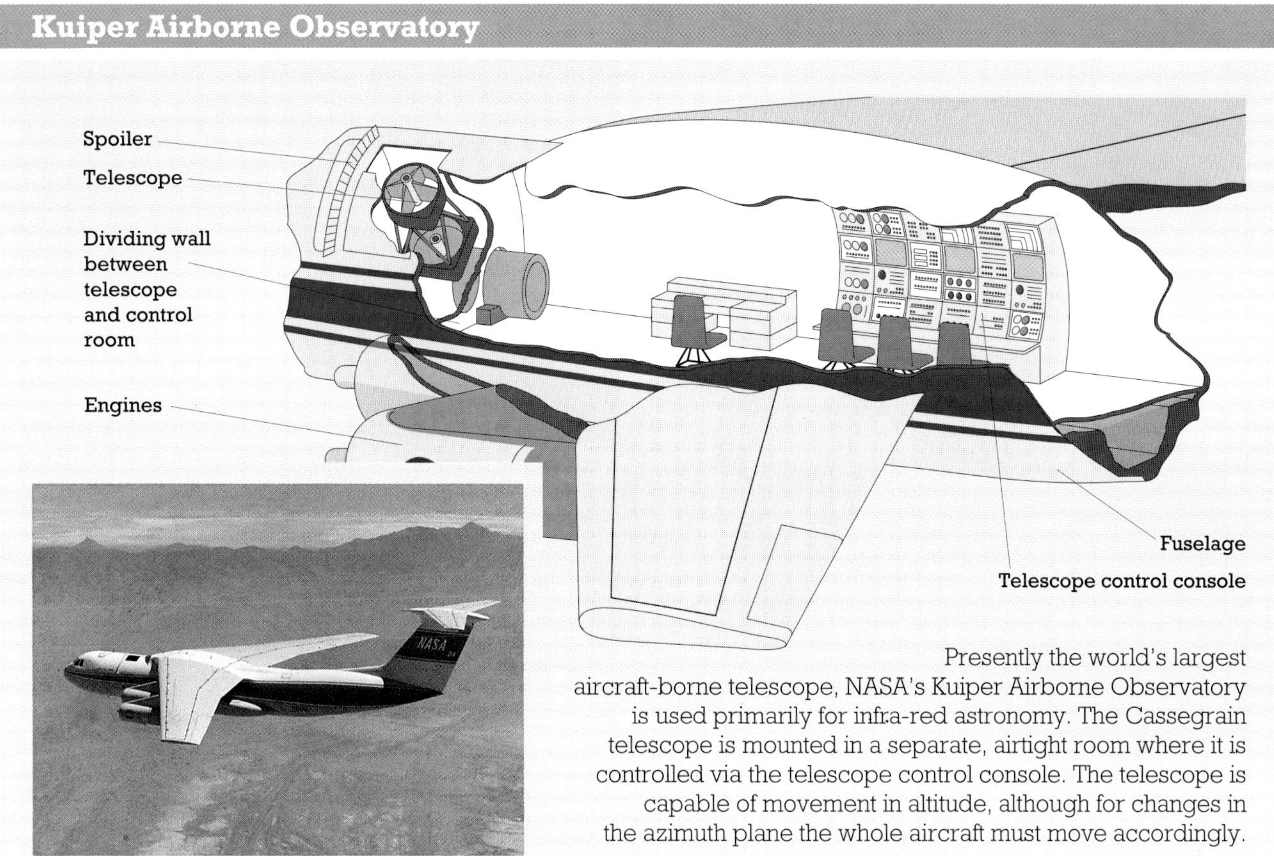

Presently the world's largest aircraft-borne telescope, NASA's Kuiper Airborne Observatory is used primarily for infra-red astronomy. The Cassegrain telescope is mounted in a separate, airtight room where it is controlled via the telescope control console. The telescope is capable of movement in altitude, although for changes in the azimuth plane the whole aircraft must move accordingly.

Kohoutek who was looking for the long-lost fragments of *Biela's Comet. Kohoutek's Comet was at the remarkably large distance of 4.75 astronomical units from the Sun when it was discovered. If it had brightened in the way that most comets do, it would have been spectacular when it reached its perihelion of 0.14 astronomical units. This failed to happen, but nevertheless the comet was comprehensively observed. A large bend in its tail was interpreted as being the result of a reversal in the direction of the magnetic field of the *solar wind. Ultraviolet observations indicated that the comet was surrounded by a *coma of hydrogen that was larger than the Sun and was produced by the loss of a million tonnes of water per day. For the first time infra-red observations of the light from the dust tail indicated the presence of silicates.

Krüger 60, a nearby red dwarf *binary star in the constellation of Cepheus, with a 45-year orbit. The brighter component is of tenth magnitude, and the fainter, of eleventh magnitude, is a *flare star known as DO Cephei, and is one of the least massive stars known, with about one-seventh the Sun's mass. The distance of Krüger 60 is 4 parsecs, and its combined luminosity is about 500 times less than that of the Sun.

Kuiper Airborne Observatory (KAO), a NASA aircraft observatory equipped with a reflecting telescope of 91 cm (36 inches) diameter. A converted Lockheed Starlifter, the KAO enables astronomical observations to be made in the more rarefied higher regions of the atmosphere. It is used primarily for infra-red astronomy because at an altitude of 12.5–13.7 km (41,000–45,000 feet) it flies above most of the infra-red-absorbing water vapour in the Earth's atmosphere. It is frequently used to observe the solar system and the Galaxy. Notable discoveries have included the detection of water in *Halley's Comet, the ring system of *Uranus, and identification of atomic and molecular constituents of the interstellar medium. Though normally based in California, the KAO also flies from Hawaii and New Zealand for studies of southern-sky objects such as the centre of the Galaxy and *Supernova 1987A. At high altitude the *scattering of light is reduced, and the KAO has been used for *occultation measurements, playing a major role in the discovery of the rings of Uranus. The observatory is named after the Dutch–American planetary astronomer and space pioneer Gerard P. Kuiper (1905–73).

L

Lacerta, the Lizard, a *constellation of the northern sky, introduced by the Polish astronomer Johannes Hevelius (1611–87). It lies on the edge of the Milky Way and has been the site of three novae during the 20th century. Its most celebrated object is BL Lacertae, originally thought to be a peculiar fourteenth-magnitude variable star but now recognized as one of a family of elliptical galaxies with active centres believed to be related to *quasars.

Lagoon Nebula, a *nebula in the constellation of Sagittarius. It is important because it displays small round dark areas known as Bok globules and thought by many astronomers to be stars condensing out of the dust and gas. It is bright enough to be visible in binoculars, though to see its detailed structure a telescope of 54 cm (21 inches) aperture should be used.

The **Lagoon Nebula** contains dark clouds of dust, called Bok globules, from which stars may be forming. One of these dark nebulae, which can be seen at the bottom left, is called the Dragon and is being intensively studied by astronomers who want to learn more about how stars are born.

Lagrangian points, places in the vicinity of two orbiting masses where the total acceleration due to gravitational and centrifugal forces is zero. There are five Lagrangian equilibrium points. Three are located along the line joining the centres of mass of the bodies: one in between the two masses and two outside, one on each side. They are therefore **collinear** points. The remaining two are located at the vertices of the equilateral triangles having the two orbiting masses at the other vertices. In the case of the Sun and Jupiter, these locations are occupied by the *Trojan group of asteroids.

Lalande 21185, an eighth-magnitude star in the constellation Ursa Major, and one of our nearest stellar neighbours. It is a red *dwarf star at a distance of 2.5 parsecs, and its luminosity is about 200 times less than that of the Sun.

La Palma Observatory (LPO), a major *observatory at a height of 2,369 m (7,770 feet) on La Palma in the Canary Islands. Officially known as the **Observatorio del Roque do los Muchachos** or the **Northern Hemisphere Observatory**, it is part of the Instituto de Astrofisica de Canarias. The excellent seeing conditions at such an altitude were first noted by the Astronomer Royal for Scotland, Charles Piazzi Smyth, over a century ago. Use of the observatory, and of the Observatories del Teide del IAC on the neighbouring island of Tenerife, is shared by several European countries including Spain, Norway, Denmark, Finland, Sweden, the United Kingdom, The Netherlands, the Republic of Ireland, and Germany. The La Palma Observatory has a number of large telescopes. The Isaac Newton group of telescopes comprises the 1 m (39-inch) Jacobus Kapteyn telescope, the 2.5 m (100-inch) *Isaac Newton telescope, and the 4.2 m (165-inch) William Herschel telescope, the third-largest single-mirror telescope in the world. The telescopes, together with their computerized recording equipment, are managed and maintained by the *Royal Greenwich Observatory. Other instruments on the site include the Carlsberg Meridian Circle, the Swedish solar telescope, and the Nordic 2.5 m (100-inch) telescope. See the photographs on page 83.

Lagrangian points

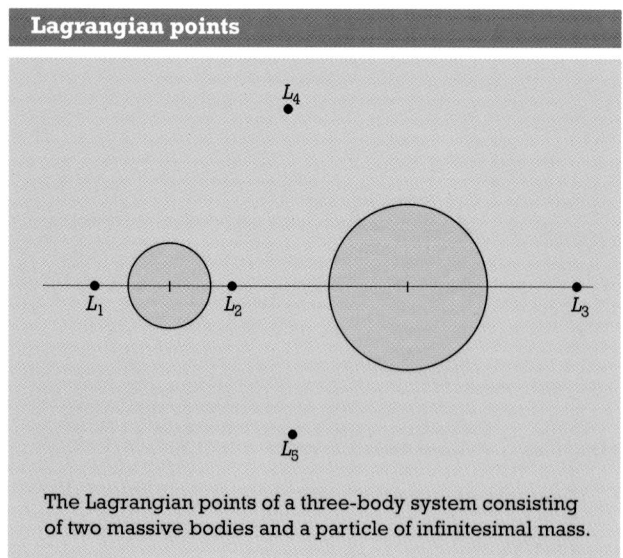

The Lagrangian points of a three-body system consisting of two massive bodies and a particle of infinitesimal mass.

Laplace, Pierre-Simon, marquis de (1749–1827), French mathematician and astronomer who demonstrated the stability of the solar system from *Newton's Law of Gravitation. He incorporated his mechanics of the solar system in his *Méchanique Celeste* (1799–1825) which included the proposal of what is now known as the *Laplace nebular hypothesis of the birth of the solar system. He also conceived the idea that a star might be so massive that particles of light could not escape from it, similar to the modern *black hole theory.

Laplace nebular hypothesis, a theory of the origin of the solar system proposed by the French scientist *Laplace in 1796. He suggested that a rotating cloud of gas, contracting under its gravitational field, would rotate faster to conserve angular momentum and so be forced to take the shape of a spinning disc. A ring of gas would be left behind by the contracting disc whenever the growing centrifugal force equalled the gravitational force. The material in each ring would collect to form a *proto-planet. This process would continue, and it was thought possible that the system of satellites of a planet would form in a similar way from the planet as it contracted. The Sun would form at the centre of the whole system.

The hypothesis certainly explained a number of properties in the solar system such as its flatness, the almost circular orbits, and the velocities of the bodies in the same direction. More recently, however, severe objections have been raised against it, including the enormous preponderance of mass within the Sun (99.9 per cent of the solar system's mass) compared to the Sun's small percentage of angular momentum (3 per cent of the solar system's). In addition, gas would not condense to form planets. Nevertheless, versions of the nebular hypothesis, such as *Weizsacker's theory utilizing modern ideas, still form the most plausible theories of the creation of the solar system.

Large Magellanic Cloud *Magellanic Clouds.

late-type star, a star of *spectral class K, M, C, or S and with a temperature of less than about 5,000 K. In an early theory of *stellar evolution, the stars were thought to spend most of their lives cooling from 'early type' (B, A) through intermediate types (F, G) to 'late types'. These terms are still in use although the theory has long been abandoned. Late-type stars of all *luminosity classes are found, including *red giants and red *dwarfs.

latitude *prime meridian.

leap year *calendar.

The **La Palma Observatory** (*below*) is located on the lip of a large volcanic crater. The telescope domes contain, from left to right in the photograph, the Royal Greenwich Observatory's 4.2 m William Herschel telescope (also shown *right*), which was commissioned in 1987, the 2.5 m Isaac Newton telescope, and the 1 m astrometric telescope. The tower houses the Swedish solar telescope.

Lemaitre model *cosmology.

lenticular galaxy, a type of *galaxy that appears lens-shaped. It shows no sign of spiral structure but does possess a central nucleus and traces of a disc. Those stars that are visible appear to be old. In the *Hubble classification such galaxies are termed So galaxies.

Leo, the Lion, a *constellation of the zodiac. In mythology it is the lion slain by Hercules as the first of his twelve labours. Leo's most distinctive feature is a sickle-shape of stars that marks the lion's head and chest. Its heart is marked by the brightest star in the constellation, first-magnitude *Regulus, while the lion's tail is marked by the star *Denebola. Gamma Leonis, known as Algieba, is a pair of yellow stars of second and third magnitudes; binoculars show a wider fifth-magnitude star. R Leonis is a red giant variable star like *Mira Ceti that varies between fourth and eleventh magnitude every 310 days. Four galaxies are visible in small telescopes under good conditions: M65 (NGC 3623), M66 (NGC 3627), M95 (NGC 3351), and M96 (NGC 3368). The Sun passes through the constellation from mid-August to mid-September. Every November the *Leonid meteors appear to radiate from Leo.

Leo Minor, the Little Lion, a *constellation of the northern sky, introduced by the Polish astronomer Johannes Hevelius (1611–87). Its brightest stars are of only fourth magnitude and there are no objects of special interest though it does contain a number of small and faint galaxies within its borders.

Leonids, a *meteor swarm that occurs approximately every 33 years on about 13 November. Leonid observations in 1799 led to the concept of the *radiant. On the night of 12 November 1833 some 240,000 meteors were seen during a period of seven hours. The Leonids were correctly predicted to return in 1866, and were well observed. The calculation of the orbit of the *meteor stream showed that it was very similar to the orbit of comet 1866 I. This led to the realization that meteor streams were the product of cometary decay.

Lepus, the Hare, a *constellation of the southern celestial hemisphere, one of the forty-eight constellations known to the ancient Greeks. It represents a hare crouched beneath the feet of Orion, the hunter. Its brightest star is the third-magnitude Alpha Leporis, known as Arneb from the Arabic meaning 'hare'. R Leporis is a red giant variable star of the *Mira Ceti type that varies between about sixth and twelfth magnitudes every 430 days. In the 19th century, the British astronomer John Russell Hind drew attention to its deep red colour which led to the name Hind's Crimson Star. Gamma Leporis is a binocular double consisting of components of fourth and sixth magnitude. The constellation contains one object from the *Messier catalogue, the globular *cluster M76.

Le Verrier, Urbain Jean-Joseph (1811–77), French astronomer whose work led to the discovery of *Neptune in 1846. Sixty years after its discovery, *Uranus was already deviating from the expected orbit based on the gravitational attraction of the Sun and the known planets. Le Verrier assumed that the irregularities were caused by an undiscovered, more distant planet lying close to the *ecliptic and obeying *Bode's Law. In an analytical *tour de force*, Le Verrier predicted where the planet should be found; it was discovered on the very first night of searching. After the discovery, an Anglo-French dispute over priority erupted because of an earlier, similar, but unpublished prediction by *Adams in Cambridge. Le Verrier subsequently embarked on a vast re-analysis of motions in the solar system, which was finally completed just weeks before his death. Planetary masses were revised, as were the size of the solar system and the *velocity of light. The revised velocity was soon confirmed in the laboratory by Léon Foucault. Mercury's orbit was found to show a minute residual motion of 38 arc seconds per century. Le Verrier tried to explain this by an undiscovered planet or asteroid belt within Mercury's orbit, but the drift was in fact a harbinger of general *relativity and the breakdown of *Newton's Law of Gravitation. In 1854 Le Verrier was appointed director of the Paris Observatory. His leadership was controversial and he was dismissed in 1870 but was reinstated in 1873. Outside astronomy, Le Verrier instituted an international network of weather reports and forecasts, and sat in the French senate.

Lexell's Comet, a long-period *comet discovered in 1770 and named after the astronomer who calculated its orbit. It approached within 2.26 million km of the Earth during that year. On its inward journey the comet actually passed between the satellites of Jupiter and remained near them for

This picture of the **Leonid** meteor swarm of 12 November 1799 was made by Andrew Ellicott off Cape Florida, USA. The Leonid swarm is probably the best known of the periodic meteor showers.

four months. The fact that these suffered no orbital change indicates that the comet had a mass that was at most 1/5,000 that of the Earth. This close passage of Jupiter was thought to have resulted in Lexell's Comet being perturbed into a short-period orbit. The comet had an inner solar system orbit with a period of 5.6 years and it was the second periodic comet to be discovered. As it moved away from the Sun it re-encountered Jupiter in the summer of 1779. The orbit changed once more and the comet has never been seen again.

Libra, the Scales, a *constellation of the zodiac. The ancient Greeks knew this area of sky as the claws of the neighbouring scorpion, Scorpius, but about 2,000 years ago the Romans made it into a separate constellation, representing a pair of scales. Sometimes it is visualized as the scales of justice held by neighbouring Virgo. The stars Alpha and Beta Librae are known as Zubenelgenubi and Zubeneschamali, Arabic names that mean 'southern claw' and 'northern claw', a reminder of the constellation's origin.

libration, the rocking motion of an orbiting body. For example, while the Moon generally keeps the same face towards the Earth, over a period of time 59 per cent of its face can be seen from the Earth, the parts near the edge of the Moon's disc becoming alternately visible and hidden. This apparent rocking motion has three main causes.

The Moon rotates on its axis at a constant rate, but its orbital velocity about the Earth varies because its orbit is elliptical. Thus, in the illustration, mountain *A* appears to move back and forth across the line from the Earth to the Moon. This results in **libration in longitude**.

The Moon's equator is inclined to the plane of its orbit. This causes the Moon to appear to nod up and down each month and is known as **libration in latitude**.

The Earth rotates faster than the Moon revolves in its orbit about the Earth so that when the Moon rises we can see slightly over the western edge of the Moon's disc and at moonset slightly over the eastern edge. This is due to differences in line of sight and is called **diurnal libration**.

Mercury exhibits librational behaviour with respect to the Sun, and *Hyperion and *Titan with respect to Saturn. Physically, libration occurs when two motions are locked in resonance.

Lick Observatory, an *observatory on the 1,280 m (4,200 feet) summit of Mount Hamilton, overlooking San Jose, California, and run by the University of California, Santa Cruz. It was the first permanent mountain-top observatory and is named after James Lick who bequeathed sufficient money for the construction of a 36-inch (91 cm) *refracting telescope, completed in 1888. The observatory also obtained the 36-inch (91 cm) Crossley *reflecting telescope and used this for spectroscopic work and the photography of nebulae. The sixth, seventh, and ninth satellites of Jupiter were discovered with this instrument, as was the existence of interstellar matter. In the 1940s a double *astrograph was added to take simultaneous blue and yellow sky photographs. The 3 m (118-inch) Shane reflector was commissioned in 1959.

life *extraterrestrial life.

light, *electromagnetic radiation to which the human eye is sensitive. It can be regarded either as a wave motion or as a stream of particles: the photon is the 'particle' of wave energy. This visible radiation region lies between about 380 and 750 nm *wavelength, the actual boundaries varying slightly from person to person. The eye is not equally sensitive to all the wavelengths in this region, the sensitivity being low at the boundaries and maximizing in the green-yellow part between around 500 and 550 nm. Light occupies only a narrow portion of the total electromagnetic spectrum, a portion sandwiched between the ultraviolet and the infra-red. The colour of the light changes with the wavelength and white light can be broken into a spectrum of colours using a prism, diffraction grating, or a rainbow-producing raindrop. *Newton categorized this spectrum in terms of seven colours: red, orange, yellow, green, blue, indigo, and violet. The Sun emits over 50 per cent of its radiated energy in the visible wavelength band. The Earth's atmosphere is also reasonably transparent in this region, so the vast majority of astronomical observations have been taken in this wave-length range. Only in recent times, with the advent of space

Libration

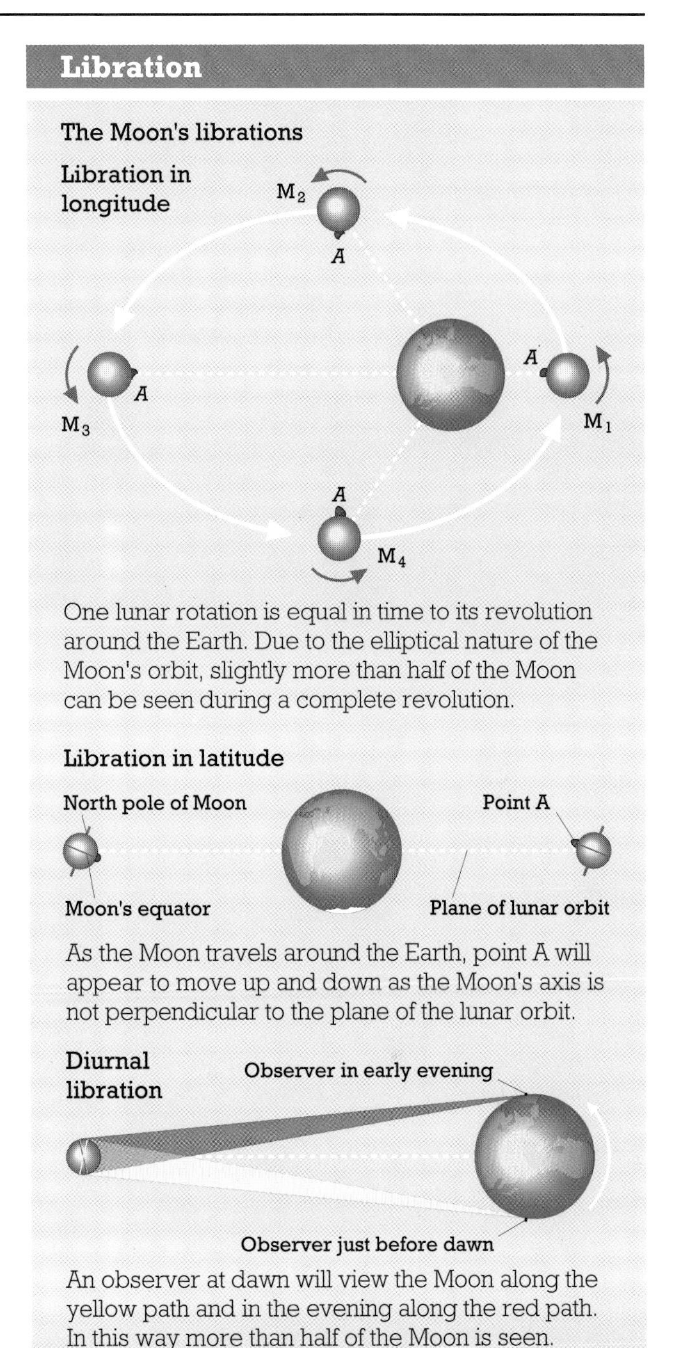

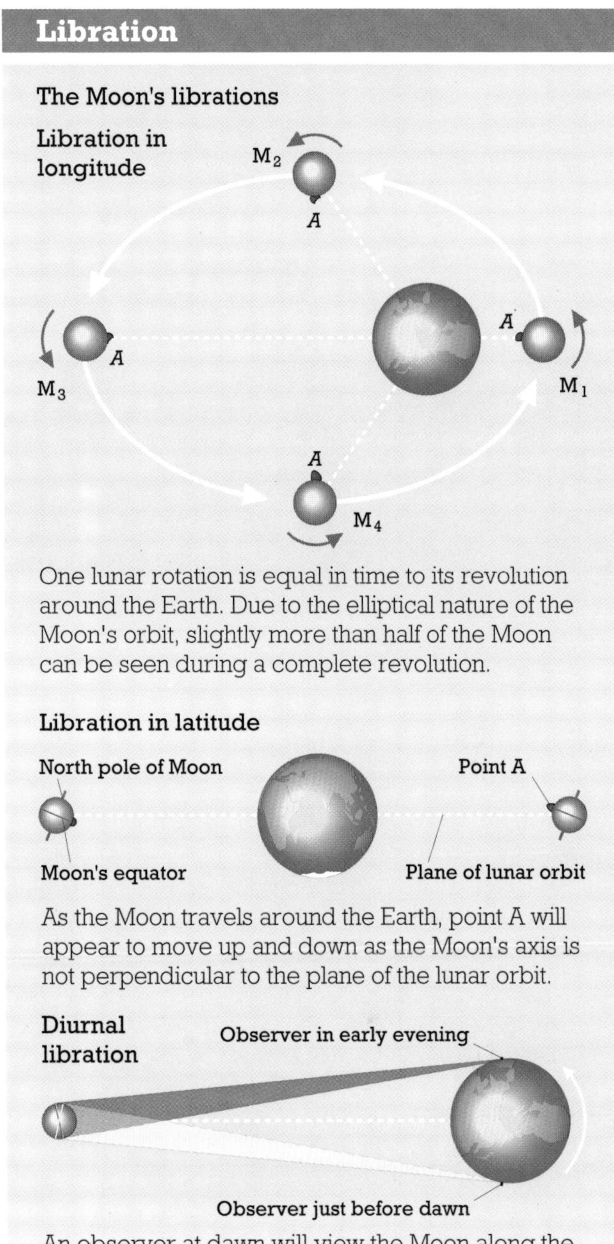
The Moon's librations
Libration in longitude
M2
A
A
M3
A
M1
A
M4
One lunar rotation is equal in time to its revolution around the Earth. Due to the elliptical nature of the Moon's orbit, slightly more than half of the Moon can be seen during a complete revolution.
Libration in latitude
North pole of Moon
Point A
Moon's equator
Plane of lunar orbit
As the Moon travels around the Earth, point A will appear to move up and down as the Moon's axis is not perpendicular to the plane of the lunar orbit.
Diurnal libration
Observer in early evening
Observer just before dawn
An observer at dawn will view the Moon along the yellow path and in the evening along the red path. In this way more than half of the Moon is seen.

Local Group of galaxies

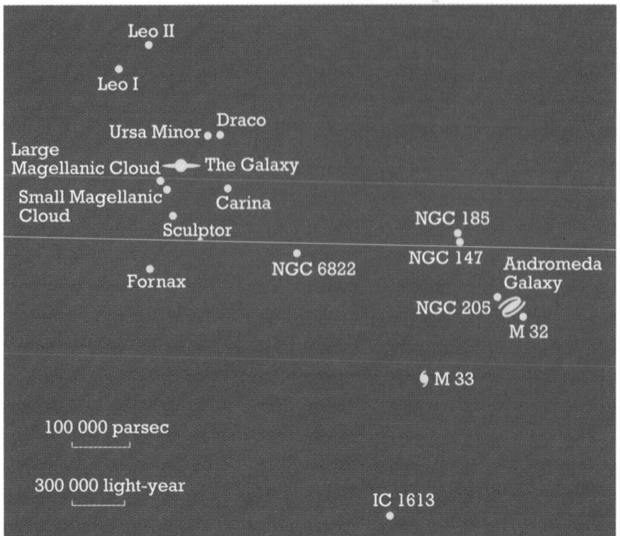

The Local Group of galaxies viewed so that our galaxy is seen edge on. The Local Group is dominated by the Milky Way and Andromeda spiral galaxies, around which smaller galaxies cluster.

travel and sophisticated instrumentation, has observational astronomy expanded to encompass ultraviolet, X-ray, gamma-ray, infra-red, millimetre, and radio regions of the electromagnetic spectrum.

light-year, the distance covered in one year by light or other electromagnetic radiation travelling in a vacuum. One light-year is equal to 9.46×10^{15} m, 63,000 *astronomical units, or 0.31 *parsecs. Although astronomers more usually use the parsec for stellar and larger distances, the light-year more easily indicates light travel times. For example, a radio message would take 2 million years to travel the 2 million light-years to the Andromeda Galaxy.

limb, in astronomy, the edge of a celestial object's visible disc. Only those objects such as the Sun, Moon, planets, and some satellites viewed in telescopes display discs to which the term may be applied. The eruptions of the Sun's surface, such as *flares and *prominences, are best viewed at the limb.

limb brightening, the increase in the strength of radio and X-ray radiation as observations in those wavelengths move from the centre of the Sun's disc to the edge or limb. Radiation in these wavelengths comes from the upper *chromosphere and lower *corona. Consequently, the longer oblique paths producing radiation nearer the limb than the centre produce radiation of higher intensity.

limb darkening, the darkening of the Sun's disc near its edge compared with its centre. The edge is darkened because light coming to the observer from the centre passes radially through the layers of the atmosphere while light from a similar *optical depth near the limb has to approach the observer obliquely through a greater thickness of the atmosphere. The observer can therefore see into the Sun's *photosphere to hotter depths near the centre of the disc.

Certain properties of the light variation from some *eclipsing binaries reveal that these stars also exhibit limb darkening. When the eclipse of one star by the other begins, the fall in brightness is initially slow, but then speeds up as the brighter central region of the eclipsed star's disc is obscured.

line of apsides, the major axis of the *ellipse described by the first of *Kepler's Laws. These laws apply to the orbit of a planet around the Sun, when the perturbations of the other planets are ignored. Both the perihelion and the aphelion of a Keplerian orbit occur along the line of apsides.

line of nodes, the intersection between the orbital plane of an orbit and a **reference plane**. For a planet orbiting around the Sun, the reference plane is usually taken to be either the *ecliptic or the *invariable plane; for a satellite orbiting the Earth, the reference plane is usually the equatorial plane.

Little Dipper *Ursa Minor.

Local Group of galaxies, the cluster of twenty or so galaxies to which the *Galaxy belongs. The Group is dominated by three *spiral galaxies: our own galaxy (2×10^{11} times the mass of the Sun), the Andromeda Galaxy (3×10^{11} times the mass of the Sun), and M33 which, at 10^{10} solar masses, is less than a tenth of the mass of the other two galaxies. Other members of the Group are the *Magellanic Clouds and small elliptical and irregular galaxies, most of which cluster around the two large spiral galaxies. The smallest known member of the Local Group is a system in the constellation of Carina and was discovered only in 1977. Its mass is comparable to those of globular *clusters, though its form is much looser. The difficulty in detecting it suggests that many other low-mass galaxies remain to be discovered in the Local Group. The diameter of the Local Group is approximately 1 million parsecs (3 million light-years). In comparison to other clusters of galaxies, such as the *Virgo Cluster, the Local Group contains relatively few galaxies, of relatively small mass.

Local Supercluster, a large group of some 50,000 galaxies containing the *Local Group of galaxies to which our galaxy belongs. Also known as the Virgo Supercluster because it contains the large *Virgo Cluster, it includes several nearby clusters. The Supercluster is roughly lens-shaped with a diameter of at least 20 megaparsecs.

Local System *Gould's Belt.

longitude *prime meridian.

Loop Nebula *Cygnus Loop.

Lorentz–Fitzgerald contraction, the shortening of the length of a moving body in the direction of its motion. This contraction was offered as an explanation for the inability of the Michelson–Morley experiment to detect the movement of the Earth through the *ether. Formulated independently by the Irish physicist George Fitzgerald (1893) and the Dutch physicist Hendrik Lorentz (1895), it suggested that the length of a moving body as measured by a stationary observer was dependent on its velocity. The effect, which becomes noticeable only at velocities near the speed of light, was incorporated into Einstein's special theory of *relativity.

Lorentz transformation, a set of mathematical relations enabling the coordinates of a point measured in space and time in one frame of reference to be changed into the coordinates of the same point measured in another frame of reference. Both frames must be inertial, that is, moving with respect to each other with a constant relative velocity v. The relations involve a factor $\beta = \sqrt{(1 - (v/c)^2)}$, where c is the velocity of light. The set, used in the special theory of *relativity, reduces to the set used in classical mechanics when v is much less than the velocity of light.

Lovell, Sir (Alfred Charles) Bernard (1913–), British radio astronomer. Lovell was educated in Bristol and later joined the University of Manchester, researching cosmic rays. In 1946 he was awarded the OBE for the work he did on radar during World War II. After the war he returned to research on cosmic rays. Lovell founded the *Jodrell Bank Experimental Station, a permanent research establishment provided by the University of Manchester, and was its director from 1951 to 1981. Jodrell Bank is also the site of the world's first large steerable radio telescope, completed in 1957. The 76.2 m (250-foot) Mark I radio telescope is now called the Lovell telescope.

Lowell, Percival (1855–1916), US businessman and astronomer. Lowell was the eldest son of an illustrious Boston family. A generous patron and an energetic organizer of astronomical research, he built a private observatory at Flagstaff, Arizona, which is now used by the astronomical community and known as the *Lowell Observatory. His fascination with astronomy started in 1877 when the Italian astronomer Giovanni Schiaparelli discovered the *canals of Mars. Lowell conducted lengthy observations at his observatory and produced a series of maps and a set of books mainly concerned with the possibility that the canals were constructed by intelligent life. In the early part of the 20th century he made a careful study of the orbits of Uranus and Neptune. He noted that these orbits exhibit irregularities which he attributed to another planet beyond Neptune. In 1905 he funded and organized a systematic search, and fourteen years after his death Pluto was finally discovered by Clyde Tombaugh.

Lowell Observatory, an *observatory at Flagstaff, Arizona, 2,210 m (7,250 feet) above sea-level. *Lowell used the 24-inch (61 cm) refracting telescope to observe the *canals of Mars. He also predicted the existence of another planet beyond Neptune by studying the irregularities in the orbits of Uranus and Neptune. The search for this planet started in 1905 using an equatorially mounted 5-inch (13 cm) photographic objective lens. Pluto was finally discovered in 1930 by Clyde Tombaugh, a young assistant at the observatory. In 1916 Vesto Slipher, using the facilities of the observatory to measure the *red shift of a sample of galaxies, found that they are receding from us at high velocities.

luminosity, the radiation output per unit time from a star or other celestial object, usually expressed as a multiple of the Sun's output. It is a measure of **brightness** and is given by the formula

$$\log (L/L_{\odot}) = 0.4\,(M_{\odot} - M)$$

where L and M are the luminosity and absolute *magnitude of the object, and $L_{\odot}$ and $M_{\odot}$ are the corresponding quantities for the Sun. The comparison is usually made in visible light, but can be made in other ranges of wavelength. For total radiation output the luminosity is either determined with a *bolometer or is calculated from multi-wavelength *photometry.

Luminosity classes are groups of stars which do not lie on the main sequence of the *Hertzsprung–Russell diagram.

lunar calendar, a *calendar based on the variations of the phases of the Moon as seen from Earth. The lunar year contains 12 *synodic months, these consisting of 29.5305882 days, the synodic month being defined as the time interval (*synodic period) between new moons. Therefore, the lunar year of 354.3672 days is about 11 days shorter than the solar year of 365.24219 days. A true lunar calendar quickly gets out of step with the seasons and is often replaced by a lunisolar calendar in which every third or fourth year contains 13 as opposed to 12 lunar months, a leap month being intercalated (added) as required. Islamic, Jewish, Hindu, Buddhist, and all Christian festivals except Christmas itself are timed according to the lunar calendar. The synodic month was probably the first exact interval of time to be calculated by ancient civilizations.

lunar exploration The *Moon is our nearest neighbour in space and at times the most prominent and brightest object in the night sky. Its nature has long fascinated the human race and it was a favourite object of study in the first centuries of telescopic observation, but as telescopes improved and astronomical photography developed, most observing time became devoted to other celestial objects and the Moon's brightness became a nuisance to astronomers. With the advent of rockets a new era in lunar exploration began, firstly with close-up photography of the Moon's hitherto unseen far side, then with the crash-landing and, later, soft-landing of instrument packages on its surface. This last phase was followed by *Project Apollo which landed astronauts on the Moon where they set up batteries of scientific equipment which operated long after the astronauts left the lunar surface. The astronauts also brought back about one-third of a tonne of lunar material which has been extensively analysed in terrestrial laboratories and which has provided an enormous amount of information about the Moon's origin, age, and geological evolution. The Soviet Union has successfully launched spacecraft carrying their Lunokhod vehicles for lunar exploration. The vehicles' movements and operations were controlled from Earth. Some small samples of lunar material have also been brought back to the Earth in such operations. Towards the end of the 20th century a new era in lunar exploration will begin if the present plans for self-sustained Moon bases providing shelter and support for human beings are realized.

lunar theory, a mathematical description of the *Moon's positions and distances from the Earth's centre over a period of time, usually some thousands of years. Lunar theory must be able to account precisely for the measured positions of the Moon. Many of the greatest mathematical astronomers such as *Newton, Peter Hansen (Denmark), Alexis-Claude Clairaut, and the mathematicians Leonhard Euler (Switzerland), Charles Delaunay (France), George Hill (USA), and Ernest Brown (UK) have worked on lunar theories, invariably spending years on the task. Lunar theory is difficult because not only must the gravitational effects of the Sun on the Moon's *trajectory (see *evection) be taken into account, but also the effects of the planets and the Earth. Furthermore, the Earth and the Moon are only approximately spherical, and their departures from true spheres have measurable effects. In addition, the tides raised by the Moon and Sun in the Earth's oceans produce a slow evolution in the lunar orbit, detectable from a study of ancient *eclipse records.

Brown's lunar theory contains 1,500 separate mathematical terms and was the most exhaustive treatment of the problem in pre-electronic computer days, taking him twenty years to complete. Successive improvements in accuracy over the past 300 years, especially in the recent development of radar and laser-ranging techniques, have rendered each new lunar theory obsolete. Today such a theory should give the Moon's distance from the Earth's centre (about 384,000 km) to an accuracy of a few centimetres.

lunation *synodic month.

luni-solar precession *precession.

Lupus, the Wolf, a *constellation of the southern hemisphere of the sky, one of the forty-eight constellations known to the ancient Greeks. It represents a wild animal impaled on a pole by neighbouring Centaurus, the Centaur. Lupus lies in a rich area of the Milky Way and contains numerous interesting double stars for small telescopes, among them Epsilon Lupi, Kappa Lupi, Mu Lupi, and Xi Lupi. NGC 5822 is a large open *cluster visible in binoculars.

Lynx, a *constellation of the northern sky, introduced by the Polish astronomer Johannes Hevelius (1611–87). He named it the lynx because, he wrote, one must have the eyes of a lynx to see its stars, which are mostly very faint. It contains several interesting double stars for small telescopes but is otherwise unremarkable.

Lyra, the Lyre, a prominent *constellation of the northern sky, one of the forty-eight constellations known to the ancient Greeks. In mythology it represents the instrument played by the musician Orpheus, although it has also been seen as an eagle or vulture. Its brightest star is *Vega. Near Vega is Epsilon Lyrae, a star popularly known as the 'Double Double'. Through binoculars it appears as a wide double star, but telescopes show that each star is itself a twin, making this a rare quadruple star. Beta Lyrae is a cream and blue double star for small telescopes; the cream component is the prototype of a class of variables known as *Beta Lyrae stars. Delta Lyrae appears as a red and blue double star through binoculars; the red component varies between fourth and fifth magnitudes. Zeta Lyrae is another double star easily seen in small telescopes. RR Lyrae, a faint star on the border with *Cygnus, is the prototype of a famous class of variables, the *RR Lyrae stars. Between Beta and Gamma Lyrae lies the famous *Ring Nebula; the globular *cluster M56 is also found in the boundaries of Lyra. Every April the *Lyrid meteors radiate from the constellation.

Lyrids, a rather weak *meteor shower that reaches its peak on about 22 April each year. It was first recorded in 687 BC in the chronicles of Chou Chuang Wang. The shower is produced by the *meteor stream resulting from the decay of comet 1861 I. The orbit of the meteor stream is inclined at a large angle to the Earth's orbit ensuring that it suffers very little from perturbation and is therefore very stable.

Lysithea, a small *satellite of Jupiter. It has a diameter of 36 km and revolves about the planet at a mean distance of 11.72 million km in a highly eccentric orbit in a period of 259 days. Discovered in 1938, Lysithea is one of a group of four revolving about the planet in orbits of much the same size.

Mach's Principle, the idea put forward by the Austrian physicist Ernst Mach that the *inertia of a body is due to the presence of the rest of the universe. If the body were alone with no other body present in the universe, the body would have no inertia. By its very nature, Mach's Principle is unverifiable but has stimulated the formation of cosmological models.

McMath solar telescope *see at Mc-.

Maffei galaxies, two galaxies discovered by the Italian astronomer Paulo Maffei in 1968. They are difficult to see visually because of their proximity to the *galactic equator and the densest regions of interstellar dust. Maffei 1 is bigger and brighter than the medium-sized Maffei 2 and is a large elliptical galaxy about ten times the mass of our galaxy.

Magellanic Clouds, two small *galaxies in orbit around our own galaxy, and of great importance because they are sufficiently close for detailed study. The first Europeans to see the Clouds were Portuguese navigators in the 16th century who named them in honour of Ferdinand Magellan. Part of the *Local Group of galaxies, to the naked eye they appear as hazy, cloud-like patches close to each other and to the south celestial pole. The **Large Magellanic Cloud**, also known as **Nubecula Major**, is at a distance of 50,000 parsecs (160,000 light-years) and its mass is believed to be some 10 billion times that of the Sun. Long-exposure photographs reveal an almost circular disc of stars, while shorter exposures are dominated by a densely starred 'bar', resembling a *barred spiral galaxy. The Cloud contains many star *clusters, *stellar associations, and diffuse *nebulae including the *Tarantula Nebula which is the largest diffuse nebula known anywhere in the universe. The **Small Magellanic Cloud** (**Nubecula Minor**) is less massive (about one billion times the mass of the Sun) and more distant (57,000 parsecs, or 185,000 light-years). Its structure is less clear with fewer clusters, associations, and nebulae. Part of the Small Cloud extends towards the Large Cloud possibly due to its gravitational attraction. The important *period–luminosity law for *Cepheid variable stars was discovered in the Small Magellanic Cloud.

Both Clouds appear less developed than our galaxy. They contain a lower proportion of heavy elements, which undoubtedly affects the course of stellar evolution. Stars are generally younger and gas and dust more plentiful than in our galaxy and the Clouds contain stars of greater luminosity. In 1987 a blue (not red) *supergiant star in the Large Cloud exploded to become the first naked-eye supernova in almost 400 years (*Supernova 1987A). Radio observations show that the Clouds are surrounded by a common envel-

The Large **Magellanic Cloud**, a companion galaxy to our own Galaxy, is close enough for detailed study. It is a small irregular galaxy spanned by a 'bar' of densely packed stars. The brightest areas, such as the Tarantula Nebula seen here near the bar, are probably active regions of star formation. The red-coloured filaments of glowing hydrogen are associated with very young blue stars.

ope of hydrogen, the Magellanic stream, which may extend to our galaxy. Both Clouds contain many X-ray sources.

magnetic field, a region in which magnetic substances, such as iron, and charged particles experience a force produced by a magnet, or by a conductor carrying an electric current, or by a flow of charged particles. The most common examples of this phenomenon are the orientation of a compass needle in the Earth's magnetic field, or the alignment of iron filings along the lines of force of a magnet, easily demonstrated if the filings are scattered on top of a horizontally held card under which the magnet is placed. Magnetic fields occur in a wide scale of intensity throughout the universe in planets and stars, and the interplanetary, interstellar, and intergalactic environments.

magnetograph, an instrument used for measuring the solar *magnetic field. With it a chart called a magnetogram can be obtained giving the strength, polarity, and distribution of magnetic fields across the Sun's disc. It measures the *Zeeman effect, the splitting of a *spectral line produced when atoms emit or absorb radiation in a magnetic field.

magnetohydrodynamics, a branch of physics in which properties of a *plasma or an electrically conducting fluid under the action of electric and magnetic fields is studied. Since plasmas occur in stars, especially in the processes creating a *nova or *supernova, the study of magnetohydrodynamics is an important aid in understanding such processes. *Alfvén's theory describes how magnetohydrodynamic waves can propagate in a highly conducting fluid in the presence of a magnetic field.

magnetometer, an instrument for measuring the strength and direction of a *magnetic field. Apart from those designed for terrestrial use, various kinds of magnetometers have been flown in rockets, satellites, and planetary space missions, and some of them have been taken to the surfaces of the Moon, Mars, and Venus.

magnetopause *magnetosphere.

magnetosheath *magnetosphere.

magnetosphere, a cylindrical region surrounding the Earth in which the *solar wind is accelerated by the Earth's magnetic field. In the early 1960s it was found from satellite measurements that the Earth's magnetic field does not extend indefinitely but is constrained to exist in a cavity within the solar system which is called the magnetosphere or **magnetosheath**. The cavity has the shape of a blunt-nosed cylinder and is produced by the interaction of the hot plasma escaping from the Sun (the solar wind) with the weak outer regions of the Earth's magnetic field. The nose of the cylinder is in the direction of the Sun with the boundary, known as the **magnetopause**, at about ten Earth radii. On the night side the magnetosphere is stretched out into a large tail well beyond the orbit of the Moon. Other planets with magnetic fields such as Mercury and Jupiter also have magnetospheres. In the magnetosphere, charged particles are accelerated by changing magnetic fields, and form the *Van Allen radiation belts and the *aurora. The magnetosphere protects the Earth from harmful cosmic rays.

magnifying power, a measure of the size of the telescopic image of an object compared with the apparent size of that object viewed with the naked eye. It is the ratio of the angle that the telescopically viewed image makes at the eye and the angle that the original object makes without optical aid. Its value can be calculated as the ratio of the focal lengths of the telescope's objective or primary mirror and its eyepiece.

magnitude, a measure of the brightness of a star. A star's magnitude may be illusory: a faint star which is near the Earth may appear to be brighter than a bright star which is further away because the intensity of light received from a star decreases in proportion to the square of its distance from the Earth. The **apparent magnitude** of a star describes how bright the star appears to be as observed from the Earth. In contrast, **absolute magnitude** is a measure of the actual brightness (*luminosity) of a star: it is the apparent magnitude that the star would have if it were at a distance of 10 *parsecs. *Hipparchus invented the system of (apparent) magnitudes in the 2nd century BC. Stars were assigned numbers according to their apparent brightness: the brightest stars were called first magnitude and those at the limit of visibility were called sixth magnitude. This system was rationalized in the middle of the 19th century particularly through the work of the British astronomer Norman Pogson. The present magnitude scale was established by assigning numbers to a representative group of stars near the north celestial pole. Then magnitudes could be assigned to all the other stars in the sky. This is a logarithmic scale on which a first-magnitude star is one hundred times brighter than a sixth-magnitude star. As measurements became more accurate, it was necessary to use decimal numbers between the whole numbers, one to six. The very brightest stars have magnitudes of less than one and several are even assigned negative values. As fainter and fainter objects are being recorded using powerful telescopes, the scale has had to be extended to larger and larger numbers. The limiting magnitude is now much fainter than +25.

main sequence *Hertzsprung–Russell diagram.

major axis *ellipse.

major planet, any of the largest bodies orbiting around the Sun. Mercury, Venus, Earth, Mars, Jupiter, Saturn, Uranus, Neptune, and Pluto are the major planets. Their orbits are coplanar, that is, they all lie approximately in the same plane and have small eccentricities. They are widely spaced so the mutual perturbations are small and the orbits are stable. The status of Pluto as a major planet is somewhat questionable, because of its smaller size and because its orbit comes closer to the Sun than Neptune's. The stability of the orbit of Pluto is also doubtful and may be governed by *chaos theory.

Maksutov telescope *reflecting telescope.

mantle, that region of a planetary interior that lies between the thin surface crust and the core. In the case of the Earth the mantle extends from a depth of 10–80 km (6–50 miles) to a depth of about 2,900 km (1,800 miles); the region below 1,000 km (625 miles) is known as the lower mantle. The divisions of the mantle are defined in terms of the velocities of the seismic waves that travel through them. The mantle consists of hot rock which is undergoing a very slow convective turnover. The temperature varies from 1,000 °C to 3,000 °C as the depth increases. Even at the higher of these temper-

The **maria** are large, flat, lightly cratered areas of the Moon's surface. Lunar lava flooding over the surface in the relatively recent past has covered most of the ancient craters leaving only a few low-altitude hills and bumps. The impact craters to be seen in the maria today were therefore formed since this flooding took place.

atures the rock remains solid because of the high pressures to which it is subjected. The mantle is heated by radioactive decay.

mare (pl. maria), a vast expanse of very dark, iron-rich basaltic lava that has flooded out from cracks in the lunar crust to fill large impact *craters. Most of this flooding took place 3–4 billion years ago. The maria, also known as *ring plains, are, in the main, confined to the Earth-facing side of the Moon because the lunar crust is at its thinnest on this side. The mare material is dark and reflects only about 5 per cent of light incident upon it. The maria differ slightly in colour because the chemical composition of the lava gradually changes after it has flowed to the surface. The word 'mare' is Latin for 'sea' and the smoothness of the maria led early astronomers to believe they were expanses of water. Smaller maria were described as bays, lakes, and marshes.

Mariner, a highly successful series of US spacecraft launched to explore the planets Mercury, Venus, and Mars. In December 1962 Mariner 2 made the first successful fly-past of Venus and in July 1965 Mariner 4 made the first successful fly-past of Mars. Mariner 9 was put into orbit around Mars in November 1971. Its 7,329 pictures enabled the entire Martian surface to be mapped. Close-up views were also taken of the two small moons of Mars. The *Viking spacecraft later landed on the surface of Mars. In February 1974, Mariner 10, used the gravitational field of Venus to change its trajectory, enabling it to rendezvous with Mercury. It made three close fly-pasts of Mercury and provided more than 10,000 pictures of its surface.

Markab, a second-magnitude star in the constellation of Pegasus, and also known as Alpha Pegasi. It is the brightest star in Pegasus and lies at a distance of 31 parsecs. Intrinsically its luminosity is about eighty times that of the Sun. Three degrees south of Markab there is a *barred spiral galaxy to be seen almost face-on.

Markarian galaxy, a type of galaxy, discovered by the Soviet astronomer Beniamin Markarian, which emits strongly at ultraviolet wavelengths. The source of the ultraviolet radiation often resides in the nucleus but may lie in a more extended region. Markarian galaxies with ultraviolet sources in their nuclei are further subdivided into *Seyfert galaxies, N-type galaxies, and *quasars. Those where the ultraviolet radiation originates from an extended source

The **Mariner** 6 spacecraft was launched by an Atlas-Centaur rocket on 24 February 1969. Mariner 6 successfully flew past the planet Mars, transmitting seventy-five photographs of the Martian surface back to Earth.

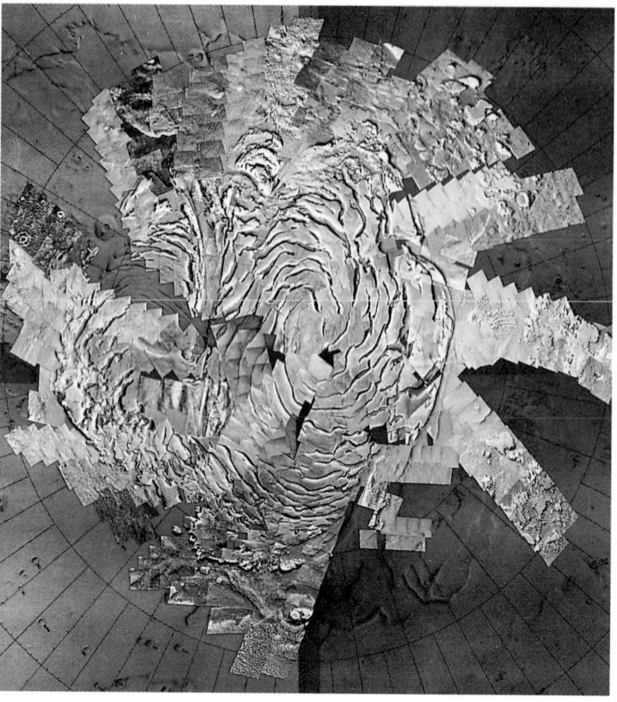

This mosaic of the north polar ice cap of **Mars** was composed using 391 pictures taken by Viking orbiter 2. They were taken during the Martian summer when the ice cap was at its smallest. The ice is water ice, rather than carbon dioxide, as had been previously conjectured.

resemble galaxies where star formation has recently taken place.

Mars, the fourth *planet from the Sun. Mars moves in a markedly elliptic orbit, its mean distance from the Sun being 1.52 times that of the Earth's. Its diameter is approximately 6,800 km, a little more than half that of the Earth. Its *period of revolution is 1.88 years, giving it a *synodic period of 2.14 years. At *opposition its distance from the Earth can vary from 100 million km at aphelion to about 50 million km at perihelion, very close oppositions occurring every fifteen or seventeen years on average. At such times Mars is a bright, reddish, star-like object and the subject of extensive telescopic study. By study of the orbits of its two moons, *Phobos and *Deimos, discovered by the US astronomer Asaph Hall in 1877, the mass of Mars can be measured to be one-tenth that of the Earth. The pictures from the Mariner and Viking spacecraft show a solid surface with a wide variety of features: thousands of impact craters; many huge volcanic craters, none of which are now active; a great equatorial valley running eastwards for 3,000 km before turning northwards; and many smaller valleys and chasms with branching canyons. There also exist eroded plateaux, collapsed regions, and two white polar caps which change size as the Martian seasons come and go. However, it has no canals; the network of waterways thought by *Lowell and others in the late 19th century to cover the surface does not exist. Bombardment by meteors, and erosion, weathering, and volcanic activity have created complex surface conditions on Mars. It is possible that orbital and precessional changes over long periods of time may release water in large quantities, creating torrents which erode out valleys. Among the prominent surface features are *Argyre Planitia, *Chryse Planitia, *Elysium Planitia, *Hellas Planitia, *Nix Olympica, *Pavonis Mons, *Syrtis Major, *Tharsis Ridge, *Utopia Planitia, and *Valles Marineris.

The planet's thin atmosphere (at the surface it is only 1 per cent of the density of Earth's atmosphere) is mainly composed of carbon dioxide. The polar caps are made of snow, ice, and carbon dioxide and it is possible that much water may be permanently locked up as ice in permafrost conditions. Clouds and sandstorms are often seen. At the equator the temperature can vary from 10 °C to below −75 °C. The automatic laboratories of the Viking landers detected no definite signs of life. Nevertheless Mars remains the planet least inhospitable to life as we know it and future missions to it will undoubtedly put a high priority on the search for life there. See the figure on pages 94–5.

mascon, a region of the Moon's surface where the rock has a higher density than the other material below or around it. A mascon, which is short for 'mass concentration', is produced by layers of basalt lava rising up through the Moon's surface crust and solidifying on top of the low-density crust. Mascons were first detected by spacecraft orbiting the Moon when they felt the slight changes in the Moon's gravity over the mascons. The *Imbrium Basin is a mascon.

maser (microwave amplification by stimulated emission of radiation), a region of gas containing high-energy molecules in which radio emission from some molecules stimulates other molecules to emit more radiation of the same wavelength, resulting in amplification of the radio emission, analogous to a laser. Natural masers in the *interstellar medium generate very intense radio *spectral lines which can be investigated by *radio astronomy. The first cosmic maser sources to be discovered were *hydroxyl radical (OH) masers associated with the formation of young stars. Thousands of cosmic masers are now known.

mass, the ratio of the *force acting upon a body and its acceleration, according to *Newton's Laws of Motion. The equivalence principle states that this inertial mass is exactly proportional to the passive gravitational mass which is related to the quantity of material contained in the body; therefore the orbit of a particle under the effect of gravitation is independent of the size of the particle. The equival-

In 1976 the two Viking landers took many photographs of the dusty, rock-strewn surface of **Mars**. Part of the spacecraft is visible in the foreground. Samples of the surface were scooped up for analysis by the landers which then transmitted their data back to Earth.

The Keck telescope, housed in this dome at **Mauna Kea** in Hawaii, is a multiple-mirror telescope with an effective diameter of 10 m. This huge size is achieved by segmenting the mirror into thirty-six hexagonal mirrors, each 1.8 m wide. These are co-ordinated a hundred times per second by computer so that the light from each is combined to form a single image. This technique is necessary because it is not at present possible to construct a single mirror whose diameter is greater than about 8 m.

ence principle has been verified to an extremely high degree of accuracy. There is no way to measure the inertial mass of a freely falling body. For example, only the ratios of the masses of the planets can be determined from their mutual perturbations. What are directly measured from the orbits of the planets and satellites are the active gravitational masses; by *Newton's Law of Gravitation, the latter are also proportional to the inertial masses, the proportionality constant being the universal gravitational constant G. As a result, the product of G multiplied by the mass is known with extreme accuracy for the main bodies of our solar system, while the inertial mass is poorly known. At relativistic speeds, comparable to the speed of light, the inertial mass is increased with respect to the mass at rest.

mass–luminosity relation, a relationship, found by *Eddington in 1924, between bolometric luminosity L and the *stellar mass M. The relation is approximately of the form

$$L/L_\odot = (M/M_\odot)^a$$

where $L_\odot$ and $M_\odot$ are the luminosity and mass of the Sun, and a is typically in the range 3–5. The relation is a consequence of the fact that the observable properties of normal stars are determined primarily by their masses. It can be derived theoretically for *dwarf stars and in this case it is in good agreement with observations. It is thought to hold approximately also for *supergiant and *giant stars, though their masses are poorly known from direct measurements. It does not apply to *white dwarfs, because it overestimates their luminosities.

Mauna Kea, a dormant volcano on the island of Hawaii on whose 4,200 m (13,800 feet) summit there is a collection of telescopes, some designed to take advantage of the special

Mass–luminosity relation

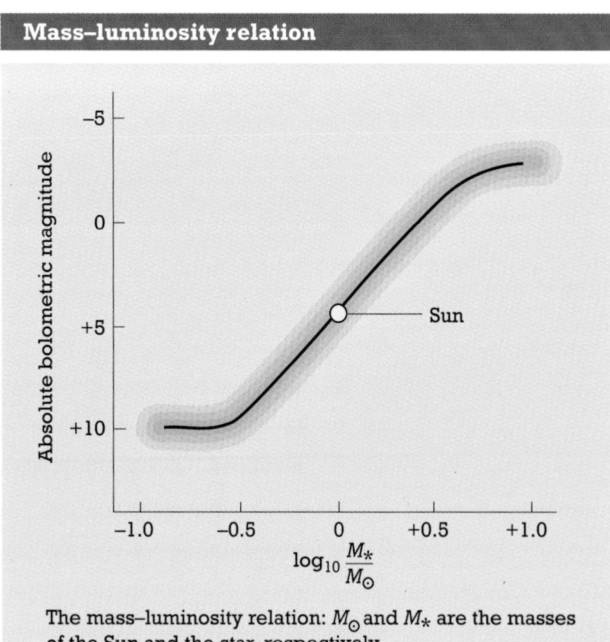

The mass–luminosity relation: $M_\odot$ and M_* are the masses of the Sun and the star, respectively.

Mars

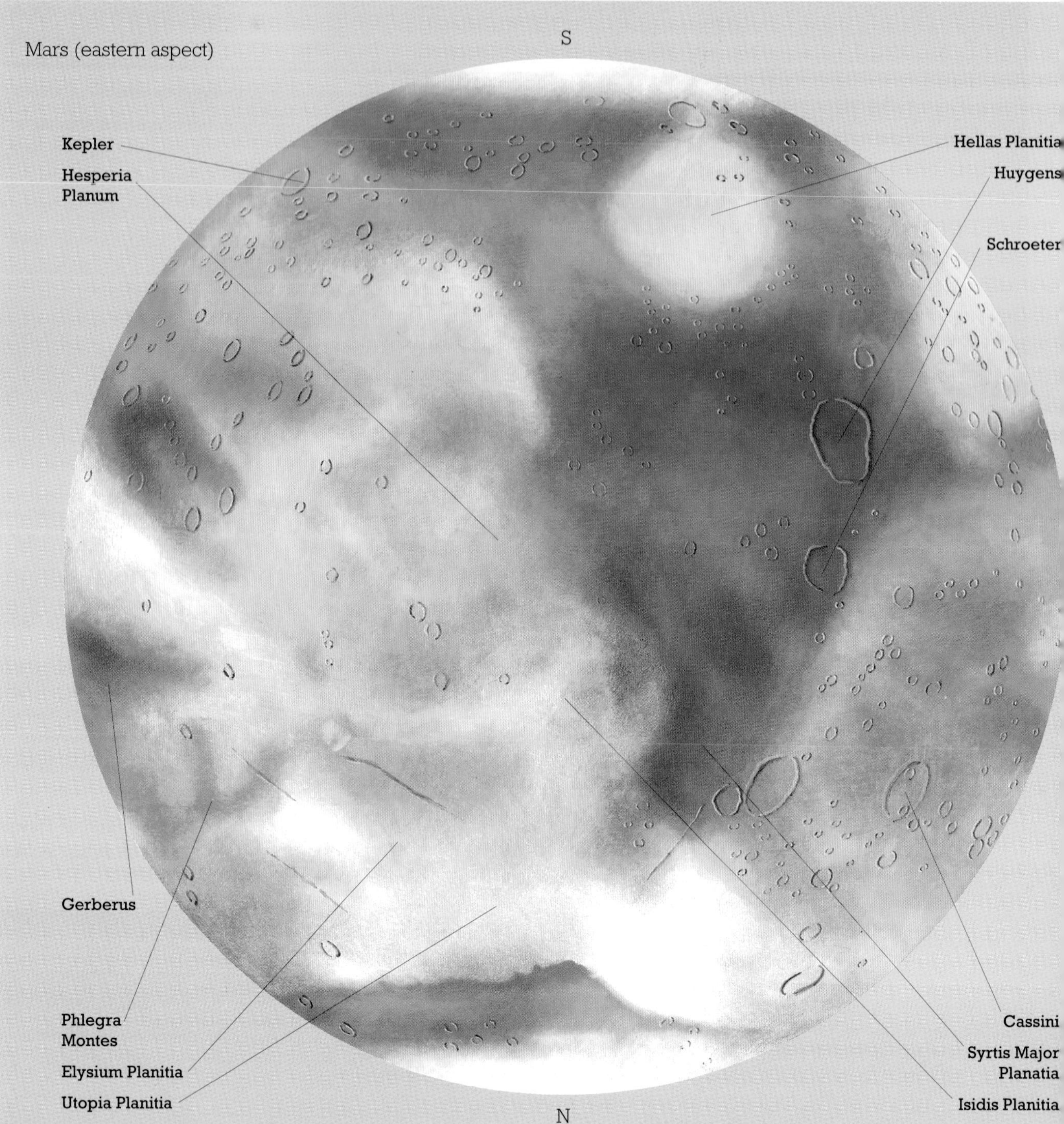

Mars is the fourth planet from the Sun and the second smallest in size (after Mercury) of the inner planets.

Its northern hemisphere has, in contrast to its south, relatively few impact craters, but is occupied by younger plains which appear as lighter areas. The surface of the south illustrates the primitive topography of the planet that has been intensely bombarded by meteorites since its formation 4.5 billion years ago, and which has since been encroached upon by the flooding effect of the north. Both poles contain ice caps which when seen from directly above have a spiral appearance. The caps themselves show little evidence of meteorite bombardment and are therefore thought to be relatively recent additions to the topography of Mars.

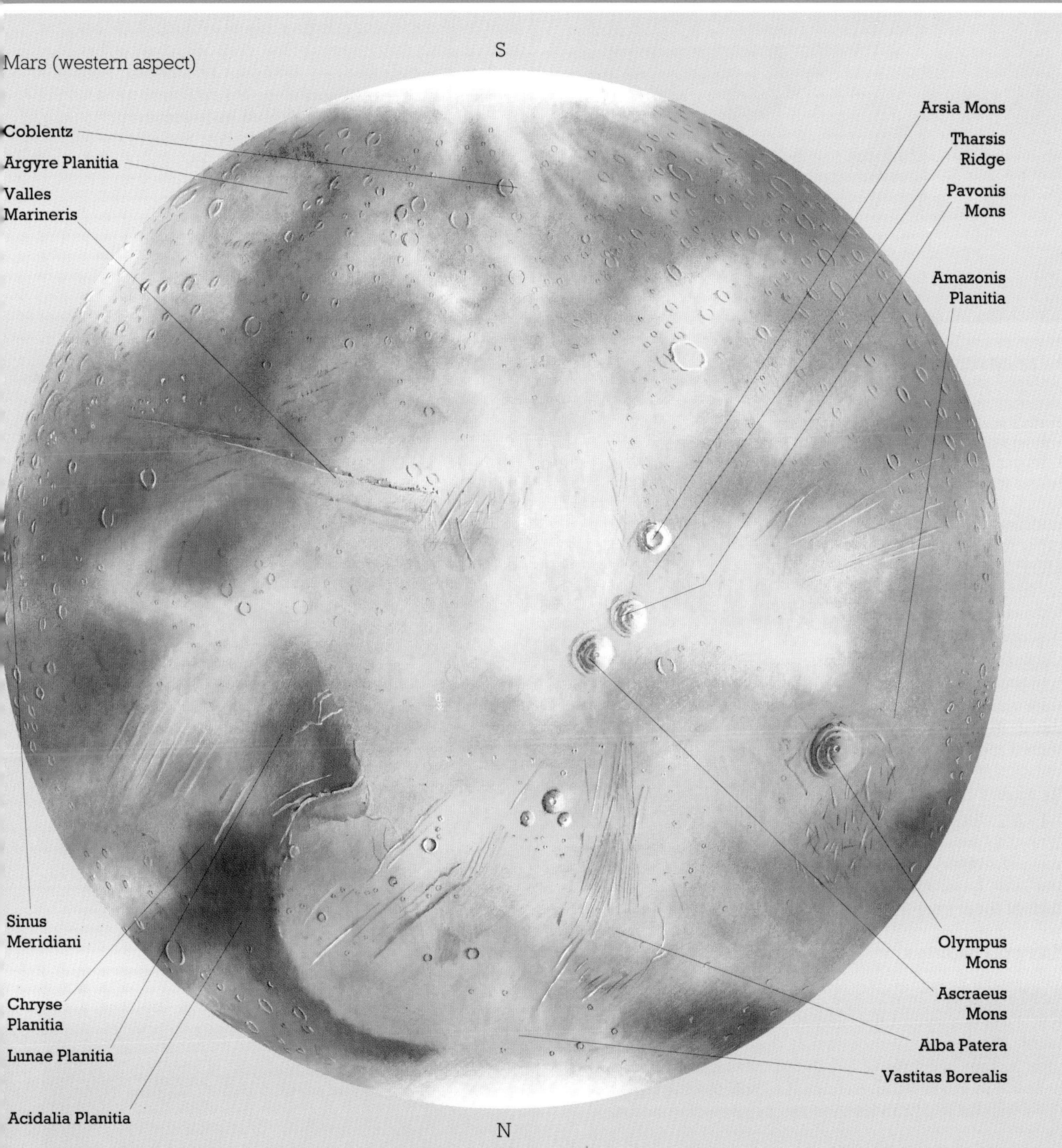

Before the 1960s the best images of Mars were those obtained through Earth-based astronomical telescopes. However, even through the most powerful telescopes, its surface features cannot be observed with any clarity: although the north and south polar regions can be distinguished, the rest of the surface appears to be reddish with indistinct dark and light areas. It was not until the 1960s and 1970s that martian space probes enabled detailed maps to be constructed. Modern maps are based on thousands of photographs taken mainly by Mariner and Viking spacecraft as they flew past or orbited the planet. This map is drawn with the south pole at the top because that is how the image of Mars would appear when viewed through an astronomical telescope based in the Earth's northern hemisphere.

The **McMath solar telescope** is housed in a tall tower so that light reflected into it from the Sun by a heliostat on top of the tower is unaffected by the hot, turbulent air present at ground-level. For greater air stability, a uniform temperature is maintained inside the building and the instrument itself is underground.

conditions found there. The site has extremely low humidity and is ideal for infra-red observations. The *Royal Observatory, Edinburgh, operates the 3.8 m (150-inch) UK infra-red telescope (UKIRT) and the 15 m (49 feet) UK–Dutch millimetre telescope known as the James Clerk Maxwell telescope (JCMT). There is also the 3.6 m (142-inch) *Canada–France–Hawaii telescope (CFHT), the 3 m (118-inch) infra-red telescope facility (IRTF) operated by NASA, and a 2.2 m (87-inch) telescope for the University of Hawaii. The 10 m (33 feet) Keck telescope, constructed with a segmented mirror, was built by the University of California.

Maunder minimum, the period of time between 1645 and 1715 when very few *sunspots appeared. According to *Spörer's Law, sunspots normally appear in an eleven-year cycle. However, longer-term variations are superimposed upon this basic cycle and the Maunder minimum was a result of these long-term variations.

Maxwell Montes *Venus.

McMath solar telescope, a *reflecting telescope at the *Kitt Peak National Observatory in Arizona, USA. It was specifically designed for observing the Sun's disc. The telescope has an aperture of 1.5 m (59 inches) and employs a *heliostat to guide the light from the Sun into the telescope. It provides an image of the Sun 0.8 m (31 inches) across.

mean solar day *day.

mean solar time *time.

mean sun *day.

Mensa, the Table Mountain, a *constellation near the south celestial pole. It was introduced by the 18th-century French astronomer Nicolas Louis de Lacaille in honour of Table Mountain in southern Africa from where he observed the southern stars in 1751–2. The Large *Magellanic Cloud, part of which lies in the constellation, represents the cloud that caps the real Table Mountain. Mensa is a faint constellation, with no stars brighter than fifth magnitude.

Mercury, the innermost *planet of the solar system. Mercury is named after the winged messenger of the Gods in early Greek mythology. It is visible as a *magnitude 0 star at approximately two-monthly intervals alternating between the evening and morning twilight. The planet follows a remarkably elliptical orbit moving from 46 to 70 million kilometres from the Sun and completes one orbit every 88 days. In a small telescope it shows *phases like those of the Moon, due to its varying position relative to the Earth and Sun. Very occasionally, the Earth, Mercury, and the Sun line up precisely and the planet is then seen to *transit the Sun as a small dark spot crossing the solar disc. This last occurred on 12 November 1986 and will occur again on 14 November 1999. Observations of faint dusky markings seen on Mercury through large telescopes around 1900 led astronomers to conclude that the planet rotated on its axis and revolved round the Sun at the same 88-day rate, keeping the same face pointing sunwards. After being accepted for eighty years, the 88-day rotation period was finally replaced by the correct 58.5-day period deduced from radar observations.

Mercury is 4,878 km in diameter, midway in size between Mars and the Moon and its daytime surface is baked by the proximity of the Sun. The midday equatorial temperature varies, due to the elliptical orbit, from 415 °C at the closest point to the Sun to 285 °C at the farthest. In contrast, the lack of an atmosphere allows the temperature to plummet to −175 °C at night. The spacecraft *Mariner 10, which passed close to Mercury on three occasions in 1974 and 1975, revealed the planet as a barren airless world covered with impact craters similar to those on the Moon. The largest structural feature is the *Caloris Basin, a ring basin 1,300 km in diameter, whose floor has been intensely disrupted by fractures and ridges. Caloris was probably caused by the impact of a body some tens of kilometres in diameter. Some craters show ray systems.

The perihelion advance of Mercury has been used to confirm predictions made by the general theory of *relativity.

This composite photograph of the cratered surface of **Mercury** was constructed from ten images taken by Mariner 10 as it left the planet shortly after its fly-past on 29 March 1974. The north pole is at the top and the equator is about two-thirds down from the top.

meridian, any vertical great circle across the sky that passes from the observer's south celestial pole to the observer's north celestial pole. The particular meridian that passes through the observer's *zenith is called the observer's meridian. All meridians intersect the *equator at right angles.

MERLIN *multi-element radio linked interferometer network.

mesopause, the upper boundary of the *mesosphere at a height of around 75 km (47 miles). This is the coldest part of the atmosphere and the temperature is about −90 °C. The mesopause is above the *ozone layer and below the *ionosphere, both of which are relatively warm. The atmospheric pressure in this region is about one millionth of that at sea-level and the molecules move about a centimetre before striking other molecules.

mesosphere, or middle atmosphere, the region in the Earth's atmosphere that extends between the heights of 60 and 75 km (38–47 miles) between the *stratopause and the *mesopause. The temperature decreases from about 10 °C at the base to about −90 °C at the top of the region. As in all the regions up to a height of about 100 km (60 miles) the mesospheric composition remains constant and equal to the ground-level composition of 78 per cent nitrogen, 21 per cent oxygen, with the remaining 1 per cent being made up of mainly carbon dioxide, water vapour, and argon. The small amounts of water contained in the region quickly become dissociated. Because of the rapid decrease of pressure and density with height, only about one part in a thousand of the Earth's total atmospheric mass lies above the height of 60 km (38 miles).

Messier catalogue, the first significant catalogue of *nebulae and star *clusters, compiled by the French astronomer Charles Messier (1730–1817). The first version, published in 1774, listed forty-five nebulae or clusters that had mostly been discovered by Messier during his assiduous search for new comets. A later version listed 103 objects, incorporating discoveries by another French astronomer Pierre Méchain and others. The total has been increased further through examination of Messier's papers by modern astronomers. Messier and his co-discoverers worked with telescopes of small aperture, and the catalogue is thus a list of the brightest extended objects visible from mid-northern latitudes. As such, it contains a wide variety of bodies, and Messier designations are still extensively used. Examples include: M1, the Crab Nebula, a supernova remnant (the first of Messier's discoveries, in 1758); M13, a globular cluster in Hercules; M31, the Andromeda Galaxy; M42, the Great Nebula in Orion; M45, the Pleiades, an open cluster and reflection nebula; and M87, a giant elliptical galaxy in the Virgo Cluster.

meteor, the streak of light that is seen in a clear night sky when a *meteoroid burns itself out in the Earth's upper atmosphere. The meteoroid has a velocity of between 74 km/s, if it hits the Earth head-on, and 11 km/s, if it is catching the Earth up. At these speeds the friction between the meteoroid and the atmosphere usually creates enough heat to vaporize it entirely and it burns up in the region between 75 and 115 km (50 to 70 miles) above the Earth's surface. The friction also causes the evaporating meteoroid to leave behind a thin stream of glowing atmospheric gas and a trail of ionized atoms and molecules which can reflect radar pulses. Meteoroids that are visible or that can be detected by radar have masses of between a microgram and a thousand kilograms. Meteoroids smaller than this are not visible and, after losing speed in the atmosphere, merely float down to the Earth's surface. Very large meteoroids that survive their journey through the atmosphere and reach the ground are called *meteorites. A typical visual meteor, also known as a **shooting star**, is produced by a meteoroid about a centimetre in size with a mass of about a tenth of a gram and a density of only about one-third that of water. Much larger meteoroids occasionally enter the Earth's atmosphere producing a brilliant meteor known as a **bolide**. An observer can see about 120,000 square kilometres of the atmospheric layers in which meteoroids burn up. On an average day between the hours of 2 a.m. and 6 a.m., when the observer is on the leading side of the Earth as it orbits the Sun, about fifteen meteors per hour can be seen. The rate drops to about five per hour by 6 p.m. Many of these are **sporadic meteors**, that is, lone meteors not associated with a *meteor shower. However, on certain nights of the year, a meteor shower can yield fifty to eighty meteors per hour.

meteorite, the remnant of a *meteoroid that has survived its passage through the Earth's atmosphere as a *meteor. For a meteoroid to reach the Earth's surface its mass must be at least several grams and it must enter the upper atmosphere with a low velocity. Even then, only 25 per cent of the meteoroid remains to reach the ground. Meteorites are named after the geographical localities in which they are seen to fall or are found. Immediate recovery of meteorites that have been observed to fall gives a reasonably unbiased indication of the abundance of the different types. The vast majority (84 per cent) are *chondrite stones; 8 per cent are achondrites, that is, stones that have no chondrules; 1 per cent are stony-iron meteorites known as **siderolites**; and 7 per cent have an exceptionally high iron content. Meteorites that are recovered some time after they have fallen are frequently of the iron type, simply because they are easier to recognize in the ground; other types of meteorites tend to go unnoticed. About six new fallen meteorites are recovered each year. Stone meteorites have masses between about 100 grams and 1,500 kg and iron meteorites between a few grams and 20,000 kg. Meteorites smaller than a few grams are hard to find, even if they reach the Earth's surface, and those much larger than the largest recovered meteorites produce craters in the Earth's surface and are blown apart on impact. Meteorites represent the fragments of those *asteroids that have orbits with large eccentricities, and can thus cross the Earth's orbit.

meteoroid, a small body that, if it enters and burns out in the Earth's atmosphere, produces a *meteor. Meteoroids are fragments of *comets or are dust particles in the *interplanetary medium, and have masses ranging from thousands of kilograms to less than a million millionth of a gram. The larger ones are porous, friable aggregates of stony or metallic dust. Meteoroids are often grouped into a *meteor stream, the debris of a decaying comet. If the Earth passes through such a stream a *meteor shower is observed.

meteoroid stream *meteor stream.

meteor shower, a group of *meteors that appear to come from the same point, known as the *radiant, on the celestial sphere. A meteor shower occurs whenever the Earth passes through a *meteor stream, and the number of meteors per hour, known as the rate, reaches its peak when the Earth crosses the centre of the stream. The showers last for as long as the Earth is in the stream and this can vary from a day to a few weeks. During a prolonged shower the visual rate on each day is at its maximum when the shower radiant is highest above the horizon. Showers are named after the constellation in which the radiant lies. Most showers have rates that vary little from year to year and the major ones are the Quadrantids, Beta Taurids, Perseids, Orionids, and Geminids; these are summarized in the table on page 195. The Leonids and Giacobinids are exceptions and produce *meteor swarms. The orbits of some meteor streams about the Sun change from time to time because of perturbations by the planets. For this reason the Quadrantids and Geminids have been seen only since the mid-1800s. The Perseids, however, have been seen for millennia.

meteor storm *meteor swarm.

meteor stream, an elliptical ring of debris, principally dust and rubble, occupying the orbit of a *comet. As a comet approaches the Sun, *radiation pressure and the *solar wind cause dust and gas to be ejected. This debris has a velocity of a few kilometres per second with respect to the comet. Depending on the angle of emission some of the material gains on and some falls behind the cometary nucleus. Gradually the dust and gas spreads out right around the comet's orbit. The resulting ring gets broader as it ages but is thinner near perihelion than at aphelion. When the Earth's orbit intersects the meteor stream there is a chance of seeing a *meteor shower. A meteor stream is also known as a **meteoroid stream**.

meteor swarm, a particularly impressive *meteor shower. When the Earth crosses the orbit of an old *meteor stream whose particles are scattered around its orbit, a meteor shower will be seen. However, in a new stream the debris from the comet that originally produced the stream will still be concentrated in a cloud around the comet. If the Earth passes through this cloud a brilliant shower, called a meteor swarm or **meteor storm**, is seen. One of the most impressive was the *Leonid swarm of 12–13 November 1833 when the number of visible meteors exceeded fifteen per second.

Metis, a small satellite of *Jupiter, discovered by the US astronomer S. P. Synnott from *Voyager images in 1979. It orbits at about 128,000 km from the centre of the planet, just outside **Jupiter's rings** and close to the orbit of *Adrastea. It has an irregular shape with a mean radius of about 20 km, an albedo of about 0.1, and a composition that is probably rocky.

Metonic cycle, a 19-year period of time. In the 5th century BC the Athenian astronomer Meton discovered that the number of days in 19 years is almost exactly equal to the number of days in 235 *lunations. The discrepancy, less than $1\frac{2}{3}$ hours, means that if full or new moon occurs on a particular date, full or new moon will occur on the same date 19 years later. This cycle produced the first adequate framework for recording astronomical phenomena and started with the summer solstice of 432 BC which Meton observed in Athens with his associate Euctemon.

Michelson interferometer (stellar), a device for measuring the angular diameters of stars using *interferometry. Developed by the US physicist Albert Michelson (1852–1931), it originally comprised a pair of slits placed across a telescope aperture so that the interference of the light collected by the two zones could be examined as a fringe pattern in the eyepiece. The slit separation (baseline) could be adjusted and the separation corresponding to minimum visibility of the fringes allowed the angular diameter of the source to be determined. For the measurement of stars, Michelson had to provide a baseline for the slits wider than the largest telescope available and he did this by fitting a steel beam across the 100-inch (2.5 m) telescope at the *Mount Wilson Observatory. Four mirrors were used, of which two were fixed and two were adjustable; the separation of the adjustable mirrors could be varied. The light collected by the mirrors was sent into the telescope aperture and the resulting fringes were observed in the telescope eyepiece. This system operated in the 1920s and measurements were achieved for several nearby supergiant stars. The technique is now being explored again using very long baselines (telescopes separated by tens of metres) with laser links and photoelectric devices to monitor the interference patterns. It has also been used in radio astronomy with two *radio telescopes providing interference patterns that allow the positions of radio sources to be fixed.

Michelson also devised another instrument known as the **divided amplitude interferometer** within which a light beam is split and then recombined after the divided beams have travelled separate path lengths via mirrors. The resulting interference pattern depends on the path difference between the two beams and their spectral content. By changing the optical path of one of the beams and monitoring the variations of the strength of the fringes, the *spectrum of the source may be investigated. The device is capable of high spectral resolution and has been used to record spectra of the planets and stars. In its original form, it was this interferometer that was used in the Michelson–Morley experiment which proved that an *ether was not required for the propagation of electromagnetic waves.

Michelson–Morley experiment *ether.

microdensitometer, an instrument for measuring the intensity of a photographic image. Its main astronomical application is that of converting images of stellar spectra into measured quantities such as apparent magnitudes, spectral line strengths, or radial velocities. The microdensitometer measures the image strength by scanning the photographic plate to detect the fraction of incident light transmitted through the plate.

Microscopium, the Microscope, a small *constellation of the southern sky, introduced by the French astronomer Nicolas Louis de Lacaille in 1752 as one of a series of modern instruments of science and the arts. It is a faint and barren constellation with brightest stars of only fifth magnitude.

microwave background radiation, an electromagnetic radiation field that permeates the whole universe. Its existence was predicted in 1948 by the US physicists Ralph Alpher, Hans Bethe, and George Gamov, who proposed the *Big Bang theory of the universe, and it was discovered in 1965 by the US physicists Arno Penzias and Robert Wilson. It is now accepted that this background radiation is a remnant of the hot *black body radiation created in the first few moments of the expanding universe at the time of the Big Bang. This radiation, also known as the **cosmic back-**

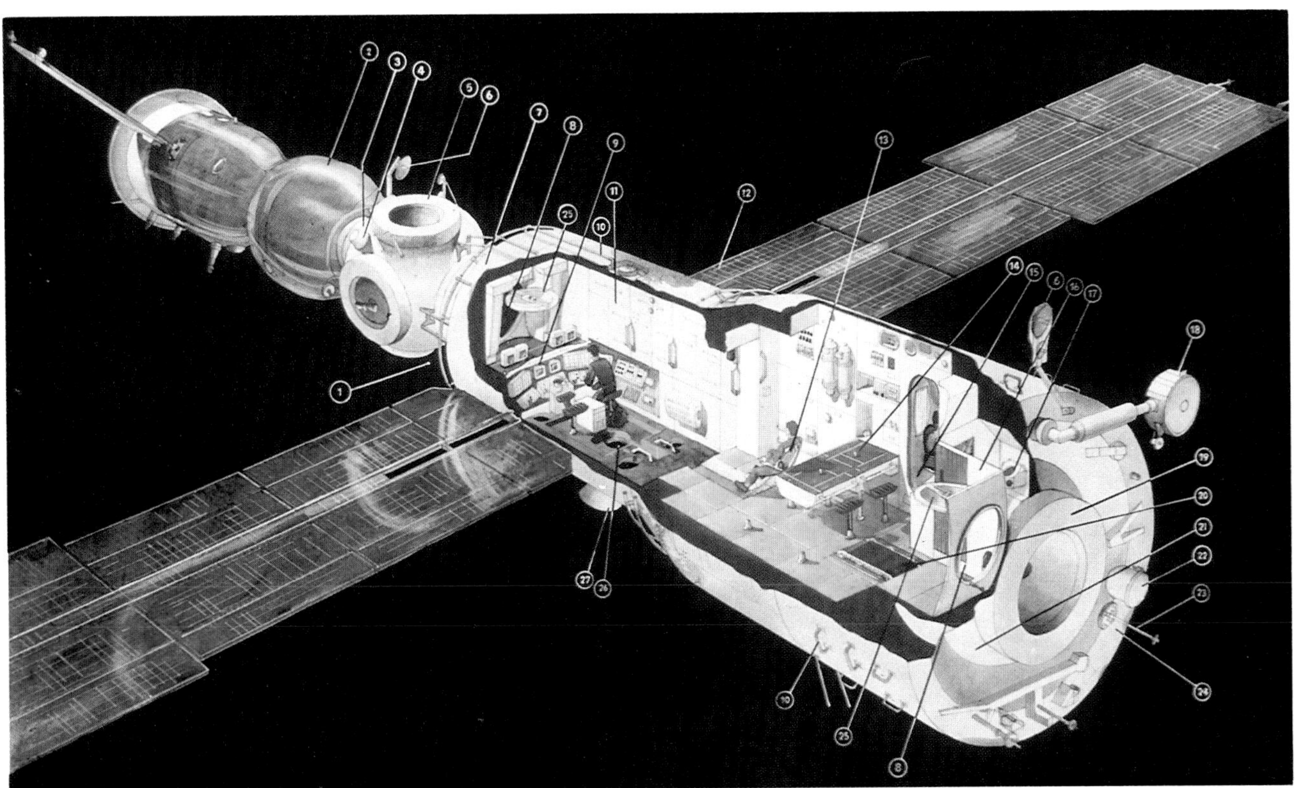

On this diagram of the original 'base' module of the **Mir** space station, number 9 points to the central control unit; 12 to the solar batteries; and 18 to the antenna for communication with Earth. Mir is increasingly being used for economic, not just scientific, purposes. For example, it now has a specialized module for remote sensing of Earth resources and for the manufacture of semiconductors in weightless conditions.

ground radiation, has a wavelength of 0.0735 m, a temperature of 2.7 K, and is *isotropic to within one part in 100,000. However, it may contain tiny irregularities that are the echoes of the galactic formation process in the early universe.

Milky Way *Galaxy.

millimetre astronomy (millimetre-wavelength radio astronomy), the observation of the extraterrestrial universe by *radio astronomy at millimetre wavelengths. The technology of radio telescopes made significant advances during the 1960s towards shorter wavelengths. This has led to the advent of millimetre astronomy which explores the sky using *electromagnetic radiation between the radio and the infra-red wavelengths. Many molecules in space emit radiation in this part of the spectrum and a study of them is giving new insights into the *interstellar medium. For many years the 11 m (36 feet) telescope at the *Kitt Peak National Observatory in Arizona dominated the field but several new telescopes throughout the world became operational in the early 1980s, in particular the 15 m (49 feet) James Clerk Maxwell telescope at *Mauna Kea, Hawaii. In addition, the International Infra-Red Astronomical *Satellite (IRAS) was successfully orbited in 1983 to give unprecedented coverage of infra-red objects.

Mimas, a satellite of Saturn, discovered by William *Herschel in 1789. Its orbit has a semi-major axis of 185,520 km and its diameter is about 390 km. Its density is 1,200 kg/m^3 which implies a predominantly water ice composition. *Voyager images have shown that, in spite of its small size, Mimas is nearly spherical. The surface is heavily cratered, with the largest crater, **Herschel**, being one-third of the size of the whole satellite with a central peak 6 km high.

minor axis *ellipse.

minor planet, an object orbiting around the Sun but of smaller size and mass than a *major planet. Equivalent to the term *asteroid introduced by William *Herschel it is not generally used for comets, although the distinction between comets and asteroids is not always clear. For example, the orbits of *Apollo–Amor objects sometimes resemble those of comets. The largest asteroid, Ceres, is less than half the size of the smallest of the major planets, Pluto. At the lower end, there is no obvious boundary between small asteroids and large *meteoroids: they are part of a continuous distribution of sizes resulting from collisions and fragmentations.

Mir, a space station put into Earth orbit by the former Soviet Union on 20 February 1986 and manned by successive crews of cosmonauts. Each crew is able to spend some months in orbit carrying out programmes of scientific and technological research before being returned to Earth. Mir has a habitable volume of 100 cubic metres. Since its launch it has been increased in size by attaching new segments, such as the Kvant astrophysics module, making it the largest continuously manned space station built so far. Astronauts from a number of nations including Austria, Britain, France, and Germany have been included in the crews. It is regularly supplied by unmanned Progress vehicles.

Mira Ceti, a *binary star in the constellation of Cetus, and also known as Omicron Ceti. The brighter component is a red giant *variable star whose magnitude varies in the range

Mira Ceti

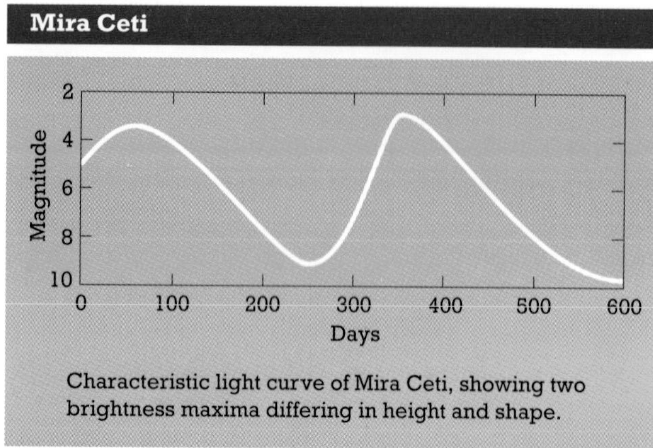

Characteristic light curve of Mira Ceti, showing two brightness maxima differing in height and shape.

3.5–9.1 over a period of 332 days. This star gives its name to the class of Mira variable stars. These are long-period pulsating stars whose mechanism resembles the *Cepheid variable stars except that it is a layer of hydrogen rather than helium that undergoes periodic ionization and recombination. 6,000 Mira stars are known. The fainter component is VZ Ceti, an irregular variable star with magnitudes in the range 9.5–12 that may be a *white dwarf with an *accretion disc. The system resembles the *symbiotic stars but has a wider separation, with a 400-year orbit that will bring the components closest together in the year 2001. Mira was discovered by David Fabricius in 1596, and its period was determined by J. P. Holwarda in 1638.

Miranda, a satellite of Uranus, discovered by the Dutch–American astronomer Gerard Kuiper in 1948. It has a nearly circular orbit at 129,390 km from the planet's centre with an inclination of 4.2 degrees to the equator of Uranus; its diameter is about 480 km. Photographed by *Voyager 2, its surface consists of two strikingly different types of terrain: an old, heavily cratered, rolling terrain, and a young, geologically very complex terrain with parallel sets of bright and dark bands, steep slopes, ridges, and troughs. These features occupy three distinct regions and the effects of ancient eruptions can still be seen. One possible explanation for these unique geological features is that Miranda was broken apart by a collision into several large pieces that subsequently reassembled into a single body.

mirror *reflecting telescope.

missing mass problem, in astronomy, the accurate determination of the masses of clusters of galaxies. This can be done by methods based on either adding up the masses of all the galaxies in the cluster, or considering the way the galaxies move within the cluster. The two methods give answers that differ by a factor of ten, with the dynamical methods giving the larger values (about 10^{15} times the mass of the Sun for the large rich clusters). For a variety of reasons, astronomers generally assume that the larger mass is correct, turning the problem into one of finding the missing mass. The most likely hiding place for the mass is now thought to be dark massive haloes round the galaxies in the cluster. Finding the true mass of the universe is important in distinguishing between models in cosmology, in particular between the *open universe and the *oscillating universe.

Mizar, a second-magnitude *multiple star in the constellation of Ursa Major, and also known as Zeta Ursae Majoris. Two components are hot *dwarf stars moving in an orbit of very long period, and forming the first double star to be discovered, by Giambattista Riccioli in 1650. The brighter star was also the first spectroscopic *binary star to be found, by the US astronomer Edward Pickering in 1889, and the fainter star is another spectroscopic binary. A distant fourth-magnitude companion, Alcor, can be seen with the naked eye. The distance of Mizar is 18 parsecs, and its combined luminosity is about forty times that of the Sun.

momentum, the product of a moving body's *mass and its *velocity. Since velocity is a directed (vector) quantity, that is, a speed in a particular direction, momentum is also a directed quantity. It is a body's momentum that gives it its *kinetic energy, equal to half the product of its mass and the square of its velocity. Doubling a body's mass without changing its speed doubles its momentum and its kinetic energy; doubling its velocity, however, doubles its momentum but quadruples its kinetic energy. This is often not appreciated by car drivers who do not realize that the car brakes will take four times as long to stop the car if the driver doubles the speed of the car. A different illustration of the effects of momentum is worrying the space organizations at the present moment. Tens of thousands of fragments from previously launched artificial satellites now encircle the Earth so that collision of new space satellites with this debris constitutes a real danger, since even a fragment of paint the size of a thumbnail colliding with a space station at a few kilometres per second would have the kinetic energy of a high-velocity bullet.

Monoceros, the Unicorn, a *constellation straddling the celestial equator, introduced by the Dutchman Petrus Plancius (1552–1622) and representing the mythical one-horned beast. Despite its faintness (the brightest stars are of fourth magnitude) it lies in the Milky Way and contains several fascinating objects including the *Rosette Nebula. The Cone Nebula is a dark dust cloud associated with the star cluster NGC 2264; the cluster's brightest member is the highly luminous supergiant S Monocerotis, estimated to lie 1,000 parsecs (3,000 light-years) away. M50 (NGC 2323) is a large star cluster visible through binoculars. Beta Monocerotis is a superb triple star whose components, of fifth and sixth magnitudes, are separable by small telescopes. Plaskett's Star is a sixth-magnitude spectroscopic binary; its component stars are each at least fifty-five solar masses, making this the most massive binary known.

month, in astronomy, a period of time related to the Moon's orbit about the Earth. Five months of different lengths can be defined:

synodic month or **lunation**, the time between successive similar geometric positions of Sun, Moon, and Earth; for example, the time between successive new moons;

sidereal month, the time it takes the Moon to rotate once round the Earth with respect to the stellar background;

anomalistic month, the time between successive passages of the Moon through perigee;

tropical month, the time between successive passages of the Moon through the vernal *equinox;

nodical or **draconitic month**, the time between successive passages of the Moon through the same *node.

The average lengths of these months are given in the table for time on page 192.

In civil usage the month is any of usually twelve periods of time into which the year is divided, or any period between the same dates in successive such portions. The primitive calendar month of ancient nations began on the day of a new moon or the day after, and thus coincided (except for fractions of a day) with the synodic month. However, the number of synodic months in a year is not a whole number, so many civilizations found it desirable to divide the calendar year into an integral number of shorter periods, and the synodic months were superseded by a series of twelve periods each having a fixed number of days. In the *Julian calendar the months in a leap year had alternately 31 and 30 days while in other years February had only 29 instead of 30. Under Augustus this symmetrical arrangement was broken up by the transference of one day from February to August, and one from September and November to October and December, respectively, producing the system of months now in use.

Moon, the Earth's only natural satellite. The Moon's surface is highly irregular, the main large-scale features being the *maria, the mountains, the *craters, and bright rays which may be *ejecta. The maria, conspicuous dark almost level plains, are depressions filled with solidified lava. The largest of them are over a thousand kilometres wide. The other three features are almost certainly the results of a colossal bombardment of the Moon's surface in the early period of the solar system, a bombardment lasting some hundreds of millions of years, while debris was being swept up by the Moon and the planets. This debris, formed of objects ranging in size from asteroids of up to a hundred kilometres across down to dust particles, created the Moon's main features, as well as producing a surface layer of rock that has been shattered, melted, vaporized, ejected, and cooled. The period of bombardment was followed by a flooding of the maria regions by basaltic lavas as a consequence of radioactive heating in the Moon's interior. Seismic activity, also known as **moonquakes**, has been recorded by the *Project Apollo instruments and the lunar samples brought back to Earth by astronauts show that the Moon must be at least 4.3 billion years old, probably the same age as the Earth. The types of rock found are made of the same chemical elements that make up the Earth, in much the same proportions. The Moon has no atmosphere and probably never had one to any marked extent, and life does not seem to have ever been present before men landed there. A recent theory suggests that the Moon's origin may have been a collision between a Mars-like planetesimal and the very young Earth. The temperature of the surface of the Moon reaches 100 °C in the lunar daytime and falls to below −200 °C during the lunar night. There is no erosion except for the slow crumbling of rock under repeated expansion and contraction and the occasional sudden localized disturbance due to the impact of a meteorite.

The Moon's mass, 1/81.38 that of the Earth, is accurately measured by tracking artificial lunar satellites, and its diameter of 3,476 km is 1/3.67 that of the Earth. It is an ellipsoid, although the three perpendicular diameters differ from each other by no more than a kilometre. The Moon rotates such that the longest axis always points in the general direction of the Earth, the Moon's period of rotation being equal to its period of revolution about the Earth so that, apart from *libration effects, it always keeps the same side to the Earth. The mean density of the Moon is 3,330 kg/m^3, very close to that of the basic rocks under the Earth's crust, and the surface gravity is only about one-sixth that at the Earth's surface, so that objects on the Moon weigh only one-sixth of what they weigh on Earth.

The Moon is the nearest celestial object to the Earth. It revolves about the Earth in an orbit with the properties described in the table on page 196. If the Earth and Moon were point masses or rigid spheres whose density varied only with distance from their centres, and no other celestial body existed, then the orbit of the Moon about the Earth would be a fixed ellipse. The Sun, however, and to a far lesser extent the planets, exert their gravitational influence upon the Moon in its orbit perturbing the *elements of orbit in a highly complicated manner so that the semi-major axis, eccentricity, and inclination suffer a myriad of cyclical changes or **inequalities**, oscillating about their mean values. The other three orbital elements, the longitude of the ascending *node, the argument of perigee (the angle between the ascending node and perigee, which specifies the location of the perigee with respect to the equatorial plane), and the time of perigee passage not only alter in a periodic manner but also change secularly, that is, increase or decrease steadily over time. Thus the orbital plane precesses backwards, the *line of nodes making a complete revolution of the *ecliptic in 18 years 219 days, while the argument of perigee advances, its period of revolution being 8 years 310 days. The shapes of the Earth and Moon contribute to these *perturbations of the lunar orbit but the Sun's gravitational effect (see *evection) dominates. From the time of *Newton, many astronomers, such as Peter Hansen, Charles Delaunay, and Ernest Brown, have attempted to create a mathematical description of the Moon's behaviour capable of predicting its position and speed to the accuracy with which it can be observed. Computers have recently been used, as in André Deprit's *lunar theory, to make much more accurate predictions, a requirement arising from the increased

These four photographs of the **Moon**, the Earth's only natural satellite, show its two principal surface features. In contrast to the heavily cratered regions, the maria appear from Earth to be flat and featureless. However, photographs from Lunar spacecraft have revealed much detail in the maria surfaces.

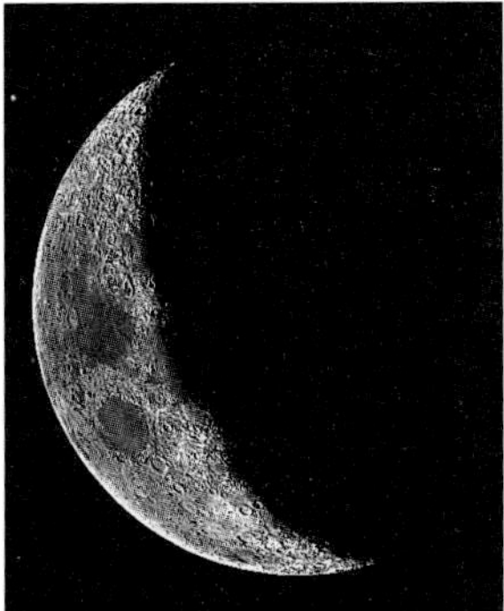

Moon

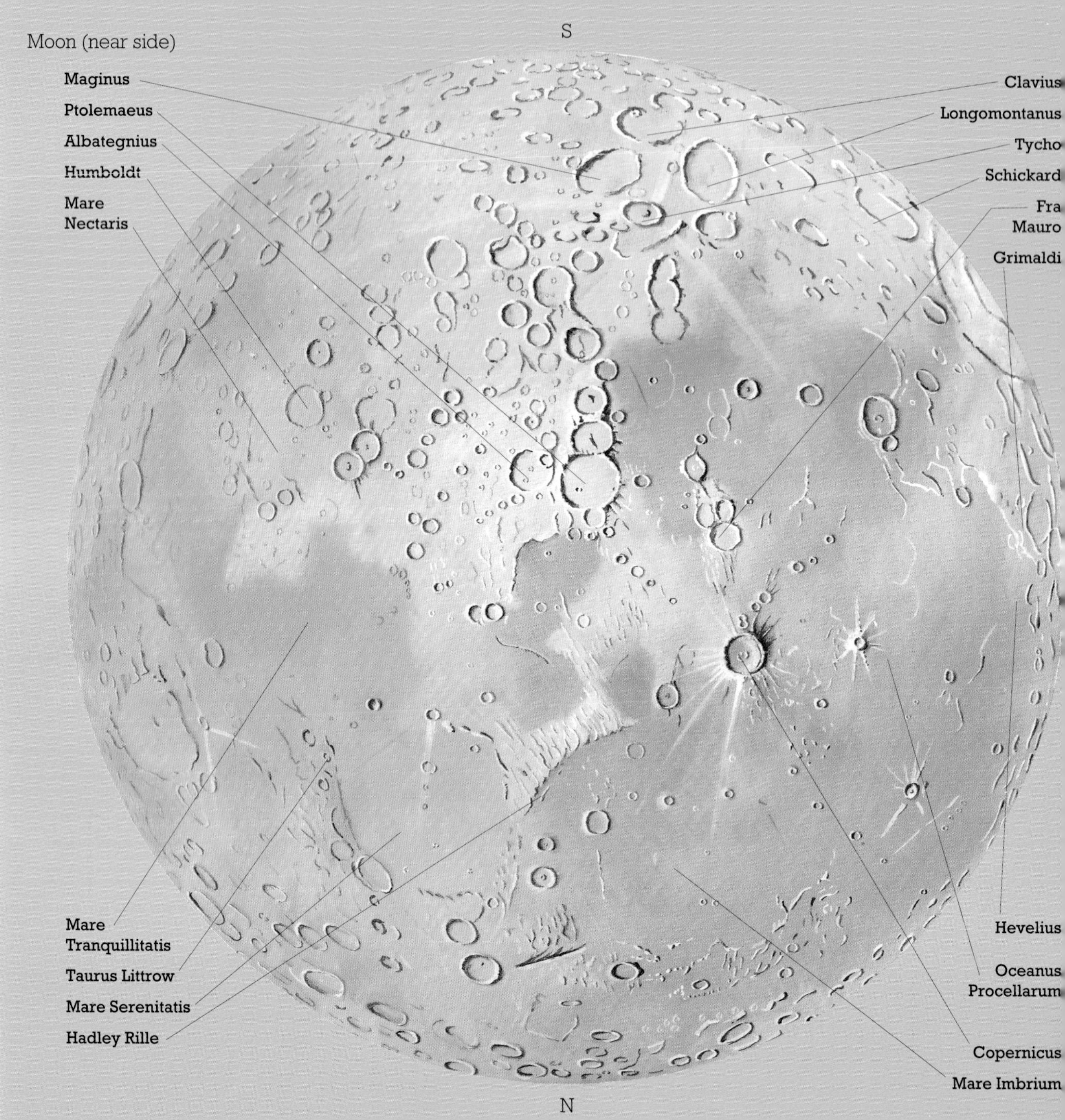

As the Moon orbits the Earth, it always presents the same face, called the near side, to Earth-based observers. The darker, smooth areas of the Moon, called maria, are only evident to any extent on the near side. These maria which cover some 35 per cent of the near side's surface are lava flows produced by volcanic action. Being far less cratered they are thought to be much younger than the surrounding highlands.

In this map of the near side, the south pole is shown at the top as it appears when seen through an astronomical telescope from the northern hemisphere. However, when seen from the northern hemisphere with the naked eye or through binoculars the south pole is at the bottom. The image in a telescope based in the southern hemisphere has the north at the top.

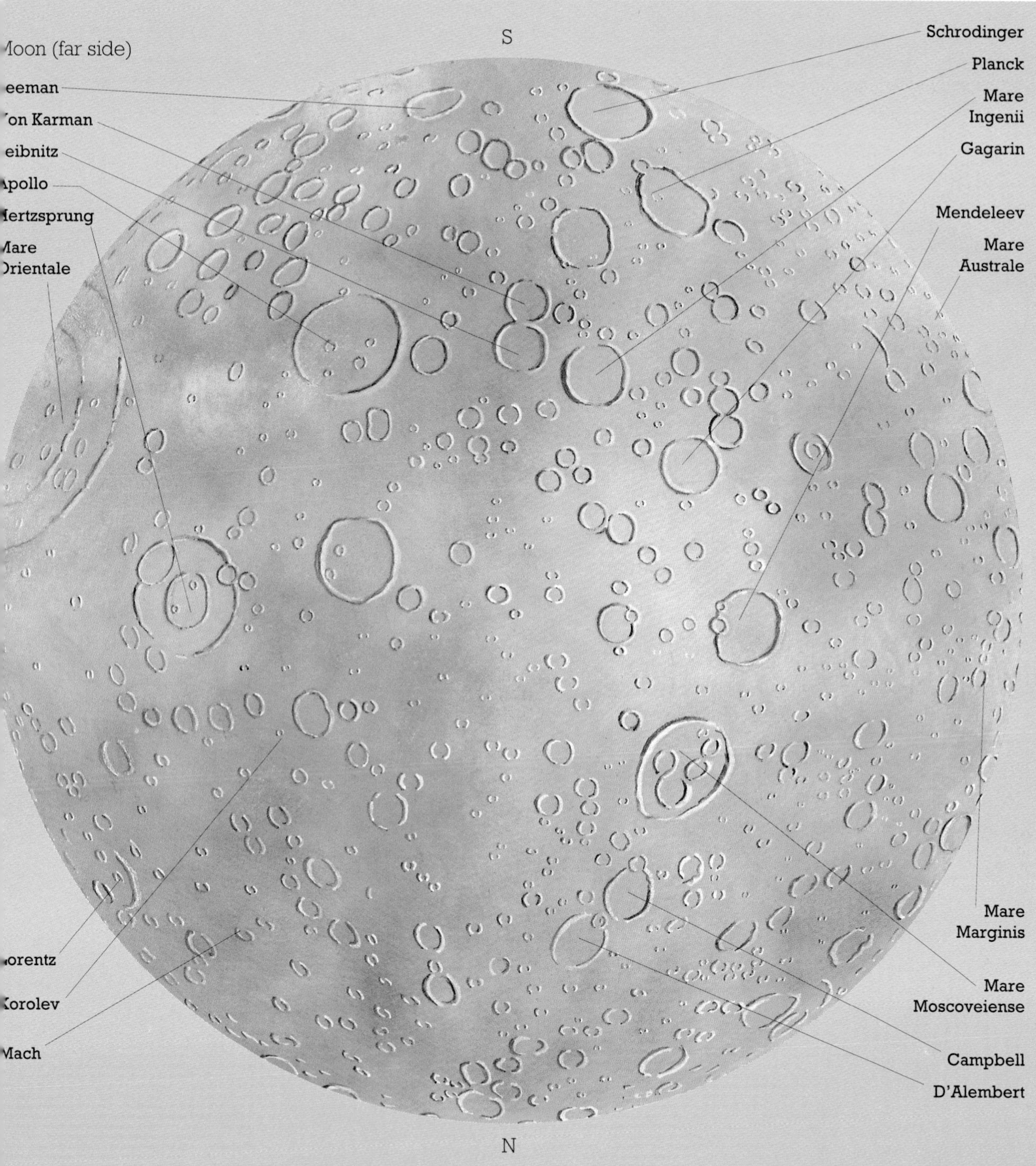

By contrast, the far side of the Moon can never be seen from the Earth, apart from a small fraction which is visible because of libration effects. Therefore maps of the hidden side have been constructed from thousands of photographs taken over many years by spacecraft orbiting or flying past the Moon. The first photographs of the far side of the Moon were taken in October 1959 by the Soviet spacecraft Luna 3.

These photographs showed that the far side of the Moon was almost completely devoid of maria. Today it is believed that this has arisen as a result of the lunar crust being thicker on the far side and therefore not allowing basalt to leak through impact craters as has happened on the thinner crusted near side.

Mounting, telescope

Astronomical telescopes are usually mounted in one of two ways, the alt-azimuth or the equatorial mounting. As the alt-azimuth mounting has a vertical axis, if the telescope tube is aligned perpendicular to this axis, it will rotate along the horizon. In the equatorial system, the equivalent of the vertical axis is aligned with the celestial pole so that if the telescope tube is aligned perpendicular to this axis, it will rotate along the celestial equator.

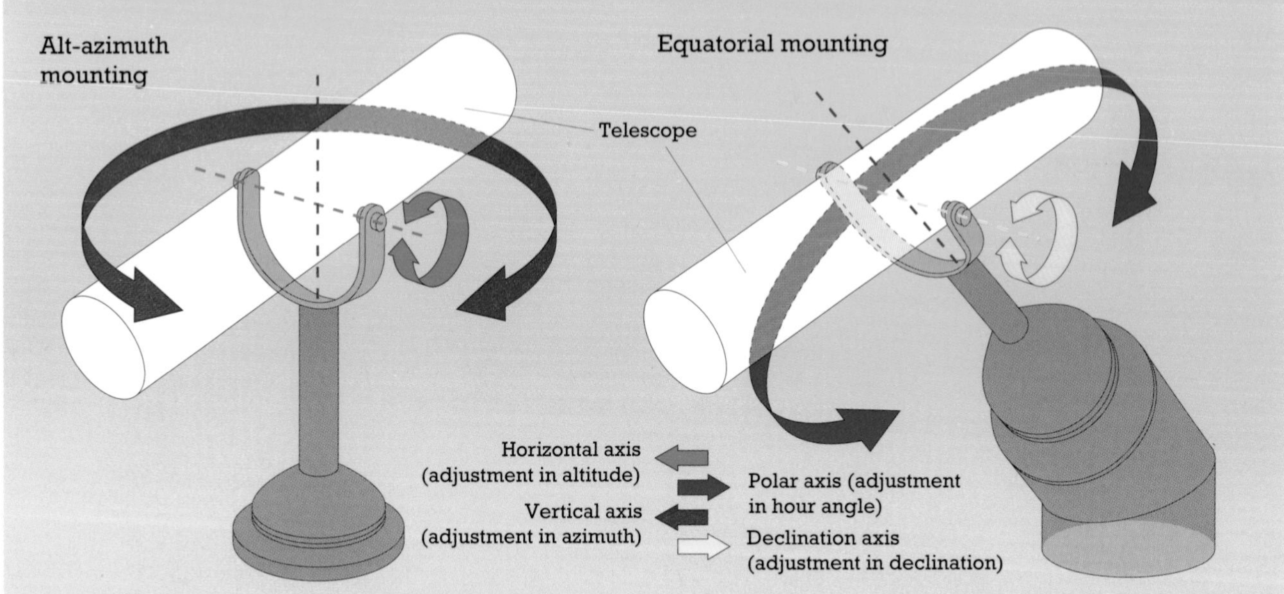

precision with which the Moon's distance and movements can be measured by lunar laser-range and range-rate methods. Such methods can give the Moon's distance to an accuracy of a few centimetres.

Man has long been fascinated by three interrelated phenomena: the Moon's *phases, the *tides, and *eclipses of the Sun and Moon. Spring tides occur at new and full moon, neap tides at first and third quarter. Lunar and solar eclipses occur respectively at full and new moon but not at every full and new moon.

moonquake *Moon.

morning star, an almost obsolete term referring principally to the planet Venus when it appears in the eastern sky before sunrise, although it may be used to refer to Mercury when it rises before sunrise. Venus and Mercury are morning stars while they move between the *planetary configurations of inferior conjunction and superior conjunction.

mounting, the structure that supports a *telescope. Because of the Earth's rotation about its axis, the celestial sphere appears to rotate about its polar axis, which extends from the north celestial pole to the south. The telescope mounting should not only enable the telescope to be trained on the celestial object of interest but also, if a lengthy observation is required, to keep it within the telescope's *field of view. Therefore, the telescope needs to be moved continuously and its direction must be adjusted about two axes. This is usually done in one of two ways. A telescope with an **equatorial mounting** has one axis (the polar axis) aligned parallel to the Earth's axis of rotation. This axis is tilted to the observer's horizon at an angle equal to the observer's latitude. The other axis (the declination axis) is at right angles to the polar axis. Once the correct angle has been set about this second axis, a motor, often with a speed adjustment facility, drives the telescope about the polar axis at the same rate as the Earth's rotation, so that the object can be tracked exactly. The equatorial mounting has a number of variations such as the German, the fork, and the English mountings. By contrast, the **alt-azimuth** mounting allows the telescope to be freely moved up and down (by varying the altitude angle) and left to right (by varying the azimuth angle). The tracking is under computer control, with both angles being adjusted at the appropriate rate according to the position of the object in the sky.

Mount Palomar Observatory, an *observatory near Pasadena, California, USA, run by the California Institute of Technology. At an altitude of 1,700 m (5,580 feet) it is the home of the 200-inch (5.1 m) *Hale telescope, an 18-inch (46 cm) Schmidt *reflecting telescope, as well as a 48-inch (1.2 m) and a 60-inch (1.5 m) reflector. The dome is 41 m (135 feet) high and weighs 1,000 tonnes. The Hale telescope was the first telescope to allow an astronomer to sit in a cage at the prime focus. The telescope was mounted so that it could easily observe the stars around the north celestial pole, a region not accessible to the 100-inch (2.5 m) telescope at the *Mount Wilson Observatory. One of the first discoveries, in 1952, was that Population I Cepheid variable stars are 1.5 magnitudes brighter than Population II Cepheids, a result that meant that the estimates of the size and age of the universe had to be doubled. The telescope was used to obtain the first quasar spectrum and has also been used to observe white dwarfs and supernovae, and to measure the velocities of stars in galaxies. Today most work is done using *solid state cameras at the Cassegrain focus.

Mount Wilson Observatory, an *observatory in the Sierra Madre mountains near Pasadena, California. The first telescope was a solar *coelostat, purchased by Helen Snow. This was followed by two solar tower telescopes, 18.3

m (60 feet) and 45.7 m (150 feet) high, that had their associated spectrographs underground where the temperature was constant. Using these telescopes *Hale discovered that sunspots were cooler than the rest of the Sun's surface and that the spots were regions of high magnetic field. He also found that their magnetic polarity varied with a twenty-two-year period. A 60-inch (1.5 m) reflector started operation in December 1908. The 100-inch (2.5 m) Hooker telescope was first used in November 1917 and for the next thirty years was the most significant telescope in the world. A *Michelson interferometer mounted at the top of this telescope provided the first direct measurements of stellar diameters.

moving cluster *open cluster.

Mullard Radio Astronomy Observatory, a radio *observatory situated at Lord's Bridge near Cambridge, UK, and operated by the University of Cambridge. The technique of *interferometry was originally developed by the Cambridge radio astronomers to measure accurate positions of radio sources and was later extended to arrays of telescopes to produce the first detailed maps of radio sources by *aperture synthesis. The discovery of pulsars was also made using an interferometer, the 3.6 hectare array. In recognition of these achievements the Nobel Prize for Physics was awarded to Martin Ryle and Antony Hewish in 1974. Martin Ryle also had the distinction of being the first *Astronomer Royal who was not an optical astronomer.

multi-element radio linked interferometer network (MERLIN), a large *radio telescope network in the United Kingdom spread over 130 km (80 miles) and employing *interferometry to map in detail celestial objects emitting radio waves. Eight radio telescopes of various sizes operate within the network though the number employed in a particular observation programme depends on the wavelength selected.

The **multiple-mirror telescope** on Mount Hopkins, Arizona, USA, consists of six separate mirrors, each 1.8 m in diameter. The movement of these mirrors is co-ordinated by computer to give the power of a single mirror with an effective diameter of 4.5 m.

multiple-mirror telescope (MMT), a *reflecting telescope containing a number of small mirrors put together to behave as one large aperture. To collect as much light as possible from celestial objects there is a need for telescopes of very large size, but telescopes with a single large aperture are costly to construct. The multiple-mirror telescope, also known as the **segmented mirror telescope**, offers a cheaper solution and, although many construction and control problems have to be overcome, successful MMTs have been built. One, on Mount Hopkins in Arizona, consists of six 1.8 m (71-inch) diameter mirrors operating together.

multiple star, a group of three or more stars that are so close together that they appear as one to the unaided eye. Normally the components are bound together by gravitation, two of them forming a close *binary star, with a third star (which may itself be a binary) moving in a larger orbit about the system's centre of mass. In some apparent multiple stars, however, one component is an unconnected star nearly in the same line of sight, as in the case of an *optical double star. Most known multiples are triple stars. Some well-known systems with more than three components are Castor and the Trapezium, which each have at least six, and Mizar, which has at least four. The components in a multiple may be too close all to be resolved in a telescope, but they may be detected as spectroscopic binaries, as in the two resolvable components of Mizar, or as *eclipsing binaries as in two of the four resolvable components of the Trapezium.

mural quadrant, a large *quadrant fixed vertically to a wall. Its greater size made it more accurate than a hand-held quadrant. The wall needed to be set north to south so that the projection of the rotating arm swept out part of the observer's *meridian on the celestial sphere. The arm was lined up on the celestial object as it crossed the meridian and the altitude was read from the position of the arm on the graduated arc.

Musca, the Fly, a small *constellation of the southern sky, introduced by the Dutch navigators Pieter Dirkszoon Keyser and Frederick de Houtman at the end of the 16th century. Its brightest star is of third magnitude but it contains little of note for the casual observer except that part of the *Coalsack Nebula intrudes across the border from neighbouring *Crux.

N

nadir, that point on the *celestial sphere opposite the *zenith. It is 90 degrees away from the horizon and is approximately the direction in which a plumb-bob hangs. Because the Earth is not spherical the direction of the plumb-bob nadir would not in general pass through the Earth's centre. Furthermore, the plumb-bob can be deviated slightly by large local masses such as mountains so that for precise astronomical measurements corrections have to be made to clarify which nadir is being used.

NASA (National Aeronautics and Space Administration), a civilian agency formed in 1958 to co-ordinate and direct the whole of aeronautical and space research in the USA. Its headquarters are in Washington, DC. As well as collaborating with industry and universities, NASA also maintains and develops its own installations. Among the latter are the Ames Research Center, California; the Goddard Space Flight Center, near Washington, DC; the Houston Manned Space Flight Center, and the *Jet Propulsion Laboratories (JPL). JPL were responsible for the *Mariner and *Voyager missions.

National Radio Astronomy Observatory (NRAO), an *observatory at Green Bank, West Virginia, specializing in *radio astronomy and equipped with a variety of *radio telescopes and a 1.6 km (1 mile) baseline interferometer. The observatory also uses instrumentation at the *Kitt Peak National Observatory.

Nautical Almanac *Royal Greenwich Observatory.

N-body problem, in *celestial mechanics, the motion in space of *N* bodies under the action of their mutual gravitational attractions. The distances between the bodies are so large that they can be considered to be points. Though the planets have a finite size, and there are other forces acting upon them, the *N*-body problem with *N* of the order of 10 is an extremely accurate model for the orbits of the planets. In 1889, *Poincaré showed that the problem cannot be solved exactly. However, solutions can be approximately computed, either analytically (with methods started by *Laplace) or numerically with modern digital computers. As a result, while it is comparatively easy to compute orbits for time-spans of a few centuries and with accuracies adequate for comparison with the astronomical observations, when the dynamical evolution of the solar system has to be studied for time-spans comparable to the age of the Earth, the problem becomes exceedingly difficult. Only recently, problems such as the stability of the orbits of the planets, the origin of asteroid families, and the source of meteorites, have become tractable, but the chaotic behaviour of the orbits (described by *chaos theory) may ultimately limit our knowledge.

neap tide *tides.

nebula, an extended patch of hazy light or darkness in a fixed position with respect to the stellar background. Nebulae are usually classified as diffuse (condensations of gas and dust), planetary, or supernova remnants. **Diffuse nebulae** can be bright or dark. Bright nebulae can occur as fairly compact structured regions or they can extend in delicate filamentary wisps over several degrees of the sky. The brightness of the nebular material is due either to light reflected from it (a *reflection nebula) or, if a very hot star is nearby, as in a *Strömgren sphere, to direct emission of *electromagnetic radiation from atoms (mainly hydrogen) ionized by the star (an **emission nebula**). The most famous example of an emission nebula is the *Orion Nebula. Information regarding the gas density and its composition may be derived from the spectra of such objects. **Dark nebulae** contain thick clouds of dust and gas which obscure and absorb light from stars beyond or within them. *Planetary nebulae, of which about a thousand are known, have nothing to do with planets but are hazy expanding shells of gas surrounding a central very hot star, probably ejected by the star. *Supernova remnants, assumed to be material ejected from supernova explosions, are also bright emission nebulae. The most famous is the *Crab Nebula, the result of a supernova explosion witnessed by the Chinese in 1054. Others, like those observed by *Tycho Brahe and *Kepler in 1572 and 1604 have similar features. *Radio astronomy can be used to investigate the *synchrotron radiation at radio wavelengths which is emitted by these nebulae. This is an indication of the existence of electrons spiralling at near the speed of light in the twisted skein of magnetic fields within the nebula.

nebular hypothesis *Laplace nebular hypothesis.

Neptune, the eighth *planet from the Sun, and the third most massive in the solar system. Neptune was discovered in 1846 by the German astronomers Johann Galle and Heinrich d'Arrest whilst trying to verify the computations of *Le Verrier who predicted that an undiscovered outer planet was perturbing the motion of Uranus. *Adams had inde-

This photograph of **Neptune** was obtained by Voyager 2 during its encounter in August 1989. As well as bands and zones reminiscent of those on Jupiter and Saturn, bright wispy clouds are here seen overlying the Great Dark Spot. This spot must therefore lie lower in the atmosphere than these clouds.

pendently arrived at the same result. Following the discovery, many observations of Neptune were found in the records of the preceding centuries, but the planet moves so slowly that it was previously thought to be a star. It was therefore possible to obtain a good knowledge of its orbit quite quickly. However, deviations of the observed motion of the planet from the computed one started to appear, and this led astronomers to search for another more remote planet. This planet, *Pluto, was found in 1930 not far from the predicted position but it does not fully account for the variations in Neptune's orbit. Observations by *Voyager 2 have greatly expanded our knowledge of Neptune, which until recently was somewhat uncertain. Neptune emits 2.7 times more energy than it receives from the Sun and, like the other *giant planets, its atmosphere is composed mainly of hydrogen. The line joining the magnetic poles of Neptune is inclined at about 47 degrees to the rotation axis of the planet. Spacecraft observations have led to the discovery of several rings, one of which is characterized by an irregular distribution of material: three arcs in this ring appear to be thicker than the rest of the ring as if the particles there were more densely packed than elsewhere. The satellite system is peculiar. As well as the large satellite *Triton and an outer satellite, *Nereid, there are also six more satellites of small to medium size, all very close to the planet, which were discovered by Voyager. Triton orbits Neptune in the opposite direction to the planet's orbital motion and was probably captured by Neptune after its formation. Nereid has a very eccentric orbit.

Neptunian inner satellites, six satellites lying inside the orbit of *Triton and discovered in the images taken during the 1989 *Voyager 2 encounter with Neptune. They move in the same direction as Neptune's axial rotation in nearly circular orbits. Only the innermost one, Naiad, has a significant inclination (about 5 degrees) to the planet's equator. These satellites have irregular shapes with dark surfaces and their sizes range from about 50 to 400 km. Two of them (Galatea and Despina) orbit just inside two narrow Neptunian rings, and may be *shepherding satellites preventing the ring material from spiralling inwards.

Nereid, an outer satellite of Neptune, discovered by the Dutch–American astronomer Gerard Kuiper in 1949. It has the highest orbital eccentricity (0.75) of all natural satellites and a semi-major axis of about 5.5 million km. Photographed at poor resolution by *Voyager 2 in 1989, it has an irregular shape, a diameter of about 300 km, and an albedo of 0.14.

neutrino, an elementary particle whose existence was predicted by Wolfgang Pauli in 1930 to avoid the violation of the laws of conservation of energy and momentum in a nuclear process called beta decay. In this radioactive process the electrons or positrons produced are ejected with an observed distribution of energies. Pauli supposed that the remaining energy was given to neutrinos. The existence of the neutrino and its antiparticle (the antineutrino) is now accepted as experimentally confirmed. Three types of neutrino exist and recently it has been postulated that some of the so-called 'missing mass' in the universe may be accounted for if neutrinos have some non-zero rest mass. The difficulty in detecting neutrinos is realized when it is known that a column of liquid chlorine one *parsec long would be required before a neutrino travelling along it would be absorbed by one of the chlorine atoms.

neutrino astronomy, the observation of numbers of neutrinos coming from celestial objects. According to the *quantum theory the Sun should emit neutrinos as a result of the nuclear processes occurring in the Sun's core where hydrogen is turned into helium. Calculations have shown that sufficient numbers of neutrinos are created in the Sun to be detected at the Earth by neutrino 'telescopes'. Since 1968 at the bottom of a 1.5 km (0.9 miles) deep disused gold mine, a tank containing 600 tonnes of tetrachloroethylene, a liquid rich in an isotope of chlorine, has been used to detect the very occasional reaction of a neutrino with a chlorine isotope atom. The result is the production of argon which can be detected. The depth below ground of the tank shields its contents from other less penetrating *cosmic rays. The rate of detection of high-energy neutrinos is a factor of three below that expected from solar nuclear reaction theory, a major discrepancy which has not been explained. Again, the supernova that appeared in the Large *Magellanic Cloud created neutrinos, a tiny fraction of which were detected by neutrino telescopes.

neutron star, a very hot star of exceptionally high density. The density is typically 10^{13}–10^{15} times that of water, and about a million times greater than in a white dwarf. The original electrons and protons of the star's constituent matter have been crushed together, so that its core is composed primarily of neutrons. Neutron stars are thought to be only about 10–20 km in diameter. Consequently they have very low luminosity and are normally impossible to detect. Rotating magnetic neutron stars can, however, be observable as *pulsars. Neutron stars are also thought to be present as components of several *X-ray binaries. They are thought to be the closing stage of *stellar evolution in most stars with masses between the *Chandrasekhar limit of 1.44 times that of the Sun and about three times the mass of the Sun. Stars of greater mass than this are thought to become *black holes. Neutron stars are formed by the collapse of the star's core when its nuclear fuel is exhausted, an event that gives rise to a Type II *supernova outburst.

Newton, Sir Isaac (1642–1727), British physicist and mathematician, most famous for his work on differential and integral calculus, gravitation, and the theory of light and colour. *Newton's Law of Gravitation is central to his work on astronomy. It states that the force between any two bodies in the universe is proportional to the product of the masses of the two bodies divided by the square of the distance separating them. This applies equally to apples falling from trees and planets on their elliptical orbits around the Sun. He also proved that the gravitational effect of a three-dimensional body such as a planet is equivalent to that of its total mass concentrated at a point at its centre. He used his theory to account for the polar flattening of the Earth, the *precession of the equinoxes, the revolution of the lunar *line of nodes, and also to measure the mass of the Sun and those planets that have moons. He proved that any body moving in space subject to a single central force moves on a conic, such as an ellipse, and he went on to devise a method of calculating cometary orbits. He applied his knowledge of optics to the production of the first *reflecting telescope in 1668. He became Lucasian Professor of Mathematics at Cambridge in 1669, and his book *Philosophiae Naturalis Principia Mathematica* (1687) is widely acknowledged as being the greatest science book ever written.

Newtonian telescope *reflecting telescope.

Isaac **Newton**'s scientific interests covered the whole of physics and astronomy as they were known in his day and he was the first to provide a unified framework in which to describe and explain all known physical phenomena. His development of the calculus was crucial to this achievement, and his work was not superseded until the advent of relativity and quantum theory some 200 years later.

Newton's Law of Gravitation, one of the most far-reaching and important physical laws ever formulated, stating that every particle of matter attracts every other particle of matter with a force proportional to the product of the particles' masses and inversely proportional to the square of the distance between them. Together with *Newton's Laws of Motion, the Law of Gravitation, published in 1687, laid the foundations for subsequent major advances in physics and astronomy.

Newton's Laws of Motion, three laws that laid the foundations of the science of dynamics. Though implicit in the scientific thought of the time of *Newton, these laws were first stated explicitly by him, and his exploration of their consequences in conjunction with his Law of Gravitation did more to bring into being our modern scientific age than any of his contemporaries' work. They are:

Every body continues in its state of rest or of uniform motion in a straight line unless acted upon by an external *force.

The rate of change of *momentum of a body is proportional to the impressed force and takes place in the direction in which the force acts.

To every action there is an equal and opposite reaction.

Nix Olympica, a bright feature on the surface of Mars visible through telescopes. Spacecraft have subsequently enabled astronomers to identify it as a huge volcano, **Olympus Mons**. Olympus Mons is an ancient mound-shaped volcano, 500 km wide at its base, that resembles the island of Hawaii. The summit of Olympus Mons has a crater 65 km in diameter and is 27 km above the mean surface of Mars; this is the highest point on the planet. Its gentle slopes are marked with channels and lava flows whose rounded extremities can be seen at its base. The walls of the central summit crater are steep and the crater was formed when the underlying lava withdrew beneath the surface. To the north is an extensive ring of grooved terrain which has either been produced by erosion or by lava erupting through ice sheets. Some astronomers think that Olympus Mons and other large Martian volcanoes have erupted only in the last billion years and that we are seeing the early stages of *tectonic activity which will eventually lead to continental drift and the production of large basins similar to those underlying the Earth's oceans.

node, one of the two intersections of the *line of nodes with an orbit obeying *Kepler's Laws. Nodes are the only points on the orbit lying in the reference plane, which is frequently the *ecliptic or the *invariable plane. The intersection at which the orbit goes from the southern hemisphere to the northern is the **ascending node**; at the **descending node** the orbit descends from the northern to the southern hemisphere. The specification of the nodes depends on the choice of the *coordinate system.

noon, the instant at which the Sun is due south of an observer in the northern hemisphere and due north in the southern hemisphere. The Sun crosses the observer's meridian at noon; the observer's meridian is the great circle on the *celestial sphere that is at right angles to the celestial

The extinct volcano **Nix Olympica** on the surface of Mars is the largest volcano in the solar system. Martian volcanoes are unusually large, perhaps because there seems to be little plate tectonic activity on Mars and because the atmosphere is thin, so that there is less erosion.

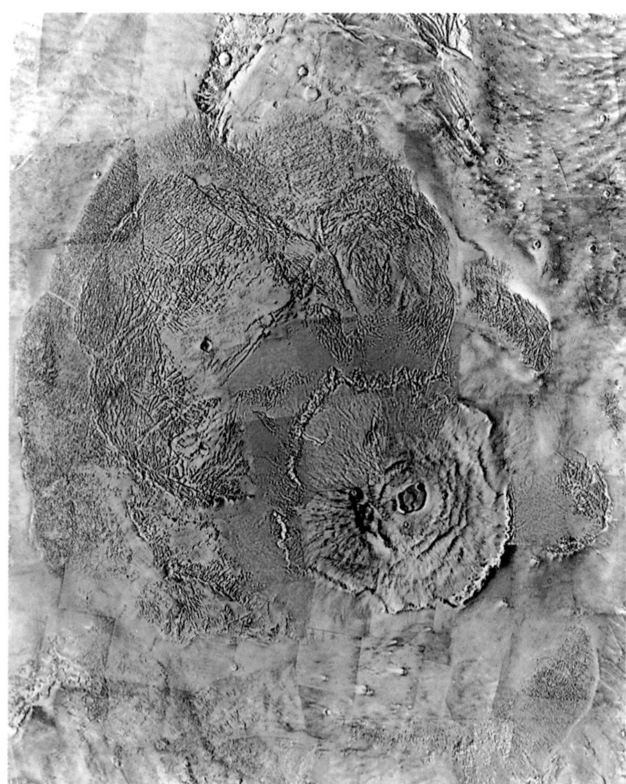

equator and passes through the north and south celestial poles and the observer's *zenith. This observed crossing is known as apparent noon. When the Sun next crosses the meridian an apparent solar day has elapsed. Because of the eccentricity of the Earth's orbit and the fact that the Earth's spin axis is not perpendicular to the orbit, the time between successive observed transits of the Sun across the meridian varies slightly from day to day. To overcome this problem the yearly average of the apparent solar day is introduced, this being known as the **mean solar day**. This is defined using a **mean sun** that moves along the celestial equator at a uniform rate. The *transit of this mean sun across the meridian defines mean noon, and at Greenwich, for example, defines 12.00 hours *Greenwich Mean Time.

Norma, the Level, a *constellation of the southern sky, introduced by the 18th-century French astronomer Nicolas Louis de Lacaille. It represents a surveyor's level, although Lacaille originally drew it as a set square and ruler. It lies in the Milky Way but its brightest stars are of only fourth magnitude and there are no outstanding objects of interest.

North America Nebula, a widespread distribution, 2 degrees across, of stars and nebular material in the constellation of Cygnus. The *nebula closely resembles the shape of the North American continent. In good seeing conditions it is visible in binoculars, though its structure is best seen by photography.

north celestial pole *celestial sphere.

Northern Hemisphere Observatory *La Palma Observatory.

north galactic pole *galactic pole.

north galactic spur, a giant arc-shaped *radio source. It runs out of the plane of the Milky Way towards the north *galactic pole, where it loops back towards the plane. The radio emission comes from high-energy electrons which spiral in the galactic magnetic field and emit *synchrotron radiation. The radio loop is part of a gigantic shell nearly 120 parsecs (400 light-years) across which is believed to be the remnant of a *supernova explosion that occurred several hundred thousand years ago. The shell has slowed down almost to rest, but its gentle expansion can still be measured through the *Doppler effect on the *twenty-one centimetre line emission from hydrogen gas it has swept up.

north polar distance, the angular distance of an object on the *celestial sphere from the north celestial pole, measured along the meridian from the pole to the object. Its value is 90 degrees minus the object's declination. If the north polar distance of a star is less than the observer's latitude, the object is a *circumpolar star.

nova, a *cataclysmic variable star that suddenly increases in brightness by 7–19 *magnitudes but returns to its previous magnitude after some months. Novae are usually binary stars in which one member is a *white dwarf with a very close companion. Gas is drawn from its companion on to the surface of the white dwarf where it erupts in a *nuclear fusion explosion throwing off a shell of gas. A nova outburst in a given system may erupt several times at intervals of perhaps tens of thousands of years but in a few **recurrent novae** (subtype NR) between two and six eruptions have

The shape of the **North America Nebula** resembles that of the North American continent. The nebula glows with the red light of hydrogen which is ionized by hot massive stars. The dark area of the 'Gulf of Mexico' is a cloud of dust obscuring the stars and the glowing hydrogen gas behind it.

Nova

Light curves of typical novae.

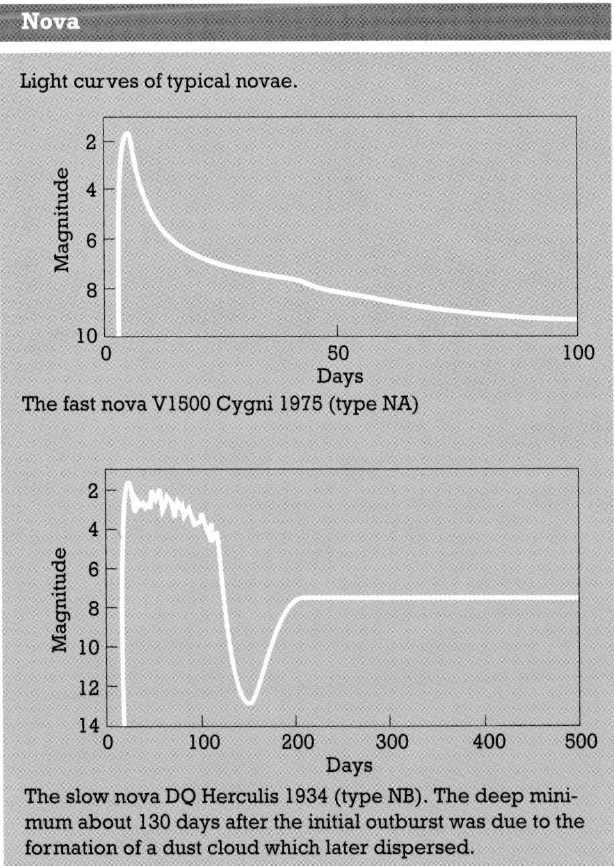

The fast nova V1500 Cygni 1975 (type NA)

The slow nova DQ Herculis 1934 (type NB). The deep minimum about 130 days after the initial outburst was due to the formation of a dust cloud which later dispersed.

When **Nova** Cygni erupted in 1975, the star, which was until then too faint to be observed, reached second magnitude; for a very short time it was approximately a million times more luminous than the Sun. Nova Cygni is therefore one of the most dramatic novae ever recorded.

been detected within a few decades. Unlike a *supernova, a nova eruption does not greatly damage the star. About thirty nova outbursts are thought to occur each year in the Galaxy, but most are either missed or concealed by interstellar dust clouds in the plane of the Galaxy. However, their high luminosity makes them visible in nearby galaxies. At its brightest a nova can have an absolute magnitude of −8. The brightness of a **fast nova** (subtype NA) fades from its maximum by three magnitudes in 100 days or less; a **slow nova** (subtype NB) takes 150 days or more. The very slow nova-like outbursts of certain *symbiotic stars are sometimes included as subtype NC.

Nubecula Major *Magellanic Clouds.

Nubecula Minor *Magellanic Clouds.

nuclear fission, any nuclear reaction that involves the break-up of atoms with the release of energy. Most nuclear power stations create electricity using turbines and dynamos driven by the energy released by the fission of uranium-235. Other elements undergoing nuclear fission are thorium and plutonium. Stars, however, derive their energy from nuclear fusion.

nuclear fusion, any nuclear reaction in which atomic nuclei collide and coalesce to form a different kind of nucleus. In stars the two nuclear reactions operating in main sequence stars are the *carbon–nitrogen cycle and the *proton–proton reaction, both being fusion reactions which change hydrogen into helium, four atoms of hydrogen fusing into one atom of helium. Producing the same type of fusion reaction (hydrogen into helium) is a long-sought goal for the world's nuclear physicists who see in its successful achievement the solution to the world's energy problems, since a by-product of the reaction is the release of energy.

nuclear synthesis, the building of the chemical elements in nuclear reactions. Suitable conditions for such reactions existed in the very early life of the universe and exist to the present day in the centres of evolving stars. In the first few seconds after the *Big Bang the temperature dropped to a level low enough for particles such as protons and neutrons to exist. Deuterium nuclei formed and in their turn produced helium. Indeed most of the helium in the universe was created at this early era. Successively heavier elements up to iron's mass number have subsequently been synthesized in the cores of stars in successive nuclear reactions. Hydrogen is converted into helium by stars in the *proton–proton reaction or the *carbon–nitrogen cycle when the core temperature reaches about 10^7 K. Thereafter, the so-called *triple-alpha process takes over. This happens when the hydrogen-to-helium phase ends, the star core collapses, the envelope, that is, the outer region of a star between the core and the atmosphere, expands, and the core temperature reaches 10^8 K. Succeeding this are other reactions, each in turn using the product of the previous reaction until iron is synthesized. Heavier nuclei can now be formed in the star by slow neutron capture. If the star then enters the catastrophic *supernova stage, the heaviest possible nuclei are formed and are ejected into the interstellar medium. New generations of stars subsequently forming from the enriched medium have an observably higher heavy element content than old stars.

nucleus, a name given to (1) the central solid region of a comet, (2) the core of a galaxy, and (3) the central part of an atom consisting of massive particles such as neutrons and protons, held together by strong nuclear forces and representing almost all of the atom's mass.

Nuffield Radio Astronomy Laboratories *Jodrell Bank.

nutation, a small oscillation superimposed on the *precession of the Earth's axis. The orbit of the Moon is inclined to the *ecliptic by 5.1 degrees so that, as the Moon orbits the Earth, it exerts a small torque on the equatorial bulge of the Earth in addition to the main torque from the Sun and Moon that produces luni-solar precession. This results in an additional oscillation, called nutation, being superimposed on the precession, and makes the Earth's north and south poles each trace out a wavy circular path. It amounts to a few arc seconds and the main period of nutation is about 13 days.

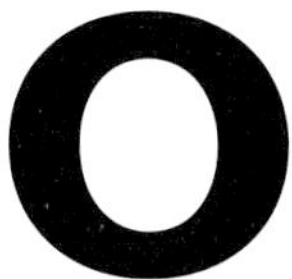

Oberon, the outermost known satellite of Uranus, discovered by William *Herschel in 1787. It has a nearly circular orbit above the equator of Uranus at 583,520 km from the planet's centre. It is about 1,500 km in diameter and its density of 1,500 kg/m^3 suggests that the interior contains a mixture of ice and rock. Photographed by *Voyager 2, its surface has some large craters, with bright rays emanating from them and with dark patches on their floors, possibly due to eruptions of dirty water. Oberon also has large mountains and steep curved slopes.

objective *refracting telescope.

objective prism, a large prism that is placed in front of the objective lens of a telescope. The incoming light is split, by the prism, into a *spectrum before entering the telescope. Photographs of star fields taken with such an arrangement record the spectrum of each star so that it can be classified.

oblateness, a measure of the degree to which the *ellipsoid shape of a planet or other celestial body departs from a sphere. It is the fraction obtained by dividing the difference between the equatorial and polar diameters by the equatorial diameter. The oblateness of the Earth is 1/297 while that of Saturn is 1/9.5.

obliquity of the ecliptic *ecliptic.

observatory, in astronomy, any site from which astronomical measurements are made. Instruments at most observatories detect *electromagnetic radiation, and each site tends to specialize in observations at a particular band of wavelengths. The main instruments at most optical observatories are housed in a protective domed building. Even before the telescope was invented, the first observatories measured the positions of the Sun, Moon, stars, and planets in the sky. Later, observatories such as the *Royal Greenwich Observatory and the *Paris Observatory used optical telescopes to measure precisely the position and motion of the stars and therefore enabled time to be reckoned more accurately. Since then the role of observatories has expanded to include all aspects of astronomical investigation. The *Mount Palomar Observatory and the *Mount Wilson Observatory are notable examples. The best *seeing conditions are to be found in remote areas away from artificial light and at high altitudes where the atmosphere is clearer and there is less turbulence. For this reason, intensive on-site testing was carried out before setting up the *Cerro Tololo Inter-American Observatory and the *European Southern Observatory, both in Chile, the *La Palma Observatory, the *Mauna Kea facility in Hawaii, and the *Siding Spring Observatory in Australia. The *Kuiper Airborne Observatory is installed in an aircraft which carries instruments above the worst of the atmospheric disturbances.

Radio observatories are those that detect radio waves from extraterrestrial objects. They comprise collections of *radio telescopes and are less troubled by atmospheric conditions. The *Arecibo Radio Observatory in Puerto Rico, the *Effelsberg facility in Germany, and *Jodrell Bank in the UK all employ *dish telescopes. Large *interferometry networks are operated at the *Australia Telescope National Facility, the *Mullard Radio Astronomy Observatory in the UK, and the *Very Large Array in the USA. Most radiation whose wavelength is shorter than that of visible light is absorbed by the Earth's atmosphere, so in order to observe at these wavelengths it is necessary to mount a specially designed telescope on an artificial satellite or balloon. Satellites carrying such telescopes can therefore also be classed as observatories. A series of Orbiting Astronomical Observatories (OAOs) was launched in the early 1970s. For example, the satellite OAO-3, renamed Copernicus after its launch in 1972, has provided a catalogue of ultraviolet stellar spectra. Valuable results have also been gathered from the International Ultraviolet Explorer (IUE) and the Infra-Red Astronomical Satellite (IRAS). The *Hubble space telescope is an infra-red, optical, and ultraviolet telescope in Earth orbit.

*Cosmic rays can be detected using special-purpose equipment. For example, a *neutrino detector operates at the Homestake Gold Mine in South Dakota. Other experimental systems have since been established and have detected cosmic rays from Supernova 1987A in the Large Magellanic Cloud.

occultation, the interruption of the light from a star or other celestial object when a nearer, solar system body

Occultation

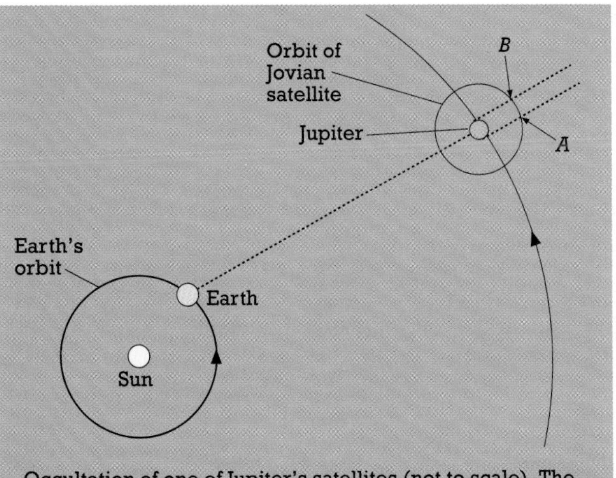

Occultation of one of Jupiter's satellites (not to scale). The satellite is occulted (or is in occultation) between *A* and *B* and therefore hidden from the Earth. Observation of these occultations enabled Ole Roemer to calculate the velocity of light in 1675.

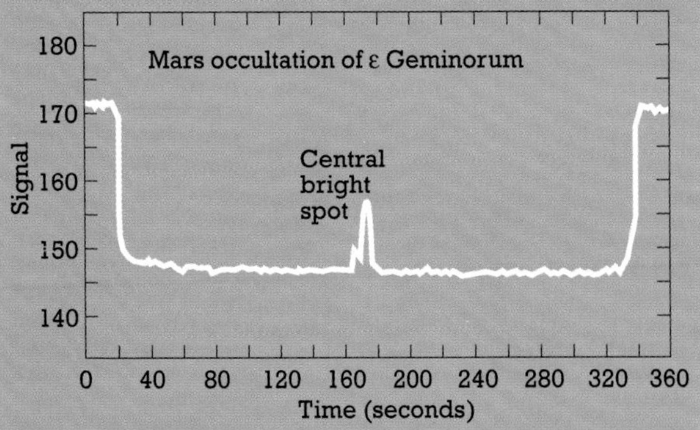

The central bright spot in this occultation light curve is due to diffraction effects.

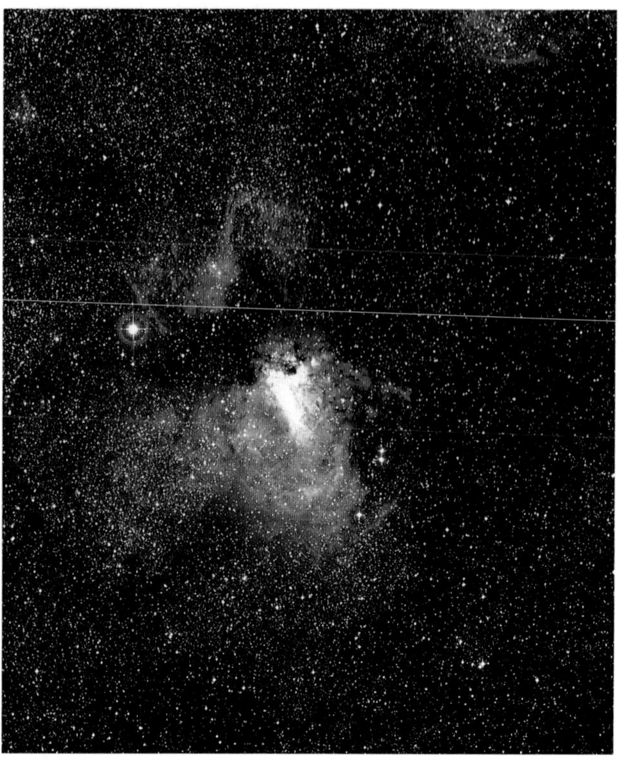

The **Omega Nebula** glows by the light emitted from hot hydrogen gas. Massive stars are thought to be forming from the nebular material. The nearby dark dust clouds are being slowly burnt away by the energetic radiation from the young stars at the heart of the nebula. The emissions from several masers have been detected from this nebula.

passes in front. Properties of both objects can be derived from details of how the light extinguishes and restores. Occultation of a star by the Moon's dark limb can indicate the local slope of the lunar mountains, and possibly reveal the angular diameter of the occulted star and whether it is a multiple star. An occultation in 1977 indicated the presence of rings around Uranus, and another in 1988 showed that Pluto possesses an unexpectedly deep atmosphere. Occultations can also furnish cross-sections of otherwise unresolvably small asteroids, and refine knowledge of solar system orbits. They are not limited to optical wavelengths: a lunar occultation observed in 1964 showed that the Crab Nebula is an extended X-ray source. Celestial bodies can also be totally or partially obscured during an *eclipse or a *transit.

Oceanus Procellarum, a vast irregular *mare produced by lava flowing into, and flooding, the *Gargantuan Basin in the north-west quadrant of the near side of the Moon. The spacecraft Surveyors 1 and 2, Apollo 12, and Luna 9 and 13 landed there. It is not a *mascon, probably because the responsible lava flow is thought to be relatively recent (3,000 million years ago) at a time when the lava was less dense than earlier lava, and of a similar density to the surrounding rocks. In the southern regions of Procellarum the lava flow was so thin that only the minor craters disappeared altogether. The Marius Hills in the centre of Procellarum display domes, *wrinkle ridges, and *rilles and provide a classic example of a volcanic field. The large craters Aristarchus, Kepler, and Copernicus figure prominently. Kepler and Aristarchus both have widespread splash marks stretching over 150 km.

Octans, the Octant, a *constellation that encompasses the south celestial pole. It was introduced by the 18th-century French astronomer Nicolas Louis de Lacaille and represents the navigator's octant, a forerunner of the sextant. The closest naked-eye star to the south celestial pole is Sigma Octantis, magnitude 5.5. The brightest stars in Octans are of fourth magnitude.

Olbers' paradox, the question formulated by the German philosopher Heinrich Olbers (1758–1840) who asked why the sky is dark at night if the universe is infinite with luminous bodies to be seen at all distances in every direction. The paradox was resolved in the 1920s with the birth of *cosmology. It is, in fact, the *recession of the galaxies that weakens the light of the stars of distant galaxies and keeps the night dark. This provides direct evidence for the expansion of the universe.

Olympus Mons *Nix Olympica.

Omega Nebula, a bright emission *nebula in the constellation of Sagittarius. Visible in binoculars, it is also a double *radio source. It is the patch on the eastern side, resembling a swan's neck and head, that gives the nebula its alternative name of the **Swan Nebula**.

Oort–Öpik Cloud, a cloud of *comets that is thought to encircle the edge of the solar system. Long-period comets seem to originate in the deep recesses of space. Both the Estonian astronomer Ernst Öpik and the Dutch astronomer Jan H. Oort wondered if they were either objects that were loosely bound to the solar system or interstellar visitors. Oort suggested that the Sun was surrounded by a spherical cloud of 100 billion comets orbiting the Sun every 1–30 million years. At their outermost points they were between 20,000 and 200,000 astronomical units from the Sun. As the nearest star at present is only 270,700 astronomical units away, a star approaching the Sun in the past could have penetrated this cloud and gravitationally perturbed some of the comets. About half of these would have entered the inner planetary regions. There are two suggestions as to how comets got into the cloud. The first goes back to the dawn of the solar system and proposes the existence of an enormous disc of cometary *planetesimals in the region of Uranus and Neptune. These comets were perturbed by the growing planets. Most were either thrown out of the solar system altogether or thrown in towards the Sun, where they quickly decayed. A small percentage went into the Oort–Öpik Cloud. The second theory proposes that the Sun picks up new comets as it passes through giant molecular clouds on its way around the Galaxy. In this theory the stock of comets is being continuously replenished.

Oort's constants, two parameters A and B introduced by the Dutch astronomer Jan H. Oort when he obtained formulae for the effects of stellar galactic orbits on the *radial velocities ρ and *proper motions μ of stars as measured from the Sun. Stars in the Sun's vicinity move in approximately elliptical orbits subject to *Kepler's Laws about the galactic centre so that those nearer to or farther from the centre than the Sun have respectively greater or lesser velocities than the Sun. The formulae for ρ and μ for a star at a distance r from the Sun in *galactic longitude l are

$$\rho = Ar \sin 2l, \quad \mu = A \cos 2l + B.$$

A and B are usually measured in km/s per kiloparsec.

opacity, in astrophysics, a quantity that characterizes the absorption or scattering of *electromagnetic radiation as it passes through a gas or *plasma. In high-opacity material a *photon travels only a short distance (its **free path**) before it is absorbed or scattered; in contrast the free path is long if the opacity is low. The opacity of the plasma in the interiors of stars is one of several factors that influence the structure and evolution of a star, because the ease with which photons can carry away energy affects the star's core temperature and thus its rate of thermonuclear energy generation. The *spectrum of the light emitted by a star is influenced by the opacity of its outer layers, or atmosphere. Opacity, which is related to *optical depth, can vary with the wavelength of the radiation so that more power is radiated at wavelengths for which the opacity is low.

open cluster, a star *cluster found mainly in or near the plane of the Galaxy and therefore often called a **galactic cluster**. The members of an open cluster have very similar velocities through space but ultimately, although the stars in the cluster had a common origin, the individual stars will be dispersed by the gravitational attractions of the galactic centre and interstellar clouds. If the cluster is near enough to the solar system the stars' *proper motions can be determined. For this reason, a nearby open cluster is also called a **moving cluster**. If these motions are plotted on the *celestial sphere they are directed at a point known as the convergent point. From a knowledge of the stars' angular distances from the convergent point and their *radial velocities, the distance of the cluster can be calculated. An open cluster usually consists of no more than a few hundred stars. Well-known examples are the Pleiades and the Hyades. Most open clusters are made up of young stars only a few million years old though in the thousand or so open clusters known, there are some whose member stars are much older.

open universe, a model in *cosmology in which the universe is boundless and infinite in extent. For an open universe to exist the present expansion of the universe would have to either continue or slow down, but would not be reversed as in the *oscillating universe model. This in turn would require that there be insufficient matter to create an overall gravitational field strong enough to halt the expansion. At present it is not known whether the amount of matter in the universe is enough to constitute an oscillating universe or an open universe.

Ophiuchids, a minor *meteor shower that reaches its peak around 20 June each year. At maximum only a few meteors per hour can be detected from the *radiant (right ascension 260 degrees, declination −20 degrees), in comparison to the 60–100 meteors per hour that are seen at the maxima of the *Quadrantid, *Perseid, and *Geminid showers. Minor showers are usually produced by the decay of small comets but the perturbation of a meteoroid stream away from the Earth's orbit and the decay of the stream over time also weakens the shower.

Ophiuchus, the Serpent Holder, a *constellation on the celestial equator, one of the forty-eight sky figures known to the ancient Greeks. It represents the mythical healer Aesculapius, visualized holding a serpent (the constellation *Serpens). Its brightest star is second-magnitude Alpha Ophiuchi, known as Rasalhague from the Arabic meaning 'head of the serpent bearer'. The constellation contains several globular *clusters, M10 (NGC 6254) and M12 (NGC 6218) being the brightest of them; both are of seventh magnitude and visible in binoculars. The southernmost regions of Ophiuchus extend into rich areas of the Milky Way between *Scorpius and *Sagittarius. The Sun passes through the boundaries of the constellation in December even though Ophiuchus is not a member of the zodiac. Ophiuchus contains *Barnard's Star, the second-closest star to the Sun.

opposition *planetary configurations.

optical depth, a measure of how far light will travel through a partially transparent medium before it is absorbed or *scattered. Quantitatively, the probability p that a *photon will traverse a certain optical depth τ without being absorbed or scattered is given by $p = e^{-\tau}$, where $e = 2.71828...$ is a fundamental constant of mathematics. A car in a fog, for example, is still distinctly visible at $\tau = 1$ because $e^{-1} = 0.37$ (or 37 per cent) of photons reflected from the car reach the observer, but the car rapidly becomes indistinct at higher optical depths. A corollary of the definition of τ is that radiation escaping from a gas or plasma will on average have been emitted at an optical depth of approximately 1. The correspondence between optical depth τ and physical depth x (in, say, metres) depends on the *opacity, κ, and can be as simple as $\tau = \kappa x$. The centre of the solar disc, for example, is brighter than the edge because the physical distance x corresponding to $\tau = 1$ probes into the Sun radially deeper (and therefore to hotter and brighter regions) when the line of sight is perpendicular to the atmosphere (as at the centre) than when it is oblique (as at the edge). This is the origin of limb darkening.

optical double, two stars that, by chance, appear to be close together because they lie nearly along the same line of sight. An optical double is not a true *binary star: the two stars may in reality be very far apart so that they are not under each other's gravitational attraction.

optical pulsar, a *pulsar that exhibits rapid variations in its light as well as in its emissions at radio and other wave-

Optical depth

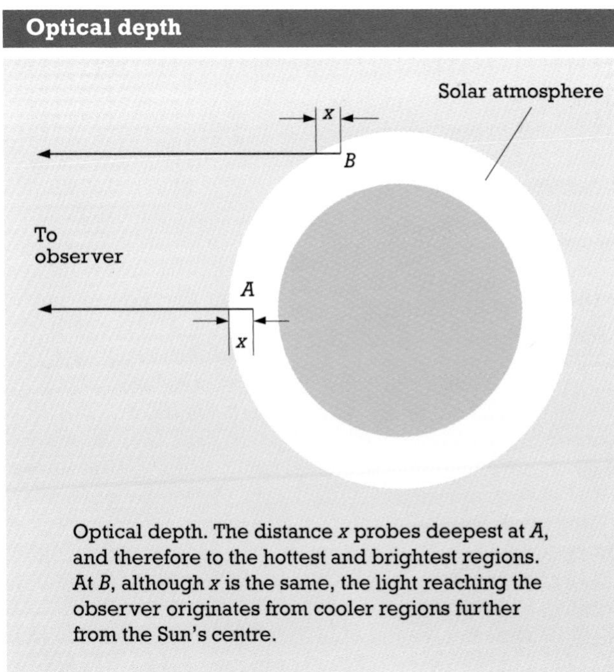

Optical depth. The distance x probes deepest at A, and therefore to the hottest and brightest regions. At B, although x is the same, the light reaching the observer originates from cooler regions further from the Sun's centre.

lengths. The first to be found was the pulsar in the *Crab Nebula. It had long been known as a sixteenth-magnitude star near the centre of the nebula, but in 1969 its light was shown to be pulsing every 0.033 seconds, the same rate as its radio pulses. Its magnitude varies between 13.9 and 17.7, with a weaker subpulse between the main pulses. The second optical pulsar to be discovered was the *Vela pulsar, and a possible third, produced by *Supernova 1987A, has been reported. Most pulsars are too faint to be seen with optical telescopes, but it is thought that rapidly spinning pulsars should emit enough light to be observed. Searches for light pulses from the most rapidly rotating pulsars and the *binary pulsar PSR 1913+16 have so far been unsuccessful.

orbit, in astronomy, the path in space followed by a planet, satellite, asteroid, or comet under the influence of gravity. The orbits of the planets are approximately elliptical and are described by *Kepler's Laws from which orbit determination can be made. However, when the mutual perturbations of the planets are taken into account by solving the *N-body problem, the accurate computation of the orbits is difficult. Indeed *Poincaré proved that it is not possible to solve the problem exactly. Approximate solutions can, however, be computed, both by perturbation methods and by numerical techniques on a computer. Such results are accurate enough to be compared with the observations of the planets, satellites, and space probes. However, over a time-span comparable to the age of the solar system (about 4.5 billion years) the computation of reliable orbits for the planets is an unsolved problem.

orbit, elements of, six quantities that describe the size, shape, and orientation of an orbit. The motion of a planet under the gravitational attraction of the Sun, when the attraction by all the other planets is neglected, follows an ellipse lying in a fixed orbital plane. To solve this *two-body problem, six quantities, the orbital elements, are required: (1) the *semi-major axis, *a*, describing the size of the orbit; (2) the eccentricity, *e*, of the ellipse; (3) the inclination, *i*, of the orbital plane with respect to some reference plane forming a *coordinate system; (4) the longitude, Ω, of the ascending *node; (5) the **argument of perihelion**, that is, the angle, ω, between the node and the direction of closest approach to the Sun; and (6) an *anomaly giving the position along the orbit at some epoch. Other sets of orbital elements can be defined, but they must be related to these six. When the perturbation arising from the other planets is taken into account, the orbit is only approximately represented by a fixed ellipse, the approximation being better for shorter time-spans. For more accurate calculations, the orbit can be described by a set of elements which varies with time.

Orbit, elements of

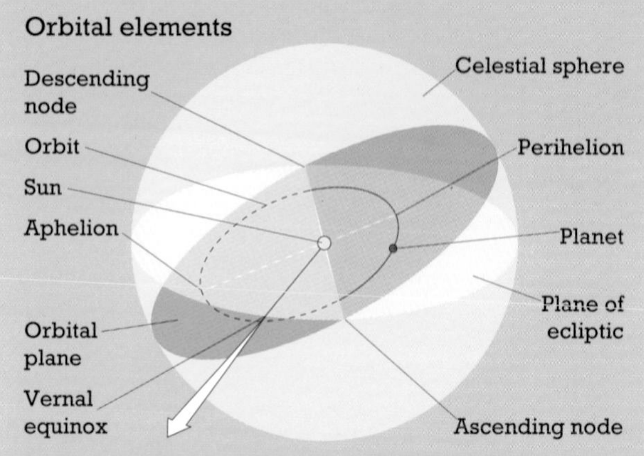

There are six elements used in determining the orbits of celestial objects, in rotation around a focus. These are described below.

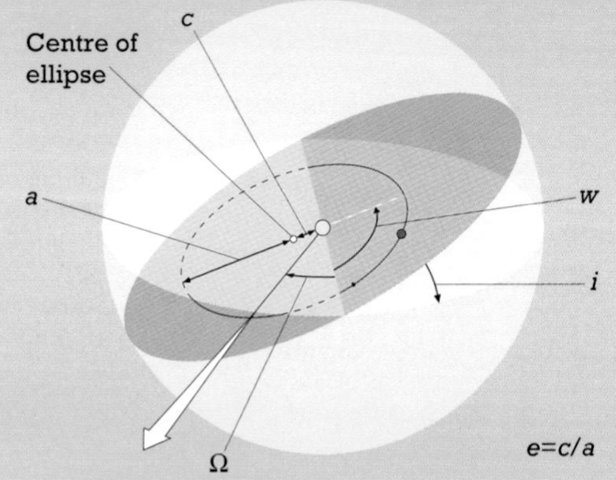

a The semi-major axis of the elliptical orbit.

e The eccentricity of the ellipse calculated by the equation $e=c/a$ where *c* is the distance between the centre of the orbital ellipse and the focus around which the body is orbiting.

i The inclination of the orbital plane to the ecliptic.

Ω The longitude of the ascending node.

w The argument of the perihelion; measured as the angle in the orbital plane between the direction of the perihelion and the nodal line.

T The last orbital element, corresponding to a time at which the object passes through the perihelion.

orbit determination, the procedure for computing the values of the elements of *orbit for some celestial body, either natural or artificial, from observations taken at different epochs. There are six independent orbital elements so, if the measurements are angular positions in the sky, observations for at least three different epochs have to be available. A procedure to compute the elements from this minimum set of observations was devised by the German scientist Karl Gauss to recover the asteroid Ceres whose position was lost immediately after its discovery in 1801. With the availability of electronic computers orbit determination is now performed numerically. For any set of observations, such as angles, distances, and radial velocities, taken at different epochs, a hypothetical orbit is computed and the differences between the real observational data and those resulting from the hypothetical orbit are computed. Then the hypothetical elements are corrected and the procedure repeated, making further corrections to the hypothetical elements, until the correct orbital elements are found with the best fit to the observational data.

Orientale Basin, a huge crater on the Moon's eastern limb that consists of at least three concentric rings of mountains. Unlike the *Imbrium Basin, the Orientale Basin is only partially flooded with basaltic lava. Orientale is 900 km across (26 per cent of the lunar diameter) and was formed about 3.9 billion years ago. The outer ring of mountains is about 6 km high. If a crater in the lunar crust is very large and deep, the stresses in the surrounding and underlying rock are greater than the strength of the rock. Concentric rings of rock then break away and slide down into the centre of the deep crater producing a mound. From above, the edges of the slipped rings form a series of circular mountain ranges. The Apollo astronauts reported seeing a light flash from the Orientale Basin and it is thought that this was due to the impact of a small meteorite.

Orion, the Hunter, a prominent *constellation on the celestial equator, one of the forty-eight constellations recognized by the ancient Greeks. In mythology Orion was a hunter, son of the sea god Neptune, who was stung to death by a scorpion for his boastfulness; in the sky, Orion sets as his slayer in the form of the constellation *Scorpius rises. The brightest stars in Orion are *Betelgeuse and *Rigel. A line of three stars marks *Orion's Belt, and beneath the belt lies *Orion's Sword and the *Great Nebula in Orion (M42), containing the *Trapezium star group. South of Zeta Orionis lies the *Horsehead Nebula. The Orionid *meteor shower radiates from the constellation every October.

Orion arm, the *spiral arm of the Galaxy in which the Sun is placed. The arm's existence has been traced by measuring the distribution of bright stars of *spectral classes O and B and the emission nebulae contained in the arm. Studies of the distribution of hydrogen by mapping the emission of radio waves also confirms the Orion arm's existence.

Orion association, a group of hot young stars in the constellation of Orion, including the stars of *Orion's Belt and the bright star *Rigel. These stars are probably only a few million years old and condensed out of the vast amounts of dust and gas still present in Orion. Because they are so hot (of *spectral class B) they excite much of the gas in their neighbourhood to become emission *nebulae.

Orion Nebula *Great Nebula in Orion.

Orion's Belt, three bright stars in the constellation of *Orion depicting the belt of the figure. The brightness of Orion enables it to be used as a convenient marker leading to other stars. In particular the belt points in one direction towards *Aldebaran and the *Pleiades and in the opposite direction towards *Sirius.

Orion's Sword, three bright stars in the constellation of *Orion depicting the sword of the figure. The *Great Nebula in Orion lies close to the middle star in the sword and can be examined in a telescope of moderate aperture. The bright star Iota Orionis, the lowest star in the sword, lies about half a degree south of the nebula.

orrery, a large complex, usually clockwork, model of the solar system. It is named after Charles Boyle, Fourth Earl of Orrery, who commissioned one from John Rowley in 1712. Usually the Sun is represented by a large central brass ball and the ivory planets and their attendant satellites are supported on radiating pivots. The clockwork mechanism drives the planets around the Sun at the correct rate and the system is adjusted to show the celestial positions of the planets at a given time. Smaller geared orreries became popular in the 18th and 19th centuries. On some orreries the planets can be replaced by a Sun–Earth–Moon system which shows the form of the Moon's orbit and the reasons behind the lunar *phase variation.

oscillating universe, a model in *cosmology in which the universe alternately expands and contracts, each cycle beginning with a *Big Bang and ending with a *Big Crunch. For a closed universe to exist the present expansion of the universe would have to be arrested and reversed. This in turn would require enough matter to create an overall gravitational field strong enough to prevent indefinite expansion. At present it is not known whether the amount of matter in the universe is sufficient to constitute an oscillating universe, also known as a **pulsating universe**, or whether we inhabit an *open universe.

outer planets, any of the *major planets beyond the asteroid belt, that is, Jupiter, Saturn, Uranus, Neptune, and Pluto. Their composition is very different from the *inner planets and, with the exception of Pluto, they are gas giants retaining a significant fraction of the hydrogen and helium of the original nebula from which all the planets were formed. Apart from Pluto, they all have complex natural satellite systems.

Owl Nebula, a *planetary nebula, also known as M97, in the constellation of Ursa Major. It takes its name from its resemblance to an owl's face, and requires a 30 cm (12-inch) aperture telescope to view it clearly. Under good seeing conditions a faint central star and two dark regions (the owl's eyes) are visible.

Ozma Project, an attempt to detect artificial radio signals from two nearby stars, Tau Ceti and Epsilon Eridani, possibly having planets suitable for life. The project was initiated by Giuseppe Cocconi and Philip Morrison of Cornell University in 1959, and was led by the radio astronomers Frank Drake and William Waltman at the *National Radio Astronomy Observatory at Green Bank, West Virginia, in 1960. Though unsuccessful, Ozma stimulated many ideas on *SETI (the search for extraterrestrial intelligence), and on the encoding of messages. The proposed *Cyclops Project, comprising a much larger telescope array, would, with recent techniques, be 4×10^6 times more sensitive. To be detected by Ozma, the source would need to emit 10^6 megawatts, but a mere 250 kilowatts would be needed for Cyclops.

ozone, a variety of oxygen in which its molecule contains three oxygen atoms in contrast to the ordinary oxygen molecule that contains only two. The ozone layer, also known as the **ozonosphere**, is part of the *stratosphere that prevents most of the harmful solar ultraviolet radiation from reaching the surface of the Earth. In recent years damage by aerosol gases to the ozonosphere has allowed more ultraviolet radiation to penetrate to the Earth's surface. At ground-level the gas is a pollutant and a health hazard.

ozonosphere *ozone.

Pallas, the second-largest *asteroid, discovered in 1802 by the German astronomer Heinrich Olbers. Pallas interacts gravitationally with *Ceres and this has enabled the mass of Pallas to be measured as $(2.15 \pm 0.43) \times 10^{20}$ kg. In 1978 an occultation of star SAO 85009 by Pallas led to a diameter measurement of 538 ± 12 km. The orbit of Pallas has a semi-major axis of 2.77 astronomical units and an eccentricity of 0.2334, putting it close to the centre of the asteroid belt. Its inclination of 34.8 degrees to the *ecliptic is unusually large.

parallax The direction of a distant object appears to shift if the observer changes his or her viewpoint. The angular shift in direction is called the parallactic angle and the distance of the object can be measured if a suitable baseline of known length is used, the directions of the object being measured from each end of the baseline. This is essentially the surveyor's method of triangulation. Within the solar system a suitably long baseline is the radius of the Earth, about 6,378 km. The angle subtended at a solar system object by the Earth's radius at right angles to the direction of the object is the **geocentric parallax** of the object and is illustrated in the figure. Most parallaxes are measured in minutes or seconds of arc, so great are the distances to solar system objects. Even the Moon, the Earth's nearest natural neighbour, has a geocentric parallax of only about 1 degree.

For objects outside the solar system, distances to even the nearest star are such that the Earth's radius is too small to act as a suitable baseline. The Earth's orbital radius of one *astronomical unit (a.u.) is used as shown in the figure. The **stellar parallax** *P* of an object is then defined to be the angle subtended at the object by 1 a.u. at right angles to the object's direction. The usual unit of *angular measure for stellar parallax is the arc second (1/3,600 of a degree). The distance *d* to a star in *parsecs is then given by $d = 1/P$. Parallaxes are very small; the largest, for *Proxima Centauri, is only 0.76 arc seconds, equivalent to the angle subtended by a coin at several kilometres. Parallaxes are in general very difficult to measure and in antiquity their apparent absence was an argument for a stationary Earth. Parallaxes are smaller than the image smearing caused by atmospheric turbulence, and certain much larger interfering effects must accurately be accounted for, principally *refraction by the Earth's atmosphere (2,000 arc seconds at the horizon) and *aberration of starlight (20 arc seconds). However, refraction and aberration corrections can be avoided if directions can be measured by simultaneous observations of angularly close but very distant stars. It was with such differential measurements of 61 Cygni that *Bessel first convincingly measured an annual parallax in 1838. Parallaxes become too small for ground-based measurements for distances greater than about 30 parsecs, though the fine guidance sensors on the *Hubble space telescope should achieve considerably greater precision. Since the Galaxy is some 30,000 parsecs in diameter, distances are only directly measurable for a very small fraction of stars, totalling less than one thousand. Parallaxes measured in this way are often termed **trigonometric parallaxes** to distinguish them from **spectroscopic parallaxes** which relate to distances determined via *spectral classes and the *Hertzsprung–Russell diagram, and **dynamical parallaxes**, which refer to distances determined from considerations of observed angular and *radial velocities.

Recently a major increase in the numbers of stars whose parallax has been measured with unprecedented accuracy has been achieved with the orbiting of the *astrometry satellite Giotto.

Paris Observatory, an *observatory founded in 1667 by Louis XIV and his finance minister Colbert. It specialized in positional astronomy, geodesy, and *celestial mechanics until activities diversified in the late 19th century. During its first century of operation, the observatory was directed by a dynasty of four successive generations of the *Cassini family, known as Cassini I, II, III, and IV. Giovanni Domenico Cassini (Cassini I) discovered several satellites of Saturn and the division in *Saturn's rings which bears his name.

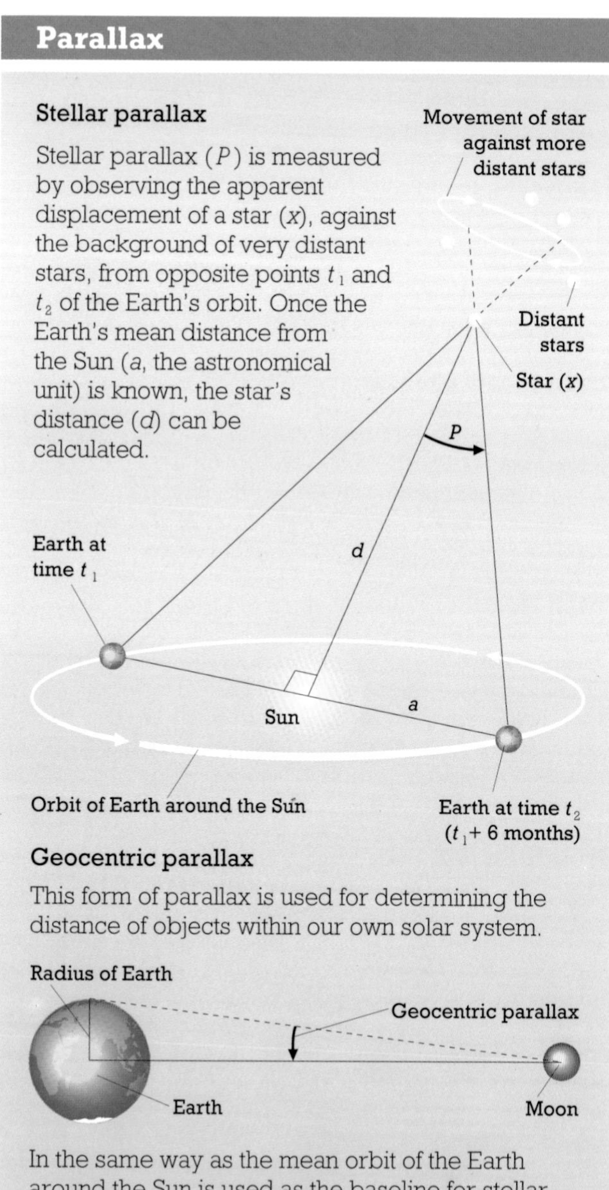

Through analysis of Cassini's tables of the motions of Jupiter's moons, *Roemer discovered that the velocity of light was finite. Cassini II initiated a series of measurements intended to determine the flattening of the Earth due to its rotation. The observatory also compiled the first tables of the orbits of Saturn's satellites. Under Cassini III and Cassini IV France was mapped at a scale of 1:86,400, the first map of this accuracy to be produced. The work of the observatory was disrupted during the French Revolution (1789), but later the observatory was involved in establishing the metre as the unit of length in the metric system. In the mid-19th century the observatory's director Francis Arago did work in physics, popularized astronomy, and was involved in republican political activities. *Le Verrier is remembered for his role in the discovery of Neptune, and for his dictatorial directorship. The enormous *Carte du Ciel* sky mapping project developed from work at the Paris Observatory. In the 20th century activities expanded into astrophysics, and in recent years have included radio astronomy and space research. The world's first speaking clock was put into service by the observatory in 1933.

parsec, a unit of distance employed in astronomy, equal to that distance at which a star would have a trigonometric *parallax of 1 arc second. The word parsec is a contraction of 'parallax–arc second'. One parsec is equivalent to 3.086×10^{16} m, 206,265 *astronomical units, or 3.26 *light-years. The nearest stars are 1–2 parsecs away; the Galaxy is about 30,000 parsecs across; and the edge of the observable universe is almost 10 billion parsecs distant.

Pasiphae, a small satellite of Jupiter. It has a diameter of 50 km and orbits the planet in the opposite direction to Jupiter's orbital motion at a mean distance of 23.5 million km. The orbit is highly eccentric and is greatly perturbed by the Sun. Discovered in 1908, it is one of a group of four small moons in orbits of similar size about Jupiter.

path of totality, the path of the Moon's shadow across the Earth's surface within which a total solar *eclipse can be observed. Also known as the **zone of totality**, it is never wider than about 300 km (200 miles). However, the path length can stretch for thousands of kilometres. The speed of the Moon's shadow is between 1,800 km/h and 8,000 km/h and depends on the relative position of the Earth, Sun, and Moon.

Pavo, the Peacock, a *constellation of the southern sky introduced by the Dutch navigators Pieter Dirkszoon Keyser and Frederick de Houtman at the end of the 16th century. Its brightest star is Alpha Pavonis, magnitude 1.9, named Peacock after the constellation itself. Kappa Pavonis is a Cepheid variable star, ranging from magnitude 3.9 to 4.8 every 9.1 days.

Pavo Cluster, a globular *cluster in the constellation of Pavo, about 15 arc minutes across. Visible in good seeing conditions with a telescope of only 7.5 cm (3 inches) aperture, it is a large condensed cluster and has been described as one of the gems of the sky.

Pavonis Mons, a large ancient volcano on the equator of Mars. Together with Arsia, Ascraeus, and Olympus Mons, it forms the main part of the *Tharsis Ridge. Ascraeus, Pavonis, and Arsia Mons lie to the south-east of Olympus Mons on a north-east to south-east line. All three are about 10 km high, 400 km wide at their bases, and are spaced about 700 km apart. They lie on top of the 9-km high Tharsis bulge and seem to be as high as the Martian crust will support. Any addition of lava will cause the underlying crust to become unstable and give way. The summit crater of Pavonis Mons is circular and about 5 km deep. The walls are terraced, indicating that it has suffered repeated collapse.

Path of totality

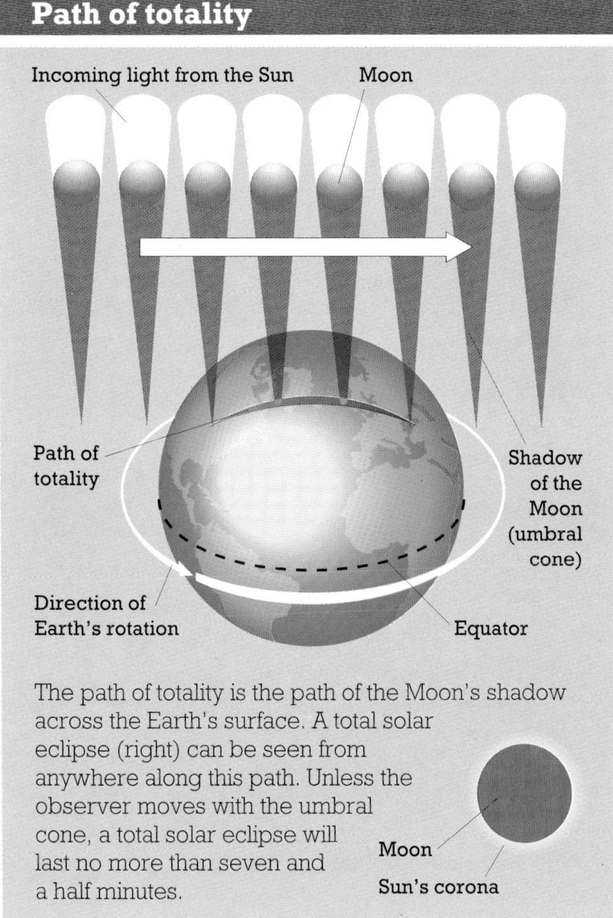

The path of totality is the path of the Moon's shadow across the Earth's surface. A total solar eclipse (right) can be seen from anywhere along this path. Unless the observer moves with the umbral cone, a total solar eclipse will last no more than seven and a half minutes.

Pegasus, the Winged Horse, a *constellation of the northern sky, one of the forty-eight constellations known to the ancient Greeks. It represents the mythical flying horse that was born from the body of Medusa the Gorgon when she was beheaded by Perseus. Only the forequarters of the horse are shown in the sky, but it is still the seventh-largest constellation. Its body is outlined by the four stars of the **Square of Pegasus**: Alpha, Beta, and Gamma Pegasi and Alpha Andromedae. Alpha Andromedae was once shared with Pegasus but is now assigned exclusively to *Andromeda. Beta Pegasi (called Scheat, from the Arabic for 'shin') is a red giant that varies irregularly between second and third magnitude. Epsilon Pegasi (called Enif from the Arabic meaning 'nose') is a second-magnitude yellow supergiant with a ninth-magnitude companion visible in small telescopes.

Pegasus Cluster, a beautiful globular *cluster which is also known as M15, lying in the constellation of Pegasus, visible in small telescopes. The luminous core of the cluster emits X-rays, possibly because of the presence of a *black hole at its centre which draws stellar material into it at high accelerations.

Period–luminosity law

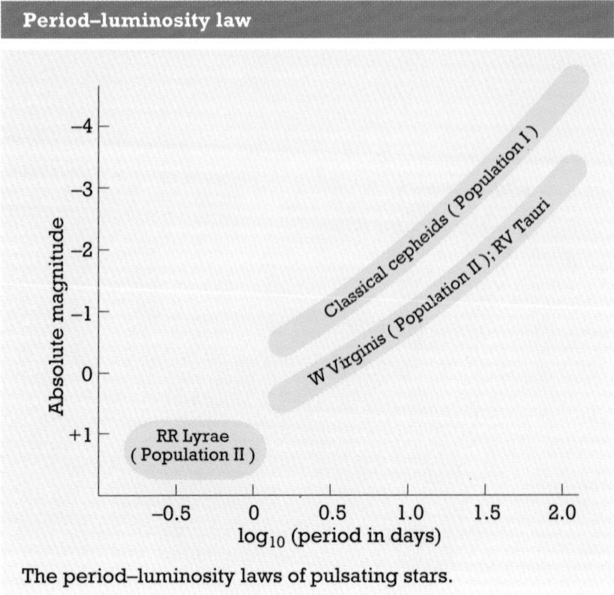

The period–luminosity laws of pulsating stars.

penumbra *eclipse.

peri-, a prefix used in astronomy to denote the point in an orbit nearest to a centre of attraction. Thus:

perihelion is that point in a planet's elliptical orbit about the Sun where the planet is nearest to the Sun;

perigee is that point in the orbit of the Moon or of an artificial Earth satellite when it is nearest to the Earth's surface;

perigalacticon is that point in a star's orbit within the galaxy nearest to the galactic centre;

periastron is that point in the orbit of one component of a binary star about the other when the two components are closest together.

The prefix used to denote the point in an orbit farthest from a centre of attraction is *ap(o)-.

perigee *peri-.

perihelion *peri-.

period, the time interval between recurrent events. The term is used very generally and periods are ascribed to rotational spins of bodies, orbital revolutions, solar magnetic phenomena, pulsations within stellar atmospheres, and so on. For example, the period of light fluctuation from the variable star Algol is 68.82 days. Some calendars are based on the Moon's *synodic period of 29.53059 days and on the *Metonic cycle of approximately 19 years.

period–luminosity law, the relation between the periods of pulsating stars and their intrinsic brightness or absolute *magnitude. The first such law was established by the US astronomer Henrietta Leavitt in 1912 by observing *Cepheid variable stars in the Magellanic Clouds. Refinements have since been made to the law to allow for there being distinct luminosity types of Cepheid. The law has been fundamental as a means of measuring the distances of stellar clusters and the nearer galaxies. If Cepheids are detected in such objects, their measured periods can be translated into intrinsic brightness via the law. Comparison of the observed brightness with the intrinsic value allows the distance to be calculated. *RV Tauri stars also follow a period–luminosity law.

Perseids, a *meteor shower that reaches its peak about 12 August each year, but its meteors can be seen for about three weeks around this date. At least a dozen historic records are known of this shower between AD 36 and 1451. The shower was known as the 'tears of St Lawrence', after the saint whose festival is near the time of maximum. In 1866 the Italian astronomer Giovanni Schiaparelli demonstrated that the Perseids had the same orbit as the comet 1862 III.

Perseus, a *constellation of the northern sky, one of the forty-eight constellations known to the ancient Greeks. It represents the mythical hero Perseus who rescued Andromeda from the jaws of the sea monster Cetus after having decapitated the Gorgon Medusa. Alpha Persei is a yellow supergiant of magnitude 1.8, surrounded by a scattered cluster of stars. Beta Persei is the famous eclipsing binary *Algol, marking the Gorgon's head carried by Perseus. Rho Persei is a red giant that varies between third and fourth magnitudes every month or two. The constellation lies in a rich area of the Milky Way and includes the *Double Cluster, which marks the sword hand of Perseus. M34 (NGC 1039) is an *open cluster well seen in binoculars. NGC 1499 is the California Nebula, named because its shape resembles that US state. Every August the *Perseid meteors radiate from the constellation.

Perseus A *Perseus Cluster.

Perseus arm, a *spiral arm of the Galaxy lying about 2,000 parsecs further out from the galactic centre than the *Orion arm in which the Sun lies. Its distribution has been mapped using the bright stars of *spectral classes O and B within it, and the radio emission from the hydrogen clouds it contains.

Perseus Cluster, a cluster of galaxies in the constellation of Perseus. The cluster is centred on NGC 1275 (**Perseus A**), an eruptive *radio galaxy which is a very intense radio source. When observed through optical telescopes, a complicated system of filaments can be seen spreading outwards from this galaxy, reminiscent of those visible in the *Crab Nebula. NGC 1275 is a supergiant elliptical galaxy surrounded by a huge cloud of hot gas which flows into it.

perturbation *celestial mechanics.

phase, in astronomy, the fraction of a body's apparent disc that is illuminated. For example, the cycle of the phases of the Moon occurs because of the changing angle between the Earth, the Moon, and the Sun, together with the fact that the Moon has no light of its own and shines by light reflected from the Sun. The illustration shows various phases within a *synodic period or lunation. At any moment, only half of the Moon's surface reflects light and only when the Moon is in opposition is the whole of the illuminated half presented towards the Earth giving a **full moon**. At other times within its lunation only a fraction of the illuminated half can be seen. When over half the illuminated face is visible the Moon is said to be **gibbous**. *Earthshine is sometimes seen near **new moon** when the Moon is in conjunction.

In general, the phase of a celestial body is zero when only the darkened side is presented to the Earth, and 100 per cent when the illuminated side is fully visible. Mercury and

Venus show similar cycles of phases, with their phase being zero at the *planetary configuration of inferior conjunction. Even Mars exhibits a changing phase, but it always shows more than 85 per cent of its illuminated face to the Earth.

Phobos, the inner satellite of Mars, discovered by the US astronomer Asaph Hall in 1877. Its orbit is 9,380 km from the planet's centre. Phobos has an irregular shape, with a mean radius of about 11 km and a density of 2,200 kg/m^3. Images from the *Viking probes showed a dark, heavily cratered and *regolith-covered surface. The biggest crater, *Stickney, is about half the size of Phobos, and a system of grooves radiates from it. Phobos is being driven inward by tidal effects.

Phoebe, a satellite of Saturn, discovered by the US astronomer William Pickering in 1898. Its orbit has a semi-major axis of about 13 million km and Phoebe moves in the opposite direction to Saturn's orbital motion. *Voyager images have revealed a nearly spherical shape of diameter 220 km and a patchy surface with low albedo (0.06). Phoebe is probably an *asteroid or *comet nucleus captured by Saturn in the early solar system environment.

Phoenicids, a southern hemisphere *meteor shower that reaches its peak on 5 December each year and was first observed in 1956 from Australia. This shower is tentatively associated with Comet Blanpain (1819 IV) and contains *meteoroids which have a very low velocity relative to the Earth. The British physicist Edward Bowen concluded that raindrops could form and grow on the dust introduced into the atmosphere by the decay of meteoroids such as the Phoenicids thus encouraging higher rainfall.

Phases of the Moon

The Moon's phases
Third quarter
Waning crescent
Waning gibbous
Quadrature
New moon
Full moon
Conjunction
Opposition
Light from the Sun
Quadrature
Waxing crescent
Waxing gibbous
First quarter

The inner ring shows the Moon as it circles the Earth, with the Sun shining on it from the left-hand side. The outer ring shows the Moon as it is seen from Earth. To see every phase of the Moon in one night the viewer would need to circumnavigate the Moon. However, by standing at one position on the Earth's surface, every phase of the Moon can be witnessed over a period of one month.

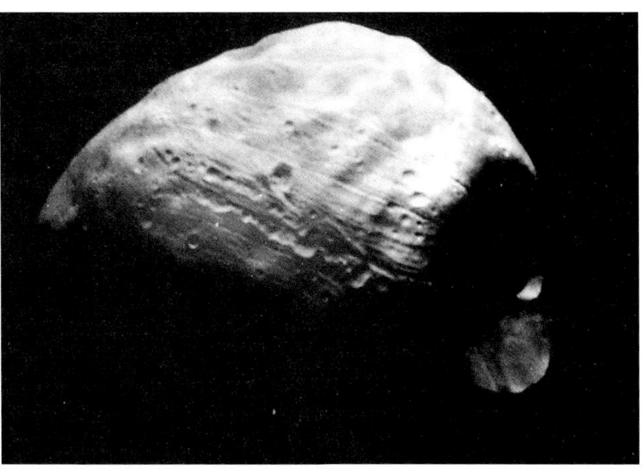

This photograph of **Phobos**, the innermost of the two natural satellites of Mars, was taken by the Viking 1 spacecraft in 1976. The crater Stickney, the largest on Phobos, is visible as the deep depression on the moon's limb.

Phoenix, a *constellation of the southern sky, introduced by the Dutch navigators Pieter Dirkszoon Keyser and Frederick de Houtman at the end of the 16th century. It represents the mythical bird that was reborn from its own ashes. Alpha Phoenicis, its brightest star, is of second magnitude. Beta Phoenicis is a close pair of fourth-magnitude stars for telescopes of moderate aperture.

Phosphorus (literally, 'light bearer'), the Greek name for the planet Venus when it appears as the morning star seen near western *elongation. Both Venus and Mercury are only visible to the naked eye as morning or evening objects. Venus as a morning star was therefore called by the Greeks Phosphorus and when seen as an evening star *Hesperus.

photographic zenith tube (PZT), a specially modified *zenith tube to which photographic recording techniques have been added to provide highly accurate determinations of time and of the observatory latitude. Because the instrument is always used in a near vertical position, errors due to flexure and atmospheric refraction are minimized. Only stars passing near the zenith are observed. The technique employed in using the PZT ensures the removal of any instrumental errors incurred in photographically recording the star.

photography, in astronomy, the production of images of celestial objects on a photosensitive surface. The first photograph ever taken was of the Sun and was produced by the Frenchman Nicéphore Niepce in 1826. Niepce's process, based on the hardening of bitumen when exposed to light, was very crude; an eight-hour exposure was required to create a picture. The first practical photographic process was developed by another Frenchman, Louis Daguerre, the details of which he published in 1839. That same year Francis Arago, the French astronomer and politician, gave a description of Niepce and Daguerre's invention to the Académie des Sciences and again that year the word 'photography' was first used by John *Herschel who had been experimenting since 1819 on the solvent action of hyposul-

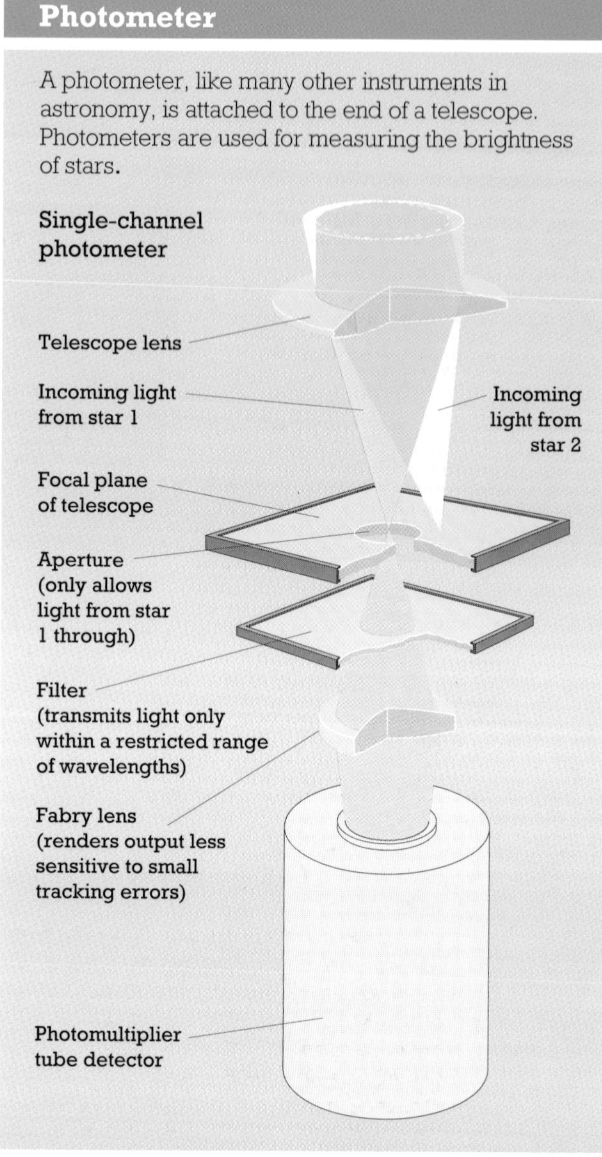

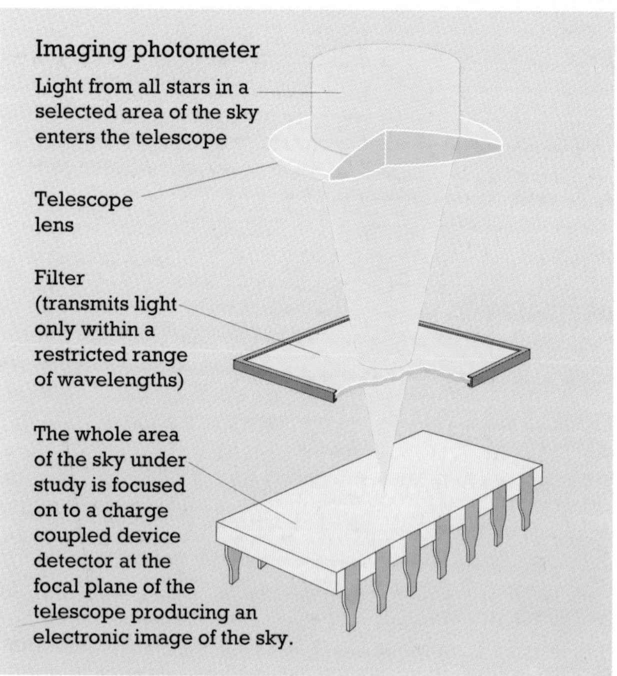

phites on silver halides. In 1840 the American John William Draper, the father of the astronomer Henry Draper, obtained a daguerreotype of the Moon with an exposure time of twenty minutes, and two years later Lerebours had obtained the first daguerreotype of the Sun. Early photography was a very complicated and time-consuming process and it was not until 1857 that stellar photography really began. Among the pioneers were George Bond of Harvard College Observatory and the photographer George Whipple. Between 1875 and 1882 the US astronomer Benjamin Apthorp Gould produced photographs of star clusters, double stars, and star fields down to the eleventh magnitude. Pictures of planets, particularly Jupiter and Saturn, were obtained by 1886, the Great Red Spot, Jupiter's Belts, and similar belts on Saturn being successfully recorded. The *Carte du Ciel*, a huge undertaking initiated at the *Paris Observatory to photograph the entire heavens, gained international support and astronomical observatories all over the world took part in the survey. However, only part of this great work was completed. Since 1839 the quality of photographs has steadily improved and a century later photographic sensitivities of some 100,000 times greater than that of the daguerreotype were being made.

Astronomical spectroscopy on photographs began with John Draper in 1843 but only really became practical with the work of the British astrophysicist William Huggins in the late 1870s. On 7 March 1882 he obtained the first photographic spectrum of the *Great Nebula in Orion. Progress thereafter depended on the development of sensitive plates, good driving mechanisms for telescopes so that long exposure times on the celestial object of interest could be used, and a clear sky free from artificial light. Nowadays, the vast majority of celestial objects such as galaxies can be viewed only on long-exposure plates because they are too faint to be seen by the eye even through the most powerful telescopes. New methods, in which light is converted into electrical signals by devices such as photoelectric detectors, image converters, photomultipliers, *photometers, and *CCD cameras, are rivalling and even surpassing the ability of the photograph to capture the images of very faint celestial objects.

photometer, in astronomy, an instrument for measuring the brightness of stars and other celestial objects. It consists of a detector that accepts some of the light from a telescope directed at the object of interest. A **single-channel photometer** is particularly suited to the *photometry of point sources such as stars or asteroids. An aperture selects the light from only a small part of the image produced by the telescope, and a filter restricts the range of wavelengths transmitted into the detector. The detector is often a photomultiplier tube that amplifies weak incident light to produce large, easily recorded electrical signals. The background light from the night sky also inevitably passes through the aperture and is corrected for by a separate measurement of the sky alone. An **imaging photometer** is conceptually simpler, comprising a filter in front of a detector such as a *CCD camera which can accept an image of an area of the sky, rather than just a point of light. The sky background is evaluated from the image itself. Despite its greater mechanical complexity, a single-channel photometer with a photomultiplier tube requires very much simpler electronic circuitry than an imaging photometer with a CCD camera. In both types of photometer a mechanism is provided for changing filters, and eyepieces are needed for directing the photometer on to the target.

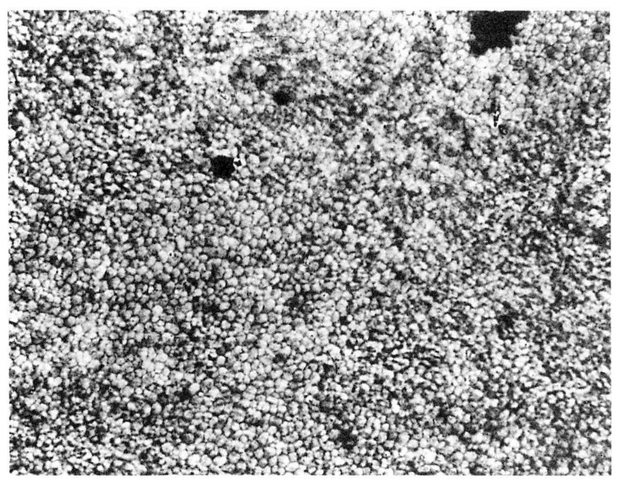

Most of the visible light emitted from the Sun originates at the **photosphere**. This close-up of the photosphere shows the solar granulation, the result of turbulence and convection in the Sun's surface.

photometry, the measurement of the brightness of light. The earliest estimates of stellar brightness were naked-eye observations by *Hipparchus and Ptolemy, which formed a basis for the logarithmic *magnitude scale invariably used in modern astronomical photometry. Most professional photometry uses *photometers with electronic detectors and the measurements are corrected for the absorbing and scattering effects of the Earth's atmosphere. A series of differently coloured filters is used to sample the light from the celestial object under study. The wavelengths (colours) passed by the filters are carefully chosen to reveal characteristic *spectral lines such as absorption lines in stars and galaxies, and emission lines and bands in gaseous nebulae and comets. Magnitudes from different filters are usually compared and interpreted as a *colour index, which is a ratio and thus independent of the actual light levels or the distance to the object. In recent years the photometry of galaxies and other extended objects has been enormously improved as *CCD cameras have displaced photography. Satellite observatories and advanced detector technology have extended astronomical photometry to include ultraviolet and infra-red wavelengths.

photon, the unit constituent or particle of *light. The energy carried by a photon is the product of Planck's constant, 6.626196×10^{-34} Js, and the frequency of the electromagnetic wave. Thus, for example, a photon at a radio frequency is of much lower energy than a gamma-ray photon. If energetic enough, a photon can ionize atoms. The particle nature of *electromagnetic radiation is deduced from its interaction with matter, such as in a detector.

photosphere, those layers of a star's atmosphere from which most of the light is emitted, and where many of the *spectral lines are formed. The photosphere is usually very thin relative to the size of a star; a representative thickness is one-thousandth of the stellar radius. The *plasma temperature drops through the photosphere as the distance from the centre of the star increases, and then rises in the overlying and more transparent *chromosphere and *corona. In the particular case of the Sun, it is the thinness of the solar photosphere that gives the Sun its apparently sharp edge. In fact the photosphere is frequently thought of as the surface of the Sun. The honeycomb appearance of the solar disc (the *granulation) is due to convection within the photosphere. The granulation consists of millions of cells of hot, rising plasma, edged with descending currents of cooled material. The Sun's *limb darkening is a consequence of the outward fall of the photospheric temperature.

Pictor, the Painter's Easel, a *constellation of the southern sky, introduced by the 18th-century French astronomer Nicolas Louis de Lacaille. Its brightest star is of only third magnitude. It contains Kapteyn's Star, a ninth-magnitude red dwarf 3.9 parsecs from us, which has the second-largest *proper motion of any star, 8.7 arc seconds per year.

Pioneer, a series of eleven US *spacecraft launched between 1958 and 1975. The most important were Pioneers 10 and 11 which reached Jupiter in 1974 and 1975 having survived passage through the asteroid belt. Many pictures of Jupiter were transmitted. Pioneer 11 reached Saturn in 1979 sending back pictures of the planet and its ring system. Both spacecraft are now leaving the solar system.

Pisces, the Fishes, a *constellation of the zodiac. In mythology it represents Venus and her son Cupid who jumped into the river Euphrates to escape from the monster Typhon and were turned into fishes; in the sky the tails of the two fishes are visualized tied together by a cord. Alpha Piscium (Alrescha, Arabic for 'cord') is a close binary star of fourth and fifth magnitudes; the stars orbit each other about every 900 years. The Sun passes through the constellation from mid-March to late April. The vernal *equinox currently lies in Pisces.

Piscis Austrinus (or Australis), the Southern Fish, a *constellation of the southern sky, one of the forty-eight constellations known to the ancient Greeks. It is depicted as a fish drinking the flow of water from the urn of Aquarius, and supposedly represents the Syrian fertility goddess Derceto who fell into a lake and was saved by (or changed into) a fish. This fish is sometimes regarded as the parent of the two smaller fishes of Pisces. Apart from first-magnitude *Fomalhaut its stars are of only fourth magnitude and fainter.

plages, bright *flocculi on the Sun's surface, in regions of higher temperature than their surroundings. They occur in regions where the Sun's magnetic field is stronger than normal and extends from the *photosphere through the *chromosphere to the solar *corona.

Planck's Law, the mathematical expression describing how much power is radiated at different *wavelengths by a *black body. The law was formulated in 1900 by the German physicist Max Planck, and signalled the foundation of quantum mechanics. The law derived from Planck's realization that energy is always exchanged in discrete packets, which Planck called 'quanta'. In particular, light is emitted as *photons whose energy depends upon their wavelength in a vacuum.

planet, a non-luminous body in space that forms from the gas and dust around a star and that shines only by reflected light. This includes all bodies from just a few kilometres across, such as the *asteroids or minor planets, up to objects with a mass of perhaps ten times that of *Jupiter. Above that mass, the body's central temperature is sufficient to commence nuclear reactions and it then becomes a *star.

Planetary configurations

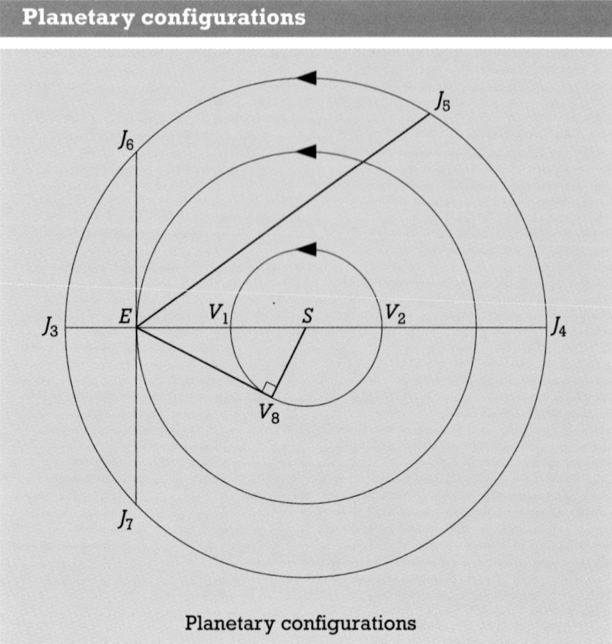

Planetary configurations

Planets may be composed mainly of either rocky and metallic materials, for example *Mercury, *Venus, *Earth, and *Mars, or gaseous materials, for example Jupiter, *Saturn, *Uranus, and *Neptune. Planetary systems are formed by the same process which forms stars. A huge cloud of cold gas and dust slowly condenses to form a disc. The heavily condensed central portion heats up to form a new star while the material in the disc condenses into a number of much smaller bodies. If the major body here exceeds a mass of ten times that of Jupiter, it too becomes a star, forming a *binary star system. However, if the disc forms a family of smaller bodies then a planetary system is produced. Those planets close to the central star are strongly heated and lose their light gaseous materials to form small rocky terrestrial planets. Those farther from the central star retain more of their light materials to produce large gaseous planets.

Our own *solar system is thought to have formed in just this way some 4.6 billion years ago, while a star system in the constellation Cygnus called MWC 349 is currently in the very early stages of this process. The central star is perhaps only 1,000 years old and is heavily obscured by dense dust and gas in which planets may now be forming. Several planets orbiting other stars have now been detected from periodic wobbles they induce in the motion of their central star. *Barnard's Star may be orbited by two giant planets about the size of Jupiter. Planetary systems are probably a common phenomenon in the universe. For tables of planetary orbital data, planetary periods and motions, and physical data for the planets, see page 196.

planetarium, a complex optical projector used to simulate the movements of celestial bodies. The word planetarium is also used to describe a large domed room or building which houses the projector. Star patterns are projected on to the inner ceiling of the dome. The projector can be geared so that it shows the daily movement of the Sun, Moon, planets, and stars in a few minutes and can therefore show how the sky appeared to an observer in the past or will appear in the future. It can also be adjusted so that it takes into account the *precession of the equinoxes and changes in the observer's latitude and longitude.

A planetarium was also the name given to a small, simplified non-mechanical form of *orrery.

planetary configurations, the positions of the planets relative to the Sun and the Earth. Certain positions of planets as seen from the Earth and measured with respect to the Sun's direction are given special names. In the figure, V denotes an *inferior planet, J a *superior planet, E the Earth, and S the Sun. When the inferior planet V lies in the same direction as the Sun it is said to be in **conjunction**, the configurations EV_1S and ESV_2 being called **inferior conjunction**, and **superior conjunction**, respectively. The superior planet J is said to be in conjunction when it also lies in the Sun's direction (ESJ_4) and in **opposition** when it lies in the opposite direction to the Sun (J_3ES). The angle between the directions of a planet and the Sun with respect to the Earth, for example (J_5ES) is termed the planet's *elongation. For an inferior planet the maximum elongation occurs when EV_8S is 90 degrees; for a superior planet all elongations between 0 degrees (ESJ_4) and 180 degrees (J_3ES) are possible. The planet is said to be in **quadrature** when the elongation is 90 degrees (J_6ES and J_7ES).

planetary exploration The collection and analysis of scientific data about the planets started when the wandering stars known as planets were magnified into a disc when seen through an astronomical telescope. As telescopes improved, the 17th, 18th, and 19th centuries saw an ever-increasing sophistication of planetary maps and the discovery of such phenomena as Saturn's rings, mountains and lava-filled craters on the Moon, and the strange and variable markings on Mars, which were called *canals. The late 17th century saw the accurate establishment of the scale of the then-known solar system. However, the perceived size of the system increased dramatically by the discovery of Uranus, Neptune, and Pluto in 1781, 1846, and 1930, respectively. The application of the *spectrometer to solar system objects enabled their rotation and composition to be measured.

Planetary exploration started in earnest when Earth-based telescopic research was enhanced by the ability to actually travel to the planets introduced by the space age. Four phases of exploration can be recognized, these being the fast fly-past, the orbiter, the surface lander, and the sample return. In the case of the Moon, our nearest space neighbour, the first of these stages was exemplified both by the Luna 3 probe that flew past the Moon in October 1959 and took the first pictures of the far side, and the Ranger 7 spacecraft that flew into the Mare Nubium region in July 1964, taking 4,308 pictures as it did so. Other famous exploratory fly-pasts were the Mariner 10 fly-past of Mercury in 1974 and 1975, the only space mission to that planet, the Mariner 2 fly-past of Venus in 1962, the Mariner 4 fly-past of Mars in 1965, the Pioneer 10 and Voyager 1 and 2 fly-pasts of Jupiter in 1973 and 1979, the Pioneer 11 and Voyager 1 and 2 fly-pasts of Saturn in 1979, 1980, and 1981, the Voyager 2 fly-pasts of Uranus in 1986 and Neptune in 1989, and the *Giotto fly-past of Halley's Comet in 1986.

There have been fewer orbiting missions, mainly because of the added spacecraft mass and the advanced technology required to reduce the craft's speed and thus allow it to drop into and stay in the planet's gravitational field. Five Orbiter spacecraft flew around the Moon between 1966 and 1967 producing a superb set of photographs. Pioneer Venus 1 orbited that planet in 1978 and a series of satellites have been producing radar maps of the cloud-shrouded surface ever since. Mariner 9 went into orbit round Mars in 1971 and obtained many images. The Galileo mission to put an orbiter around Jupiter has been launched and the Cassini mission which involves an orbiter around Saturn and the CRAF mission that will orbit a comet are planned to be launched in 1997.

The job of landing a spacecraft on a planetary surface and returning information from it is onerous. Surveyor 1 soft-landed on the Moon in 1966, Venera 7 transmitted data back from the surface of Venus in 1970, and later Venera missions returned coloured pictures and data about the soil composition. The Viking 1 and 2 missions to Mars in 1976 were searching for life. The equivalent missions to the gas giant outer planets involve a spaceprobe that slowly parachutes down through the extensive atmosphere.

The ultimate goal of planetary exploration has been realized only with the Moon. Man landed on *Tranquillitatis Mare in July 1969 and returned to Earth with a host of information and 20 kg of lunar samples. By the end of the Apollo programme in 1972 more than 380 kg of documented lunar samples had been brought back to Earth laboratories. Hopefully the future will see similar amounts of material returned from the other planets, satellites, and comets, and we could even look forward to the prospect of manned landers and permanently occupied space stations on some of these bodies.

planetary nebula, a hot tenuous gas shell surrounding certain faint but hot stars and having a planet-like appearance in the telescope. The luminous shells are found to be expanding and cooling on time-scales of 10,000 years or so, and may be produced in the final stage of a star's decay to the *white dwarf state. Similar features are shown by *Wolf–Rayet stars. At present, however, the processes which produce these shells in early and late *stellar evolution, and in close *binary stars are not clearly understood. Shells of gas are also produced by *novae. The *Dumb-bell, *Helix, and *Ring Nebulae are beautiful examples of these gaseous shells.

The bright stars of the **Pleiades** open star cluster are each embedded in a blue halo caused by their starlight being reflected from dust particles in the cold gas from which they were formed. These elongated dust grains are aligned along the interstellar magnetic field and this gives the streaky appearance to the haloes.

planetesimal, a dense aggregate formed from the cloud of gas and dust that surrounded the infant Sun. It is thought that the particles in the cloud gently collided and stuck together, gradually forming larger particles. Eventually an orbiting disc of planetesimals of up to a kilometre in diameter surrounded the Sun. Planetesimals near the Sun were rocky and metallic whereas those further out contained ice as well. Although there was a certain amount of fragmentation due to collisions or stresses in the planetesimals, the bodies gradually combined and grew in size to form the planets. The *Laplace nebular hypothesis and *Weizsacker's theory construct detailed models of this creation process.

planetology, the study of the structure and make-up of *planets and natural *satellites. The term is used in contrast to **geology** which, strictly speaking, is the study of the Earth's structure. In the past twenty-five years especially, space missions such as *Project Apollo, *Mariner, *Viking, *Voyager, and *Venera, to the solar system's bodies have produced thousands of pictures and a multitude of measurements made by a wide variety of scientific instruments. Many of the larger satellites of Jupiter, Saturn, Uranus, and Neptune were, prior to space research, star-like objects even in the largest Earth-based telescopes. Now their widely contrasting conditions are being charted and studied, providing explanations of the creation and subsequent evolution of the solar system.

planisphere, a map formed by projecting the *celestial sphere on to a plane to show the appearance of the heavens at a particular time and place. It consists of two parts. The first is a two-dimensional projection of the night sky that is centred on either the north or south celestial pole. This map is drawn to indicate the sky as seen from a particular latitude on Earth. It shows the principal stars of each constellation and the Milky Way, and is usually crossed by the celestial equator, the ecliptic, and lines of equal right ascension and declination. The second part of the planisphere is a movable overlay that can be rotated about the pole position. This overlay is equipped with a large elliptical hole through which can be seen those stars that are visible in the sky at any specific time of any night or day. The edge of the hole represents the observer's horizon and is marked out in terms of the cardinal points. By rotating this overlay one can see which stars are in the sky, and how the stars rise, set, and move. Planispheres are inexpensive, handy, and portable. They were first introduced at the beginning of the 19th century, the stellar projection being developed from a much earlier *astrolabe plate.

plasma, matter consisting of *ions and electrons. Because of its charged nature, external magnetic and electric fields can influence a plasma. In addition, the charged particles themselves can interact magnetically and electrically. Radiation acting on atoms can remove electrons so that the low-density *interstellar medium contains a plasma component. The high-temperature material in stars is also a plasma.

Pleiades, a *cluster of some 500 stars in the constellation of Taurus of which seven (the **Seven Sisters**) or more are visible to the naked eye. The Pleiades is a classic open clus-

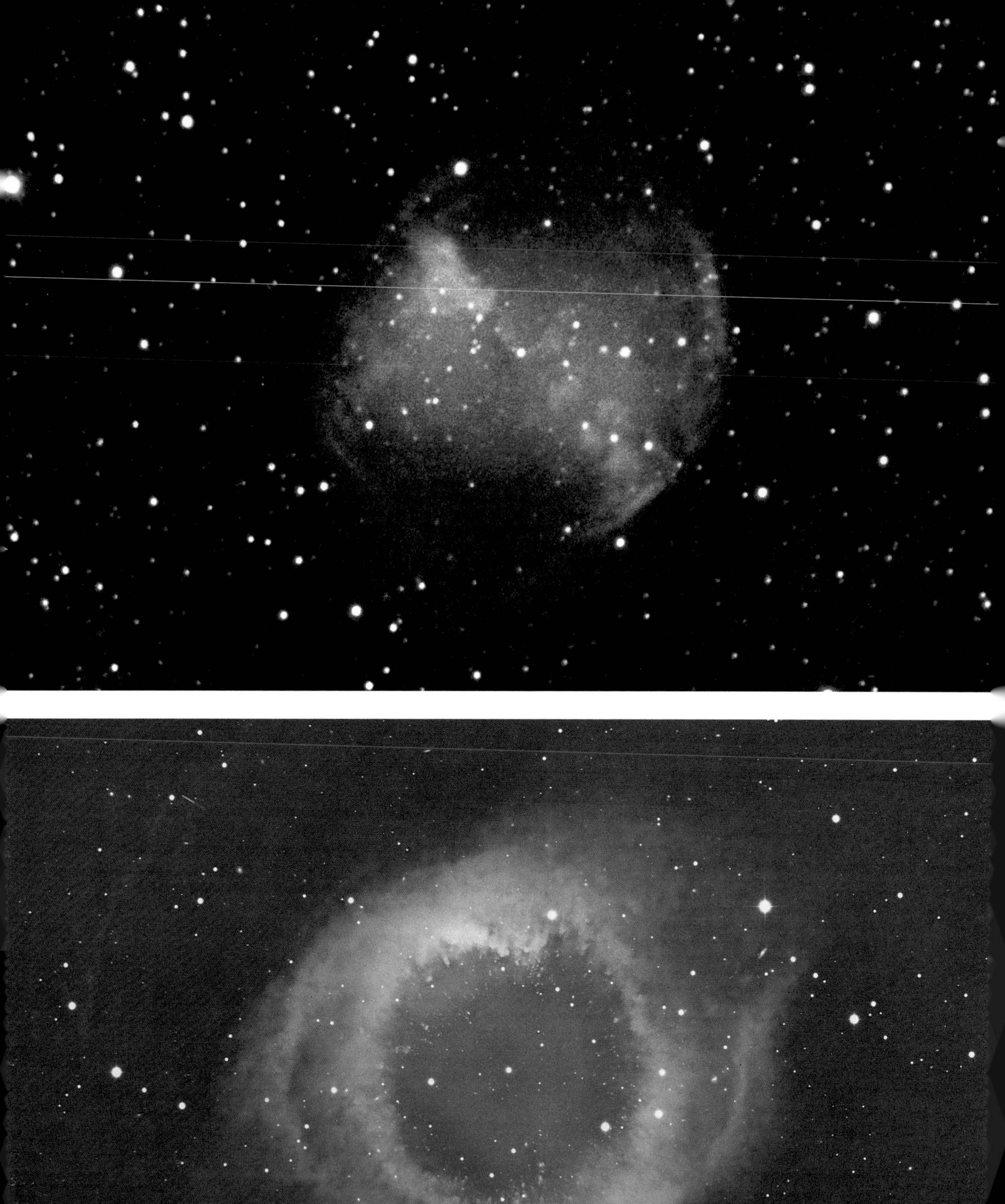

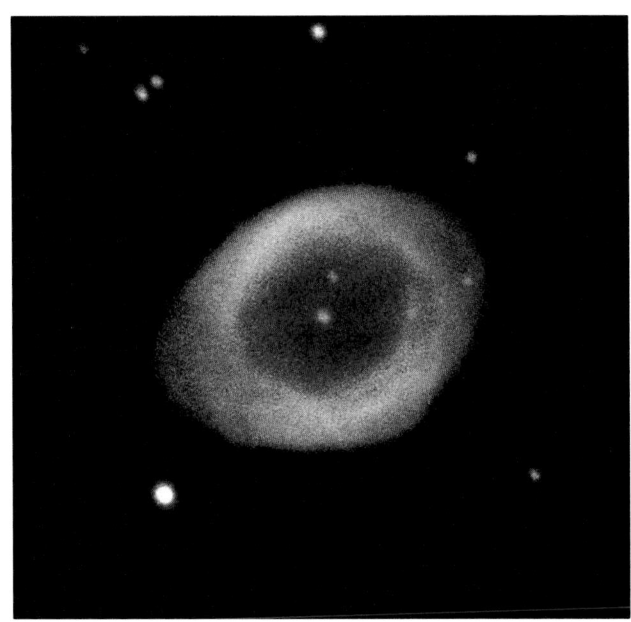

These photographs show three different **planetary nebulae.** Each nebula consists of a shell of gas thrown off by the explosion of a dying star. The Dumb-bell Nebula (*above, opposite*) is one parsec in diameter, and is expanding at a velocity of 27 km/s. The gas was ejected by the dying star (the bluish spot seen at the centre of the nebula) about 50,000 years ago. The Helix Nebula (*below, opposite*) is a shell of gas and dust ejected by the star which is now the white dwarf at the centre of the ring. The overall pink colour is due to glowing hydrogen and the blue–green inner part is due to oxygen and nitrogen. In the Ring Nebula (*this page*) the central star is probably a white dwarf. The other star that appears to be within the ring happens by chance to lie along the line of sight. The gas ring is expanding at about 19 km/s, suggesting that the central star exploded about 5,500 years ago.

ter sharing a common space velocity. There are between three and ten times as many stars per unit volume of space as there are in the vicinity of the Sun. These are *Population I stars. The remnants of the *nebula from which the stars were formed and the absence of a branch of giant stars in the cluster's *Hertzsprung–Russell diagram indicate very recent star formation; its age is about 50 million years. Located 125 parsecs (410 light-years) from us, the Pleiades cluster is 12 parsecs (40 light-years) across.

Pleione, a rapidly spinning variable *shell star in the Pleiades open cluster, and also known as BU Tauri. Three shells have been ejected in the present century, each accompanied by a fade in *magnitude followed by a brightening; the last shell episode began in 1972 and ended in 1987. Its magnitude varied between 4.8 and 5.5.

Plough, the pattern formed by seven stars of *Ursa Major. This shape is also known as the **Big Dipper**. All seven stars are bright; a good test of eyesight is to attempt to see the middle star (*Mizar) in the handle of the Plough as double. In our present era Merak and Dubhe, the two stars farthest away from the handle, point to the star *Polaris (the Pole Star).

Pluto, the outermost *planet of the solar system, discovered by the US astronomer Clyde Tombaugh in 1930. Pluto's perihelion lies inside the orbit of Neptune and this, together with its small size and mass, suggests that Pluto may have originally been a satellite of Neptune. The problem with this theory is that the orbit of Pluto avoids any close encounter with Neptune so it is difficult to see how they could have been associated in the past. There are two reasons for this avoidance. Firstly, Pluto's orbital period is about 1.5 times that of Neptune so that when Pluto is near perihelion Neptune is always in a remote part of its orbit. Secondly, Pluto always keeps far above or below Neptune's orbital plane whenever it is closer to the Sun. Pluto could thus be the innermost or largest object of an outer belt of small bodies. Also remarkable is the fact that Pluto has a comparatively large satellite *Charon, which might have split from it following a shattering impact. Pluto and Charon always keep the same face turned towards each other. Pluto's mean density of about 2,000 kg/m^3 suggests that its interior is a mixture of rocks, water ice, and methane ice; in fact, methane has been identified on the surface by *photometry and *spectrophotometry. In 1989, a stellar *occultation confirmed that the planet has a thin atmosphere, presumably composed mostly of methane, whose density probably varies significantly between aphelion and perihelion as it is heated and cooled. A haze layer may surround the solid surface. Pluto's brightness changes with its rotation rate of 6.4 days due to bright patches on the surface.

Poincaré, (Jules) Henri (1854–1912), French mathematician and theoretical astronomer, whose chief contribution to astronomy lay in the development of *celestial mechanics. In 1889 he showed that the *N-body problem could not be solved exactly. King Oscar II of Sweden had offered a prize for the first solution to the question of how long a system of several bodies could remain stable. Although Poincaré's result was only a partial answer, he was awarded the prize. He also discovered the phenomenon of *chaos, explicitly proving its existence in the three-body problem. His other contributions to astronomy included a description of the formation of planetary systems and an analysis of the rotation of the Earth, and he contributed to the discussion on the theory of *relativity. He also made many major contributions to mathematics.

Pointers *Ursa Major.

polar axis, an imaginary straight line that joins the *poles of a sphere. In a celestial context it is usually regarded as being the axis about which the object spins. The polar axis of an equatorially mounted telescope is the axis that is parallel to the Earth's spin axis.

polar caps, the circular regions of ice that have formed around the northern and southern poles of a planet. For example, the Earth and Mars have polar caps. The Earth's arctic and antarctic regions of polar water ice are reasonably constant in extent, the winter snows and subsequent glaciation compensating for the summer erosion by the breaking away of icebergs. The Martian polar caps show more seasonal variability, but both have permanent regions of ice, the north's being the larger. The south polar regions have a summer temperature of −110 °C, which is close to the frost point of carbon dioxide. The winters there are longer and colder than at the Martian north pole and about 50 cm of carbon dioxide ice is formed over the pole at that time, this frost extending up to latitudes of 60 degrees south. The northern pole has a summer temperature of −70 °C close to the temperature at which atmospheric water would con-

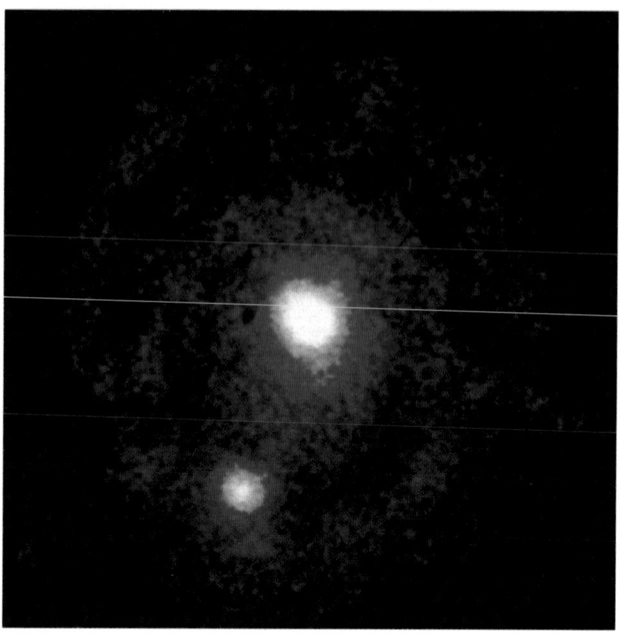

This image of the planet **Pluto** (*centre*) and its satellite Charon (*lower left*) was taken by the Hubble space telescope. Charon, whose six-day orbit around Pluto is a circle seen almost edge-on from the Earth, is shown here near its maximum angular separation from Pluto.

dense. The permanent northern polar cap is probably water ice but this is extended in the winter down to a latitude of 65 degrees north by carbon dioxide ice. The northern winter corresponds to the perihelion of the orbit of Mars, with a strengthening of the global winds. The permanent northern polar cap is much dirtier than the southern one and has a lower albedo.

polarimetry, in astronomy, the measurement of the degree and type of *polarization in light or other electromagnetic radiation coming from a celestial object. Polarization can result from reflection of the light by a surface or by the light passing through the dust in a nebula or through a *magnetic field. In polarization the wave, vibrating in a random orientation, is constrained to show a regular type of vibration. Polarimeters can measure the degree of polarization of the light which can be plane polarized vertically or horizontally. If there is a mixture of polarizations, both vertical and horizontal, the resulting effect is that the plane of polarization rotates about the direction of propagation of the light. Such an effect is called elliptical polarization. From a knowledge of physics and polarimetric measurements a great deal of information about the regions the light has passed through may be collected.

Polaris (Pole Star), a *variable star of second magnitude that lies within one degree of the north celestial pole. Found in the constellation of Ursa Minor, it is also known as Alpha Ursae Minoris. It is a *multiple star whose bright component is a *supergiant star which is itself a spectroscopic *binary star with a period of thirty years. The fainter companion is a ninth-magnitude *dwarf star moving in a very wide orbit about the spectroscopic binary; both stars are somewhat hotter than the Sun. The estimated distance of Polaris is 110 parsecs and its combined *luminosity is about 1,600 times that of the Sun.

polarization, the confinement of the waves of *electromagnetic radiation, including light and radio waves, to one plane or one direction. This property is not directly perceived by the eye but can be detected, in the case of light, by the behaviour of the light after it has interacted with various materials called polarizers. The rotation of a pair of polaroid sunglasses produces a marked change in the brightness of the light reflected off the sea, for example. Two polarizers in line, with one being rotated, can produce darkness at some angles of rotation but not at others. The first polarizer the light meets allows through only radiation vibrating in a particular direction at right angles to the direction of propagation of the light. If the second polarizer is placed so that it would allow through only light vibrating at right angles to that particular direction, then none of the light transmitted by the first polarizer is able to pass through the second. The measurement of the degree of polarization of electromagnetic radiation coming from any celestial object reveals valuable information not only about that object but also about any material lying between the object and observer.

polars *AM Herculis stars.

poles, the two points on a sphere that are 90 degrees above or below a specific great circle called the *equator. The intersecting plane that produces a great circle at the points where it intersects a sphere must pass through the centre of the sphere. A perpendicular to this plane, passing through the centre of the sphere, intersects the spherical surface at the two poles. The Earth's poles mark the intersection of the spin axis with the surface.

Pole Star *Polaris.

Pollux, a first-magnitude star in the constellation of Gemini, also known as Beta Geminorum. It is a *red giant star at a distance of about 12 parsecs, the nearest of its kind to the Earth. It is about forty times the luminosity of the Sun.

Polarization

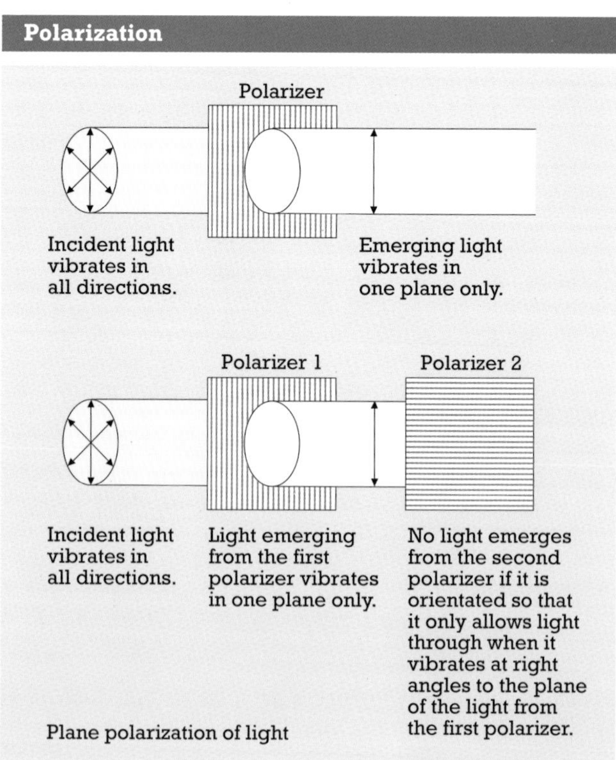

Plane polarization of light

Populations I and II *stellar populations.

pores, relatively dark markings about 1,000 km across in the *granulation of the Sun's visible surface. Indistinguishable from small *sunspots, they are of lower temperature than their surroundings and are an indication of the unceasing activity in the outer layers of the Sun.

potential, in astronomy, the *energy that a body has by virtue of its position in a gravitational field. A rock at the top of a cliff has the potential to do work if it falls over the cliff. Its potential energy, transformed into kinetic energy, can drive a nail into a plank at the foot of the cliff. A planet in orbit about the Sun has potential energy and kinetic energy, continuously transferring one into the other as its distance from the Sun varies.

Praesepe, the Beehive Cluster, a large open *cluster in the constellation of Cancer. Also known as M44, it is visible to the naked eye as a faint misty patch and is 1.5 degrees wide. Binoculars or a low-power telescope reveal its interesting features, including a number of double and triple stars and a metal-rich A0 spectral-type star, the brightest in the cluster. The cluster was first resolved into individual stars by *Galileo.

precession, the 'wobbling' motion of a rapidly spinning body or gyroscope in which the axis of rotation slowly sweeps out a cone of circular cross-section. Astronomically the term is used for the precession of the *equinoxes. Since the Earth is not exactly spherical, but has an equatorial bulge, the gravitational attraction of the Moon and Sun exert a torque on the Earth which slowly changes the direction of its rotational axis. The *average* effect of this, termed **luni-solar precession**, is shown in the *celestial sphere of the figure. The north celestial pole P describes a small circle, of radius ϵ (the **obliquity**), about K, the pole of the ecliptic, in a period of about 26,000 years. As the pole varies, so will the equator and the equinox ♈ retrogrades along the ecliptic at a rate of 50.3 arc seconds per year. As a result the right ascension and declination of any star varies, and it is necessary to specify the epoch to which these coordinates refer. The phenomenon of precession was first noted in the 2nd century BC by *Hipparchus who recognized the resulting annual increase of about 50 arc seconds in the celestial longitude ♈ of every star (see the figure).

prehistoric astronomy, the study of celestial phenomena by ancient civilizations before historical records began. Heavenly events have been visible throughout the human race's existence on Earth. Human beings of 50,000 years ago probably had the same brain capacity as at present and therefore much the same spread of intelligence as modern man. Thus, whenever intelligent humans had leisure to ponder terrestrial events such as the seasonal growth of plants, the migration of birds, and tidal phenomena, they would have noted their relationship to celestial events such as the changing rising and setting directions of the Sun, and the phases of the Moon. Keeping track of such events may have begun as far back as the Paleolithic artists who created the famed Altamira and Lascaux paintings. It has been tentatively suggested that tally-marks cut in mammoth tusks 15,000 years ago were records of the Moon's phase. According to the British engineer Alexander Thom, a pioneer of the study of astro-archaeology, one motive of the megalithic peoples of the 3rd and 2nd millennia BC in erecting vast

Precession

Cause and effect of the Earth's precession

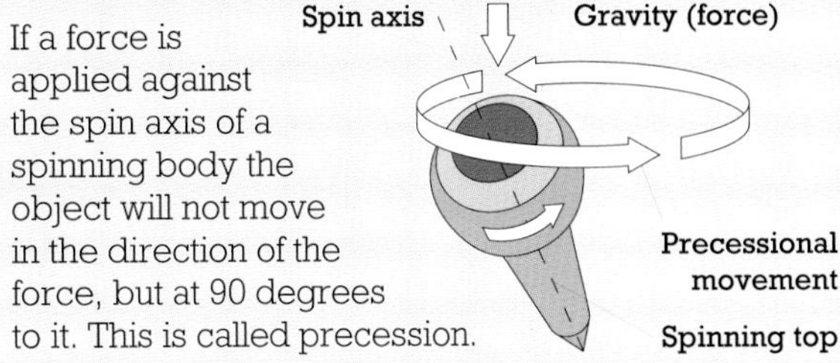

If a force is applied against the spin axis of a spinning body the object will not move in the direction of the force, but at 90 degrees to it. This is called precession.
Spin axis
Gravity (force)
Precessional movement
Spinning top

Just as gravity causes the spinning top's axis of rotation to precess (above) the attraction of the Sun on the equatorial bulge of the Earth (below) causes a similar precession of the Earth's rotation axis.

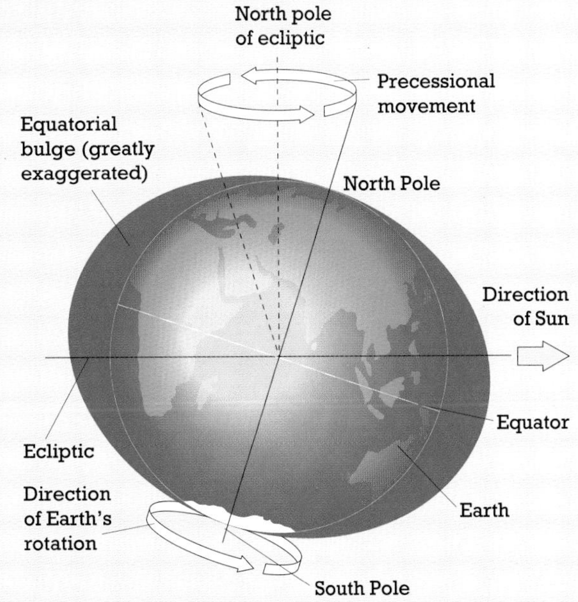

This produces a circular movement of the celestial pole about the north pole of the ecliptic, resulting in a slow regression of the intersections of the equator with the ecliptic. The equatorial coordinates of a star therefore alter with time.

Effect of Earth's precession on a star's coordinates

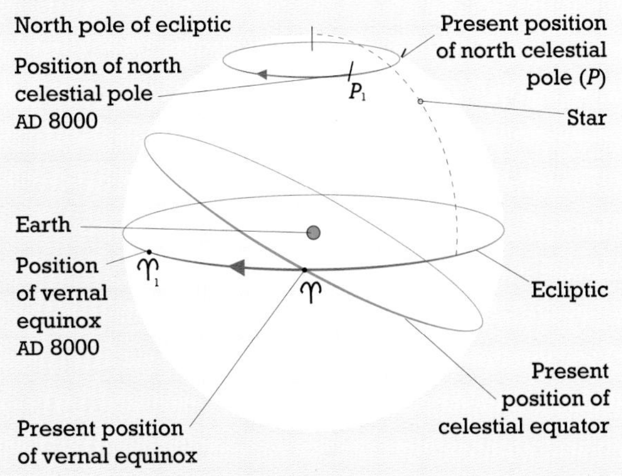

James Irwin,a member of the **Project Apollo** mission Apollo 15, checks the lunar roving vehicle near the landing site before collecting samples of lunar soil and rock. The lunar rover carried a navigation system and could travel at up to 17 km/h. Mount Hadley, in the Apennines, is in the background.

numbers of stone alignments and circles throughout western Europe, for example, at Stonehenge, Avebury, Callanish, and Carnac, was to keep track of the cyclic rising and setting points of the Sun and Moon on the horizon. In this way a practical *calendar for religious purposes could have been established and, by making ritual sacrifices at the appropriate times, the gods could be appeased and persuaded to favour the tribes. One particularly thought-provoking example of an astronomical alignment exists at the great chambered cairn of Newgrange in Ireland built during the 3rd millennium BC where the midwinter sunrise sends a shaft of light through a specific window in the stonework along the passageway deep into the heart of the cairn to strike the tomb at the far end.

prime meridian, a fixed circle, on the surface of a planet or satellite, which passes through its *poles and from which longitude can be measured east or west. Thus the **longitude** of a place is the angular distance of the place's meridian east or west from the prime meridian. The **latitude** of the place is its angular distance north or south of the body's equator measured along its meridian. The prime meridian on Earth passes through the original *Royal Greenwich Observatory. On other bodies such as the Moon or Mars, the prime meridian is defined with respect to a known surface feature.

prime vertical, the vertical great circle that passes through the observer's zenith (the point that is directly overhead) and intersects the horizon at the observer's most easterly and westerly points. This circle is at right angles to the celestial *meridian. The parts of this circle going from zenith to east and west are referred to as the **prime vertical east** and **prime vertical west**, respectively.

primordial fireball *cosmology.

Procellarum Oceanus *Oceanus Procellarum.

Procyon, a nearby first-magnitude *binary star in the constellation of Canis Minor, and also known as Alpha Canis Minoris. It comprises a subgiant or dwarf star somewhat hotter than the Sun and with about eight times its luminosity, with an eleventh-magnitude white dwarf companion in a 41-year orbit. The distance of Procyon is 3.5 parsecs.

Project Apollo, the *NASA programme resulting in the successful landing of men on the Moon. This US project involved the development of the Apollo spacecraft and the Saturn V rocket launch vehicle on which it was carried into Earth orbit. The Apollo spacecraft was designed to carry three astronauts and comprised three parts: the command module, the service module, and the lunar excursion module. These are listed in the table on page 196.

The Saturn rocket launched the Apollo spacecraft into orbit 160 km above the Earth's surface. The spacecraft was checked and put into a transfer orbit taking it to the Moon. During this transfer the lunar excursion module was detached from its storage space and attached to the nose of the command module. Once a low-altitude lunar orbit had been achieved, two of the three astronauts entered the lunar excursion module and landed it on the Moon, leaving the third crew member to orbit the Moon in the command module. The two astronauts on the surface of the Moon set up the Apollo Lunar Surface Experiments Package (ALSEP) comprising a variety of scientific instruments, and collected samples of the lunar surface. When they had completed these tasks they lifted off from the Moon in the upper half of the lunar excursion module, using the bottom half as a launch-pad, and docked with the command module. After the two astronauts had transferred to the command module, the lunar excursion module was abandoned and the three astronauts, together with their samples and scientific data, returned to Earth in the command module. Just before re-entry into the Earth's atmosphere the service module was jettisoned. Three parachutes were used to land the command module in the ocean for recovery by ships and helicopters. The table on page 197 shows the principal manned missions in the Apollo programme. The measurements and samples collected on the Moon's surface, and the information subsequently transmitted back to Earth by the ALSEP have significantly increased our knowledge of the structure, age, and history of the Moon, and of the early history of the solar system.

prominence, a glowing gas cloud apparently projecting beyond the limb of the Sun, often in a huge arc. In reality prominences are giant loops of relatively cool gas in the hot tenuous solar *corona, suspended by magnetic fields in the solar atmosphere. Prominences are usually studied through monochromatic filters and, on the disc of the Sun, they appear as dark *filaments.

proper motion, the intrinsic angular motion of a star across the sky relative to the background of more distant stars. The *velocity of a star moving through space with respect to the Sun is a vector which can be resolved into its *radial velocity along the line of sight and its transverse velocity at right angles to this across the line of sight. The proper motion is a result of the star's transverse speed through space, in contrast to its *parallax which is an apparent motion arising from the Earth's orbit about the Sun.

proton–proton reaction, the dominant *nuclear fusion reaction in *main-sequence stars whose central temperature is below 1.8×10^7 K. The proton–proton reaction creates helium-4 from hydrogen via two intermediate stages in which deuterium and helium-3 are produced together with *neutrinos and gamma rays. The Sun's central temperature is 1.6×10^7 K and so is low enough to sustain the proton–proton reaction. Hotter main-sequence stars are fuelled by the *carbon–nitrogen cycle.

proto- (planet, star), the stage in the life of a *planet or a *star when it has newly come into existence. Modern theory suggests that a proto-planet was a conglomeration of planetesimals which had impacted and stuck together. In *stellar evolution a proto-star condensed out of an interstellar cloud of dust and gas, a process that is still continuing.

Proxima Centauri, a variable eleventh-magnitude red *dwarf star, our solar system's nearest stellar neighbour, at a distance of 1.3 parsecs in the constellation of Centaurus. It was discovered by the British astronomer Robert Innes in 1915 from its large *proper motion, which it shares with *Alpha Centauri, 2.2 degrees away in the sky. Proxima is an active example of the UV Ceti stars or *flare stars: red dwarfs with an emission *spectrum and flares lasting from a few seconds to a few minutes. The flares are due to disturbances in the star's magnetic field, as on the Sun. Proxima varies by up to one magnitude in blue light. It is one of the least luminous stars known, with an absolute magnitude of 15.5, and a luminosity 19,000 times less than that of the Sun.

Ptolemaic system, a theory to describe the structure of the *solar system given in Ptolemy's **Almagest* (*c.*AD 140) and accepted by most astronomers for almost fifteen centuries. The Ptolemaic theory sought to describe the apparent movements of all the heavenly bodies and predict their future positions. It was a *geocentric system, the Earth being assumed to be the fixed centre of the universe. According to this system, the stars were attached to the surface of a transparent sphere which rotated westwards about a north–south axis making one revolution in one *sidereal day. The Sun, Moon, and planets revolved about the Earth in different periods of revolution determined from observation, the planets moving on their **epicycles**, circles whose centres in turn moved along **deferents** (circles around the Earth). This arrangement was required because the planets were occasionally seen to reverse their direction of motion against the stellar background. Because Mercury and Venus were never seen far from the Sun in direction (they were always morning or evening objects), the centres of their epicycles always lay on the line joining the Sun to the Earth. Again, to agree with observation, the radii joining Mars, Jupiter, and Saturn to the centres of their epicycles were always parallel to the Earth–Sun line. Slight tilts were given to some of the circles to improve agreement with observation. In addition, the deferents and epicycles had centres slightly displaced from the Earth's centre and the different circles, respectively. Such features were required because not only did the planets move in slightly differing planes from one another but also their angular velocities were variable.

The Arabian astronomers of the Middle Ages, accumu-

Ptolemaic system

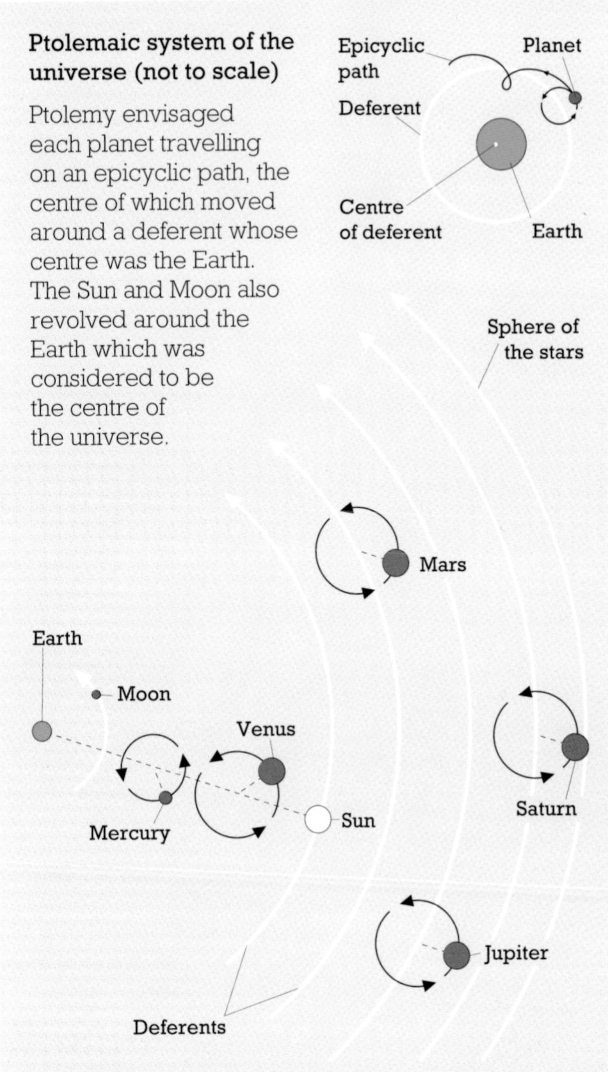

The stars were thought to be fixed to the inside of a transparent sphere which rotated westwards. The centres of Venus and Mercury's epicycles were centred on a line between the Earth and the Sun.

Prominences are huge eruptions of the Sun's surface. In this prominence, photographed in ultraviolet light by Skylab 3 in 1973, it was observed that helium erupting from the Sun does not disperse, being trapped by the Sun's magnetic field up to a million kilometres above the photosphere.

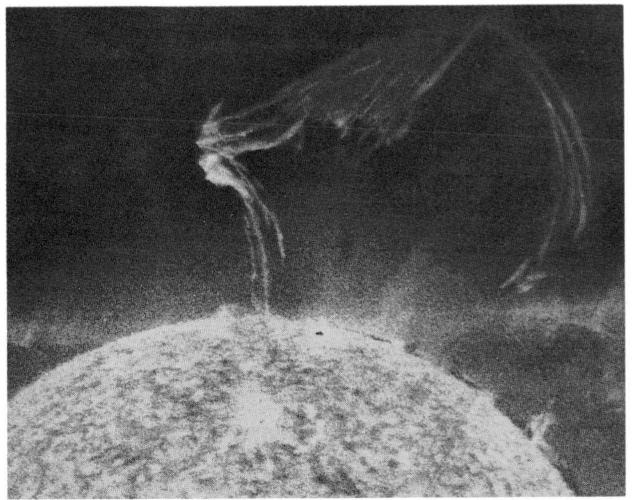

Pulsar, mechanism of

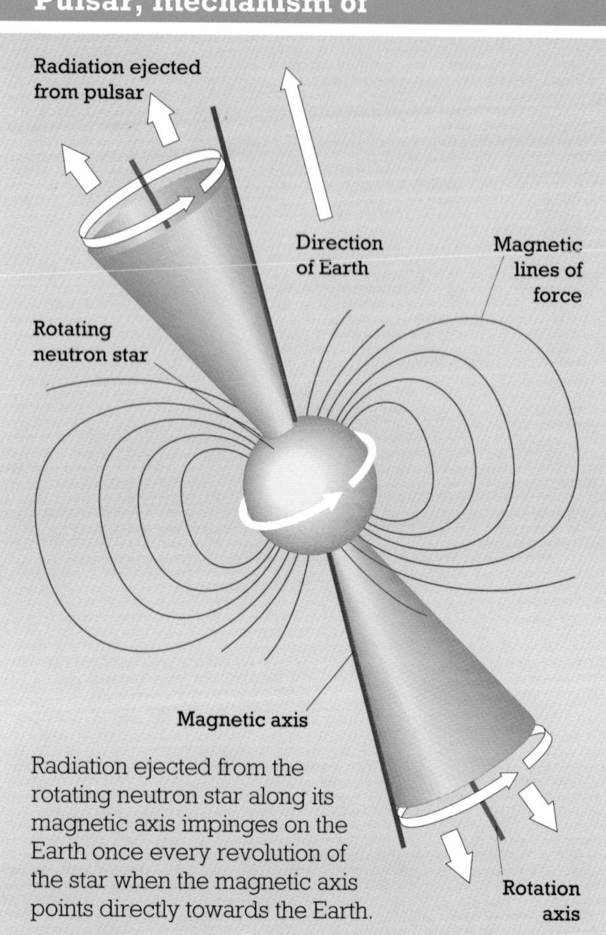

Radiation ejected from the rotating neutron star along its magnetic axis impinges on the Earth once every revolution of the star when the magnetic axis points directly towards the Earth.

lating more accurate observations of the planets, found they had to modify the Ptolemaic theory by adding epicycles to the epicycles. The theory became cumbersome and to some unconvincing, finally being replaced by the *Copernican system. See also *Greek astronomy, *Islamic astronomy.

pulsar, a rapidly spinning *neutron star emitting a narrow rotating beam of radiation, and detected as a pulsing *radio source if the beam sweeps across the Earth. A pulsar may also be a source of gamma-ray, X-ray, and optical pulses. A pulsar has a strong *magnetic field whose axis is at an angle to that of rotation. Electrons spiralling around the magnetic field lines emit *synchrotron radiation in the direction of the magnetic axis, giving rise to the observed beams. The first pulsars were discovered in 1967 and several hundred are now known. Their periods range from 0.001 seconds to over 4 seconds. Most are gradually slowing down owing to the braking effect of the Galaxy's magnetic field, although occasionally the rotation suddenly speeds up slightly, probably owing to a slight contraction of the neutron star in a 'starquake'. The youngest known pulsars, with very short periods, are *optical pulsars in *supernova remnants. These include the *Vela pulsar, the *Taurus A pulsar in the *Crab Nebula, and possibly one formed in *Supernova 1987A. However, the shortest periods, around 0.001 seconds, are found among the **millisecond pulsars** whose rotational speeds are thought to have been increased by accreting matter from a companion star in a *binary pulsar.

pulsating universe *oscillating universe.

pulsating variable *variable star.

Puppis, the Stern, a *constellation of the southern sky. It was originally part of the ancient Greek constellation Argo Navis, the ship of the Argonauts, until 1763 when the French astronomer Nicolas Louis de Lacaille divided that large constellation into three; the other parts are *Carina and *Vela. Its brightest star is Zeta Puppis, named Naos which is Greek for 'ship'. With a magnitude of 2.3, it is one of the hottest stars known. L-2 Puppis is a red giant that fluctuates semi-regularly between third and sixth magnitude every five months or so. Puppis lies in rich Milky Way star fields and contains several excellent star clusters. M47 (NGC 2422) is a naked-eye *open cluster lying next to the binocular cluster M46 (NGC 2437). NGC 2451 is a scattered cluster whose brightest star is the fourth-magnitude C Puppis. NGC 2477 is a large open cluster that resembles a globular cluster when seen through binoculars.

Puppis A, a *supernova remnant in the constellation of Puppis, resulting from the explosion of a supernova some thousands of years ago. Puppis A is an X-ray source, the radiation coming from regions where the exploding gas is in collision with interstellar dust and gas and is thus raised to temperatures measured in millions of degrees.

Pyxis, the Compass, a *constellation of the southern hemisphere, introduced by the 18th-century French astronomer Nicolas Louis de Lacaille and representing a ship's magnetic compass. Its brightest stars are of only fourth magnitude. Its most celebrated star is the recurrent *nova T Pyxidis, which can flare up unexpectedly from its normal fifteenth magnitude to as bright as sixth magnitude. Outbursts were recorded in 1890, 1902, 1920, 1944, and 1966.

quadrant, an astronomical instrument, now obsolete, which was used for measuring the altitudes of celestial objects. It is in the shape of a quarter-circle, the curved edge of which is marked off from 0 degrees to 90 degrees. The hand-held quadrant is held in a vertical plane with the right angle away from the eye and the curved edge downwards. It has a weighted cord attached to the right-angled apex. One of the straight edges is equipped with a pair of metal pinhole sights which are aligned with an astronomical object and the hanging cord reads off the altitude of this object above the horizon on the circular scale. A larger form of quadrant, known as a *mural quadrant, is mounted on a north–south wall and has a solid arm instead of a hanging cord. This obsolete instrument was superseded by the meridian, or transit, circle, a telescope free to move only in the meridian or north–south plane.

Quadrantids, an active *meteor shower that reaches its peak around 3–4 January each year. The shower is unusual because it lasts for only about a day. The comet that created the *meteor stream has yet to be identified but the orbit of the stream intersects the orbits of both the Earth and Jupiter. The *radiant is named after the obsolete constellation Quadrans Muralis, which was located between Draco and Boötes.

quadrature *planetary configurations.

quantum theory, a description of physics in which *energy exists only in discrete quantities. Quantum theory originated with the ideas of the German physicist Max Planck and *Einstein in the early years of the 20th century. In 1900 Planck suggested that *electromagnetic radiation is **quantized**, that is, it can be emitted or absorbed only in discrete amounts called quanta. Each quantum of radiation is called a *photon. This enabled Planck to explain the spectral distribution of the radiation emitted by a *black body at a given temperature and is embodied in what is now known as *Planck's Law. Einstein took the idea a step further in his paper of 1905 and proposed that all forms of energy exist as quanta. These ideas were later extended to describe the electrons surrounding the nucleus of an atom, and by the 1920s the early ideas of quantum theory had been refined into **wave mechanics**. Before wave mechanics, light was thought to be purely a wave motion, but wave mechanics showed that sometimes it was more appropriate to regard light as particles, that is, photons. Conversely, subatomic particles, such as electrons and protons, sometimes behave like light waves. Quantum theory, also known as **quantum mechanics**, is now well established and gives an accurate description of physical processes that occur within distances smaller than the size of an atom.

In the quantum theory of atoms, electrons do not circle the nucleus in precise orbits as the planets orbit the Sun in the solar system. Instead the nucleus is surrounded by a set of **orbitals**. These are in effect fuzzy orbits in which the position of an electron is inherently uncertain. At any instant the electron is spread out around the orbital and all that is known is its probability of being at each point. Each orbital corresponds to a particular energy determined mainly by the electrical charge of the nucleus, electrons in the orbitals nearer the nucleus having a lower energy than those in orbitals further away. Each electron is normally constrained to occupy a single orbital, and there is a limit to the number of electrons that can simultaneously exist in one orbital. If there is a vacancy in an orbital of low energy, an electron from a higher-energy orbital can move into the lower-energy orbital; its excess energy, E, is emitted as a single photon of electromagnetic radiation of a particular frequency ν. The frequency is given by the relation:

$$E = h\nu$$

where h is Planck's constant and has a value of 6.626196×10^{-34} Js. The sharp *spectral lines emitted by atoms can be explained by electrons transferring to lower-level orbitals. Similarly, an electron can be transferred to an orbital of higher energy if a photon whose frequency matches the energy difference between the orbitals hits the atom. Thus the photon is absorbed by the atom. If the incident photon is sufficiently energetic it can give the electron enough energy to escape from the atom altogether, creating an *ion. These processes of electron transitions between orbitals are responsible for the visible radiation emitted by *stars.

Quantum theory

Electron transitions in an atom.

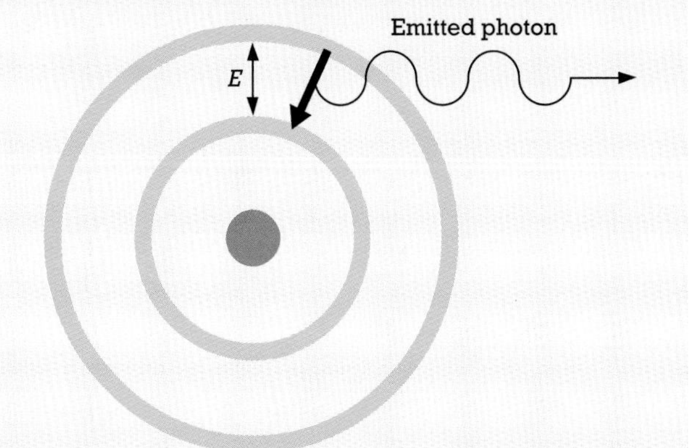

An electron transferring from a high-energy orbital to an orbital of lower energy emits a photon whose frequency is determined by the difference in energies between the two orbitals.

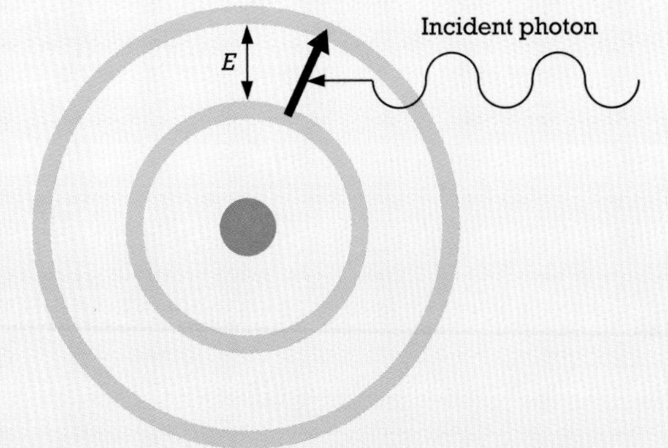

If an incident photon has a frequency corresponding to the difference between two orbitals, an electron can be transferred to an orbital of higher energy.

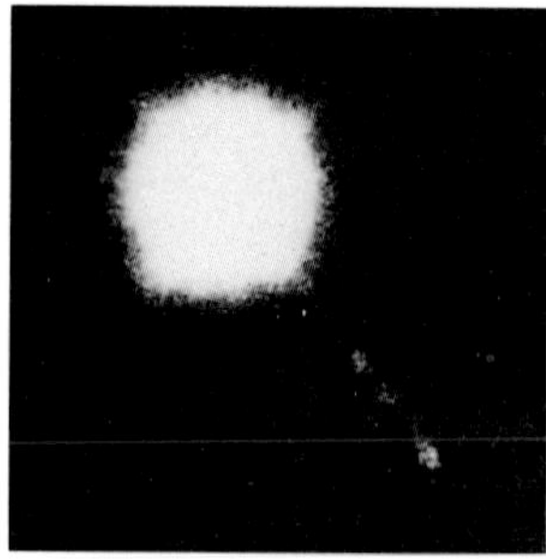

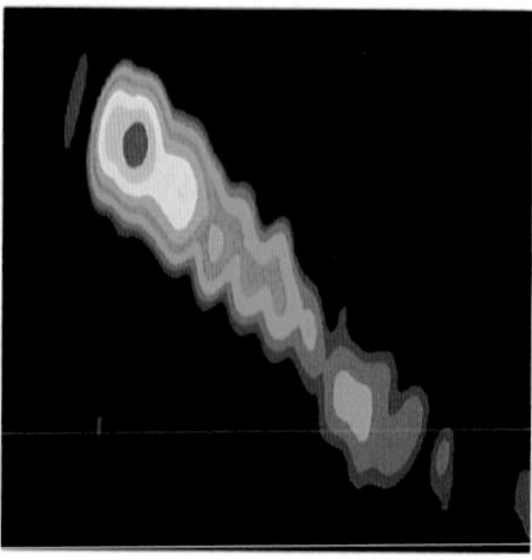

The **quasar** 3C 273 (*left*) lies in the constellation of Virgo and is the nearest known quasar, at a distance of about 700 million parsecs. One of the first to be discovered, 3C 273 is thought to be a black hole lying at the centre of a distant galaxy. Its huge but faint jet is a powerful beam of fast-moving charged particles ejected from its nucleus. The false colour image of quasar 3C 273 (*right*) shows the innermost portion of the quasar's enormous jet. The colours represent the intensity of the radio emission at a wavelength of 18 cm: red is highest intensity; blue is lowest.

Quantum theory also describes the internal behaviour of atomic nuclei and how they combine by *nuclear fusion, the source of energy in most stars.

quasar, a star-like object which emits the energy of hundreds of normal galaxies. Its original name, 'quasi-stellar object', has been shortened to 'quasar'. The nature of quasars is not well understood. They exhibit very large *red shifts of their spectral lines, so if the Hubble Law is applied, such objects must be the farthest known objects in the universe. One theory suggests they are the highly luminous centres of galaxies where *black holes drag material into them by their massive gravitational fields. Some quasars exhibit significant and rapid changes in luminosity.

R

radar astronomy, the transmission of radio waves to nearby astronomical objects and analysis of the echoes received. Radar astronomy developed from the World War II techniques of radar in which radio echoes were obtained from distant terrain or moving objects. In fact the first echoes from astronomical objects, meteors, were discovered inadvertently by radar operators on the southern coast of the UK when searching for V2 rockets in 1944. Two years later the much more difficult technical feat of obtaining echoes from the Moon was achieved. The 800,000 km round trip for the radio pulse takes 2.5 seconds, travelling at the speed of light. Meteors were a fruitful early area of radar studies; they leave a temporary ionized trail as they burn up in the atmosphere at heights of 80 to 100 km. In addition to providing data on previously undetected daytime meteor streams and accurate orbits of individual meteors, these studies gave new insights into the atmospheric winds at such heights. Since the strength of the radar echo from an astronomical object varies inversely as the fourth power of the distance, radar astronomy is restricted to investigation of the solar system. Echoes have been received from seven planets and some of their major satellites using the world's largest *radio telescopes, including the 305 m (1,000 foot) diameter telescope at the *Arecibo Radio Observatory, Puerto Rico. By observing the *Doppler effect due to the rotation of the planets, radar studies determined for the first time the rotation periods of Mercury (59 days) and Venus (244 days). *Aperture synthesis procedures have been applied as in radio astronomy to provide high angular resolution. Radar studies gave important information on the distribution of surface material (rocks and sands) of the nearby planets before the first spacecraft landings. Such studies continue to yield data on the topography of planets and on the properties of the material several metres below the surface, since echoes can usually be detected from depths of up to about ten wavelengths.

radial velocity, the component of a star's intrinsic *velocity along the line of sight. The motion of a celestial object may be separated into two components, one across the line of sight showing up as a change of position in the sky known as *proper motion, and the second along the line of sight. The second component, called the radial velocity, may be determined from the *Doppler effect by measuring the shift in optical or radio spectral lines. Any measurement of radial velocity is made relative to the Earth and so an allowance must be made for the Earth's motion about the Sun.

radiant, the point on the *celestial sphere from which a *meteor appears to come. The radiant is also the common point from which a *meteor shower seems to radiate, but in this case the effect is one of perspective, because in reality all the *meteoroids in a specific *meteor stream are moving along parallel paths. Meteor showers are named after the constellation in which the radiant lies. As the shower progresses the radiant moves eastwards by about one degree per day because of the Earth's motion in its orbit about the Sun.

radiation *electromagnetic radiation.

radiation belts, regions of ionized particles surrounding planets possessing magnetic fields. Electrically charged particles originating from the Sun become trapped in the planet's magnetic field to form one or more doughnut-shaped rings about the planet. In the Earth's case, the belts are known as the *Van Allen radiation belts.

radiation pressure *Electromagnetic radiation can be thought of as a flow of *photons which carries momentum with it. When it encounters a surface it produces a pressure on the surface. Normally this is tiny; for example, when sunlight falls on a surface it exerts a pressure per square centimetre of only 1/28,000 of the weight of a cubic centimetre of air. Radiation pressure however can be important as when the *solar wind drives dust and gas away from a comet to form the comet tail. In stellar interiors, radiation pressure can be as powerful as the hydrostatic gas pressure, supporting to a large extent the upper layers of the stars.

radioactive dating, the use of one or more radioactive elements to date the material in which they are embedded. If a rock, when it solidifies, encloses a sample of a radioactive substance then on examination, knowing the rate of decay of the substance, the proportion of it which has decayed into its stable non-radioactive product gives a measure of the time elapsed since the sample was trapped in the rock. If the so-called half-life of the substance (the time taken for one-half of the radioactive atoms in the sample to decay) is known and there is no chance of the sample or its product having been contaminated in the rock, the method gives a reliable measure of age. In this way the rocks of the Earth and those of the Moon have been dated. Relatively short-lived radioactive elements like carbon-14 have been used to date fossil and archaeological remains.

radio astronomy, the study of the extraterrestrial universe by analysing its radio-wavelength signals. Radio astronomy complements and extends knowledge obtained by conventional optical astronomy and is particularly sensitive to the emissions from objects such as the remnants of exploding stars or the explosions of very distant radio galaxies. In addition, radio waves are also characteristic of low-energy processes, particularly the generation of radio spectral lines, such as the *twenty-one centimetre line, from atoms and molecules distributed throughout the cold regions between the stars in our own Milky Way and in external galaxies. *Jansky, working for the Bell Telephone Company in New Jersey, USA, was the first to detect and recognize extraterrestrial radio waves from the Milky Way. The development of radio and radar techniques during World War II made possible a rapid expansion of radio astronomy in the immediate post-war years. (*Radar astronomy, which involves the transmission of radio waves to nearby astronomical objects such as the Moon and planets, developed in parallel, often at the same radio observatories.) Emissions from the Sun, planets, galactic nebulae, and external galaxies were soon being detected. An important class of **radio sources**, celestial sources of radio waves, was found which did not appear to be associated with any optical object. Only with the advent of the interferometer *radio telescope, which measured the position of radio sources with high precision, was it possible to begin identifying them, by the simultaneous use of the largest optical telescopes. *Cygnus A, the second-brightest radio source in the sky, was identified with a distant dumb-bell-shaped galaxy with a *red shift of 16,000 km/s which corresponds to a distance of 220 million parsecs (700 million light-years). Even relatively intense radio sources are associated with galaxies at the limit of visibility in Earth-bound optical telescopes. Even with the best optical techniques using the largest optical telescopes some 30 per cent of the brightest thousand or so radio sources remain unidentified. In 1962, *quasars were discovered; these were an entirely new type of extragalactic object which had many of the radio properties of radio galaxies but in an optical telescope they looked like stars and hence were given the name quasi-stellar objects or quasars. In fact they are intensely active galaxies with luminosities exceeding radio galaxies by factors of up to 10^4, if their distances are as indicated by their measured red shifts. Indeed, because of their high luminosity and exceptionally large red shifts, their distances have been called into question. It is argued by some astronomers that their red shift may not be cosmological, that is, caused by the expansion of the universe, but may be produced by some as yet unrecognized physical process.

Radio astronomy has made major contributions to our understanding of the Milky Way, particularly by means of the twenty-one centimetre wavelength spectral line of neutral hydrogen which has established the spiral nature of our galaxy and has produced evidence for recent violent activity at the galactic centre. Recently, interstellar molecules have been discovered in abundance in high-density clouds throughout the Galaxy. Many of these molecules, which include formaldehyde, alcohols, and ethers, may be building-blocks for biochemical molecules. It is also found that the dense molecular clouds are the birthplaces of stars. At the other end of stellar evolution, radio astronomers have found supernova remnants and more recently *pulsars, stellar embers which have collapsed into rapidly rotating *neutron stars.

No account of radio astronomy is complete without mention of its bearing on *cosmology. One of the great debates has been whether the expansion of the universe will continue with ever decreasing density (the *Big Bang theory) or whether material is being continuously replenished to preserve a constant density (the *Steady State theory). Observations of large samples of radio sources at several different observatories around the world have shown that fainter and presumably more distant sources representative of conditions earlier in the history of the universe were closer together. These radio observations therefore favour the Big Bang model.

radio galaxy, a source of radio waves identified with an optically visible galaxy and partly distinguished from a normal galaxy by its high radio power output. The output of a radio galaxy lies in the range 10^{35}–10^{38} watts, which is some 10^4–10^6 times higher than that of the average normal galaxy. A radio galaxy such as *Cygnus A often reveals a single source with two lobes or a double source with the optical object located between the two sources of radio-frequency radiation. The energy generation mechanism of radio galaxies is not understood.

radiometer, an instrument used to measure the energy coming from a source of radiation, particularly infra-red radiation. It can be used to measure the solar constant, that is, the total energy radiated by the Sun per second on to unit area at right angles to the direction of the Sun above the Earth's atmosphere.

radio source *radio astronomy.

radio telescope, in astronomy, an instrument used for the reception of radio waves from, and sometimes transmission to, extraterrestrial objects. The most common form of radio telescope consists of a steerable paraboloidal *dish that reflects the incoming radio waves on to an aerial (US antenna), called a feed, at the *focus of the paraboloid. The feed directs the signal into a radio receiver for processing. It is analogous to the optical *reflecting telescope. Large reflecting radio telescopes have great versatility; they can be used at a wide range of radio frequencies, they can be steered quickly from one radio source to another, and they can be rapidly switched from one astronomical programme to another. Such telescopes are commonly used for *radio astronomy and *radar astronomy, and are frequently used for communicating with a satellite or spacecraft. Another form of radio telescope, commonly used at long radio wavelengths or to gain very high resolution (image clarity), depends for its operation on *interferometry. It comprises an array of aerials or dish telescopes each of which receives a signal from the radio source. The resolution depends on the distance between the farthest aerials (baseline) and resolutions much higher than those obtained from existing dish or optical telescopes have been achieved by using *aperture synthesis with intercontinental baselines. The *multi-element radio linked interferometer network and the *Very Large Array have been used to produce very detailed radio maps of the sky as well as accurate maps of compact radio sources.

radius vector *Kepler's Laws.

Ramsden eyepiece, a type of eyepiece often used in an optical *telescope to magnify the image. It employs two planoconvex lenses and, like other eyepieces, it incorporates a stop which cuts out the periphery of the field of view. The **Kellner eyepiece**, a modification of the Ramsden design, replaces the lens nearer the eye with an achromatic eye lens to produce better results.

The **RATAN 600** radio telescope, at the Zelenchukskaya Observatory, has a diameter of 576 m. It is constructed from aluminium parabolic mirrors, each of which is positioned by computer to give a total area of about 10,000 m². The reflecting area may be split up and operated as four individual telescopes.

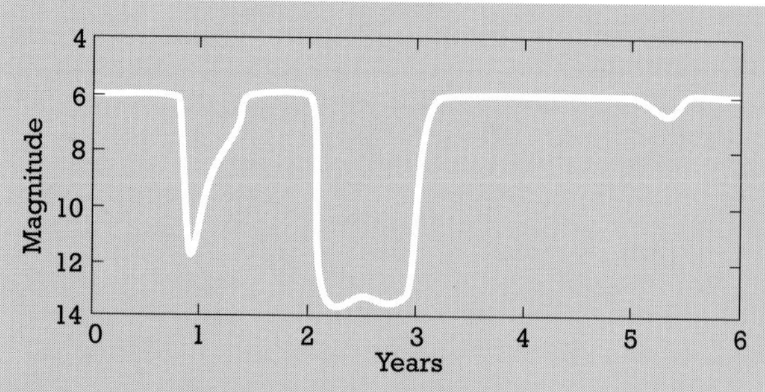

Light curve of R Coronae Borealis, showing three dips in brightness, known as extinction minima.

RATAN 600, a large *radio telescope at the *Zelenchukskaya Observatory in the Caucasus Mountains. Consisting of almost 1,000 metal sheets, 2 m (6.6 feet) wide by 7.4 m (22.4 feet) high set in a circle 576 m (1,890 feet) in diameter, it became operational in 1976. The sheets are tilted to collect and reflect radio waves to an aerial (US antenna) within the circle.

R Coronae Borealis star, a carbon-rich *supergiant *variable star that undergoes unpredictable fades. Normally it shows slight variations in magnitude as it pulsates in a period of a few weeks. According to one model, once in each pulsation cycle a cloud of gas is ejected in a random direction and condenses into a dark cloud of sooty dust, causing a sharp fade, known as an extinction minimum, if it comes towards the Earth. R Coronae Borealis stars have high luminosity but only a few dozen are known, indicating that they represent a short-lived phase in *stellar evolution. R Coronae Borealis itself varies between magnitudes 5.7 and 14.8.

recession of the galaxies, the motion of the vast majority of galaxies away from our own galaxy. It is revealed by the *red shift in their spectral lines. Their velocity of recession is related to their distance via the *Hubble constant and enables the age and size of the universe to be determined.

red giant, a *giant star that is also a *late-type star. The range of *luminosities of typical giant stars goes from fifty times that of the Sun at *spectral class G5 to a thousand times at class M5. Among the *dwarf stars, however, the luminosity range decreases from a luminosity equal to that of the Sun at G5, down to less than one hundredth of the Sun's luminosity at M5. Thus it is among the stars of late spectral class that the difference in luminosity between giant and dwarf stars becomes most apparent. Most red giants are of spectral class K or M. Highly evolved stars of spectral classes C and R, which are found only among the giant and *supergiant stars, are much rarer. Several bright naked-eye stars are red giants, including Aldebaran, Arcturus, and Pollux. Many of these stars are not truly red, but have colours ranging from orange-yellow to orange-red. Many red giants are long-period *variable stars of *Mira Ceti type, or are semi-regular variables.

red shift, a shift in *spectral lines towards the red end of the spectrum as a consequence of the *Doppler effect. Light from the more distant objects in the universe, such as galaxies and quasars, is red-shifted and this observation has to be incorporated into any theory of *cosmology. The explanation of the red shift, that it is due to the expansion of the universe and hence the distance of an object, has been the cause of considerable debate among astronomers. The debate was given a large stimulus by the realization in the 1960s that quasars have red shifts much greater than that of any galaxy known at that time. If the Hubble Law, defined in terms of the *Hubble constant, applies to these objects, they are the most distant objects seen by astronomers. It is now generally accepted that the red shift is due to quasars' very large distances from the Earth and these objects are therefore used by astronomers to give information about the universe both at great distances and at earlier epochs, since the light has taken many millions of years to reach the Earth.

reflecting telescope, in astronomy, a *telescope utilizing a large concave mirror, referred to as the primary mirror, to collect light from a celestial object and bring it to a *focus. *Newton believed there was no practical way of overcoming chromatic *aberration in the *refracting telescope of his time, and so built a reflecting telescope, now known as a **Newtonian telescope**, in 1668. Since then, as the technology of casting and grinding large mirrors developed, the size of the primary mirror has increased. In addition, a number of different ways of viewing the image have been devised. These are compared in the figure on page 136. In the Newtonian telescope, light reflected from the primary mirror, which is paraboloidal in curvature, is intercepted by a small flat mirror and sent through a hole in the side of the telescope tube to be examined by means of an eyepiece lens. The **Cassegrain telescope** has a small secondary mirror which reflects the beam of light back through a circular hole in the primary mirror before the beam comes to a focus. The position of the Cassegrain focus is useful for examination and analysis of the light by another instrument, such as a *photometer or *spectrometer. In the **coudé telescope**, a beam of light entering the telescope is ultimately reflected along the polar axis of the *mounting in a fixed direction. This is a convenient arrangement when massive pieces of equipment are being used to analyse the light because the heavy equipment can be securely fixed. The **Maksutov telescope**, named after the Russian astronomer who developed it, incorporates a correcting plate which is essentially a deeply curved meniscus lens. The primary mirror is spherical in curvature and usually has a central hole so that the telescope can be used in the Cassegrain manner. To this end, a small aluminized area on the back of the correcting plate reflects the beam of light from the primary mirror back through the hole. The **Schmidt telescope** was developed by the Estonian inventor Bernhard Schmidt in 1930. A correcting glass plate is placed in the path of the incoming light before it reaches the primary mirror. By giving the plate the correct curvature Schmidt was able to obtain undistorted stellar images over an unprecedentedly large field of view. The images are formed on a curved image surface and a correspondingly curved photographic plate placed there enables the images to be registered in a relatively short time. The Schmidt telescope has been subsequently developed and used in whole-sky surveys.

Schmidt **reflecting telescopes** are used for taking wide-angle photographs of the sky. They have a thin correcting lens at the front end of the tube, a larger concave mirror at the rear end, and a curved focal plane in the centre, where a curved photographic plate or a CCD camera is placed. Important sky surveys have been made with Schmidt telescopes, particularly the 1.2 m telescope at Mount Palomar, California, USA, and the UK Schmidt telescope at Siding Spring in Australia. The Schmidt telescope illustrated here is the 60 cm instrument at the Beijing Astronomical Observatory in China. It is used for tracking asteroids.

Reflecting telescope

The most common optical telescope employed by the professional astronomer is the reflecting telescope. Unlike refracting telescopes, a reflecting telescope uses a large concave mirror to collect and focus the light from the object being observed.

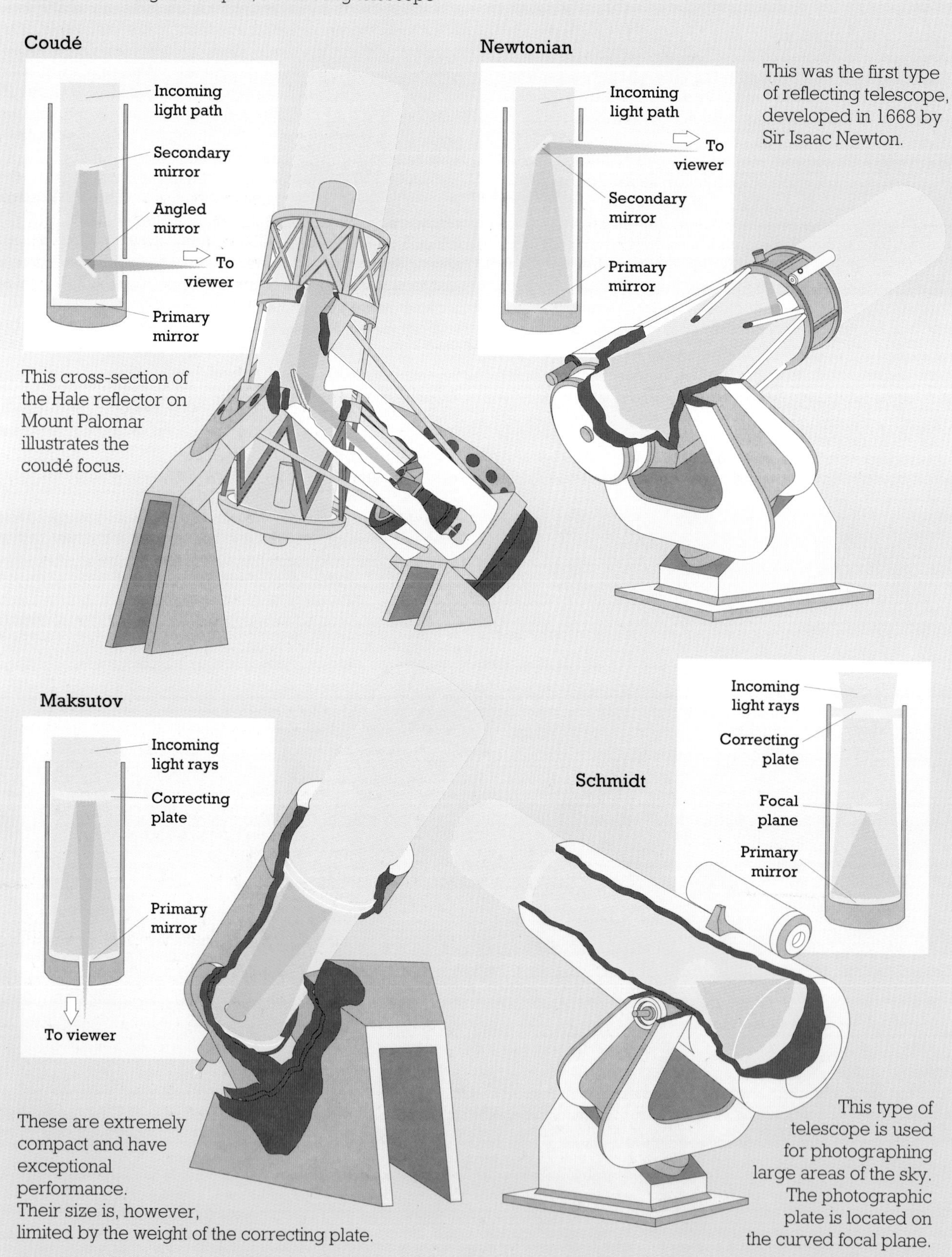

This cross-section of the Hale reflector on Mount Palomar illustrates the coudé focus.

This was the first type of reflecting telescope, developed in 1668 by Sir Isaac Newton.

These are extremely compact and have exceptional performance. Their size is, however, limited by the weight of the correcting plate.

This type of telescope is used for photographing large areas of the sky. The photographic plate is located on the curved focal plane.

reflection nebula, a luminous cloud of dust and gas which scatters light from stars in its vicinity. The light is reflected by the cloud's dust particles. Reflection nebulae are often associated with a *cluster of young stars, such as the Pleiades, which have evolved from the surrounding nebula.

refracting telescope (or **refractor**), in astronomy, a *telescope in which a large lens, known as the **objective**, acts as a light collector by focusing a beam of light from a celestial object into an image. The earliest astronomical refractor was developed by *Galileo in 1609. From 1609 to 1668, when *Newton invented the *reflecting telescope, the refractor was the only type of telescope employed in astronomy. Since then it has undergone a process of continual development. Although, in 1756, the British optician John Dolland showed that the defects of refractors, such as chromatic *aberration, could be corrected, it was eventually realized that lenses for refractors were impossible to make beyond a certain size. Thus the largest refractor in use today still remains the 40-inch (1.02 m) telescope completed in the 19th century at the *Yerkes Observatory.

refraction, the apparent bending or change in direction of *electromagnetic radiation as it moves from one transparent medium to another, for example, from air to water. The principle is used optically in the design of lenses and prisms. Light is refracted on entry into the Earth's atmosphere making stars appear higher in the sky than they actually are. The effect increases towards the horizon where the displacement is approximately $\frac{1}{2}$ degree. Allowance for refraction has to be made when making ground-based positional measurements. The *ionosphere has refractive properties similarly affecting the passage of radio waves from space.

regolith, the layer of rock-dust and rock fragments that covers most of the lunar surface and the surface of other waterless planets and satellites in the solar system. This rough soil has been produced by meteoroid bombardment, *crater formation, and the emission of *ejecta. The upper layer is readily compressible, and, on the Moon, produced fine footprints for the Project Apollo astronauts. The lunar regolith has the general mechanical properties of damp sandy soil, the average density being 1,500 kg/m^3. The porosity is around 50 per cent, and it reflects between 7 and 8 per cent of light incident upon it. The colour is a uniform medium-dark grey. The meteoric bombardment continually turns over the regolith so that it gradually increases in depth. Blocks of rocks scattered throughout the soil have been rounded by the same process.

Regulus, a first-magnitude *triple star in the constellation of Leo, and also known as Alpha Leonis. Its primary star is a hot *dwarf star at a distance of 21 parsecs, with about 100 times the luminosity of the Sun. Its distant eighth-magnitude companion, itself a *binary star of very long period, has the same proper motion as Regulus.

relativity, general theory of, a theory of gravitation published in 1915 by *Einstein. This theory reformulated the concept of a gravitational field, relating it to the *curvature of *space–time produced by the presence of matter. This is in contrast to *Newton's Law of Gravitation in which the mass of a body produces a force of attraction. A particle in the neighbourhood of a mass follows a trajectory called a geodesic which, for most practical situations, is indistinguishable from the path the particle would take under Newtonian gravitation. Einstein postulated that the effects experienced or the laws of nature deduced by an observer in a laboratory under a uniform gravitational force would be equivalent if the observer and the laboratory were instead subjected to a suitable acceleration. Einstein went on to formulate the so-called **field equations** by which the space–time curvature produced by a gravitational field can be found.

When his general theory was published three tests were available, each depending upon a different prediction made by Einstein's theory and Newton's theory of gravitation. Firstly, a known discrepancy between the rate of advance of Mercury's perihelion and that predicted by Newtonian gravitation was accounted for by Einstein's theory. Secondly, the amount by which light rays were deflected as they grazed the Sun's surface was measured using the fact that, during a total solar eclipse, the Sun is hidden and the stars are visible. The difference between the directions of a number of stars in the solar neighbourhood during the eclipse and at a different time when the Sun was elsewhere agreed with Einstein's theory. The third test, the predicted reddening of light rising through a gravitational field, was tested by measuring the wavelengths of lines in the spectrum of light from white dwarf stars. A very difficult experiment to carry out, its result nevertheless supported Einstein's theory. More recently the effect has been more clearly demonstrated by using the Mossbauer effect which describes the conditions under which gamma rays are emitted from atomic nuclei when the nuclei do not recoil. Additional support for the theory has come from the discovery of the binary pulsar PSR 1913+16. The observed change in the orbits of the components is in agreement with prediction, while the rate of loss of energy from the binary, attributed in Einstein's theory of relativity to an emission of *gravitational waves, agrees with the theory. The general theory of relativity is now firmly established, being supported by all known tests to the present date and is crucial to the understanding of *black holes.

relativity, special theory of, a theory put forward by *Einstein in 1905 in which he reassessed the laws of physics as experienced by observers in inertial frames, that is, observers moving with a fixed velocity not subject to acceleration. The theory involves the velocity of light, c, which Einstein assumed to be the same in a vacuum for all such observers. No matter what their speeds were and even if a source of light had a large velocity with respect to the observer, the velocity of light from it to the observer would be unaltered. He deduced that all physical laws and the numerical constants in those laws were the same in every inertial frame. In the theory, where events and properties in inertial frames are compared, a quantity $\beta = \sqrt{(1 - (v/c)^2)}$ appears, where v is the relative velocity between inertial frames, for example between the source of light's inertial frame and the observer's. This quantity β produces differences between classical Newtonian physics and Einsteinian physics, noticeable only when v approaches c. If the velocity of light had been 300 km/s instead of almost 300,000 km/s, the everyday speeds of objects would have produced noticeable effects on such objects. In the direction in which they travelled they would have appeared shorter; their mass also would be greater than when at rest; and a clock, moving at high speed relative to an observer, would appear to run more slowly than the observer's clock. Such effects would have been a matter of common experience; instead, because in everyday life velocities are always tiny compared to the

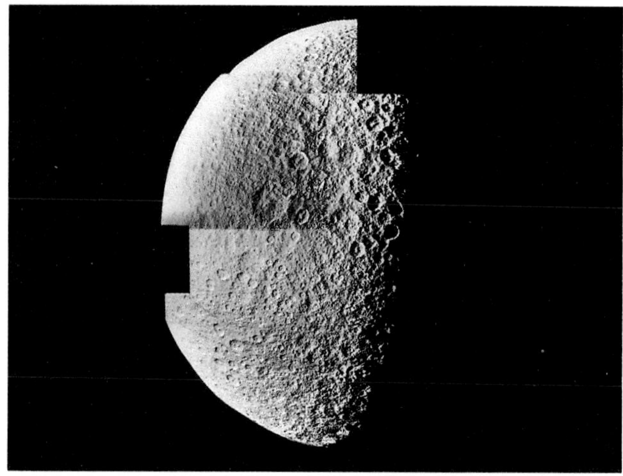

The pictures which make up this mosaic of **Rhea**, one of Saturn's moons, were taken by Voyager 1 during its fly-past in November 1980. Rhea is the most heavily cratered of Saturn's moons, the largest crater being about 300 km in diameter. Many craters have central peaks formed by the heaving of the surface after the impact of the incoming body.

velocity of light, the quantity β is very nearly equal to 1 so that Newtonian physics is indistinguishable from Einsteinian. It is only recently that observational experiments have shown that the physical world agrees with Einstein rather than Newton. For example, synchrotrons which accelerate charged particles of rest mass m_0 to velocities close to that of light have to take into account the growing mass $m = m_0 \beta$ of the particles as they spiral round. A further consequence is that no particle with a non-zero mass can ever achieve the speed of light since at that speed it would have infinite mass.

In developing his theory, Einstein also deduced that mass m and energy E were related by the equation $E = mc^2$, where c is again the velocity of light. This equation describes the rate at which energy can be extracted from nuclear reactions which take place in nuclear reactors and which cause the stars to shine. In the cores of the stars the fusion of hydrogen atoms into helium atoms involves the annihilation of mass and its transformation into radiant energy.

resolving power, the ability of a *telescope to separate two close objects, such as the components of a binary star. For a *radio telescope, the wavelengths collected are so large compared with those of visible light that even a very large instrument, such as the *Jodrell Bank Lovell telescope which has a dish of 76.2 m (250 feet) in diameter, has poor resolving power. This is overcome by coupling two or more radio dish telescopes together to employ the principle of *interferometry, as in the *multi-element radio linked interferometer network (MERLIN) and the *Very Large Array.

Reticulum, the Net, a small *constellation of the southern sky introduced by the 18th-century French astronomer Nicolas Louis de Lacaille to represent the reticle used in his telescope for measuring star positions. Zeta Reticuli is a pair of fifth-magnitude stars similar to the Sun, divisible with binoculars or even the naked eye.

retrograde motion *satellite, natural.

reversing layer *chromosphere.

Roche lobe

1 Detached system

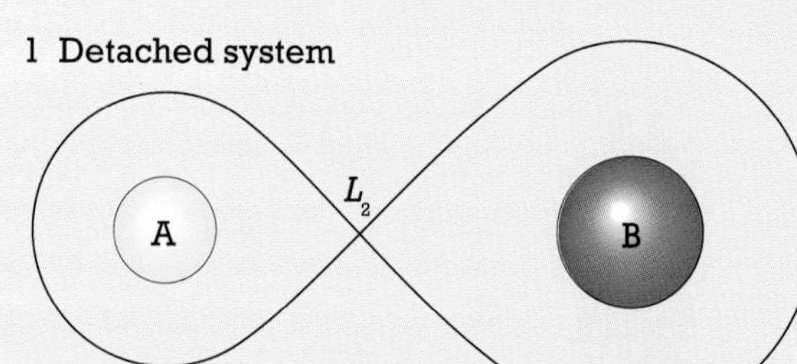

Neither star fills its Roche lobe.

2 Semi-detached system

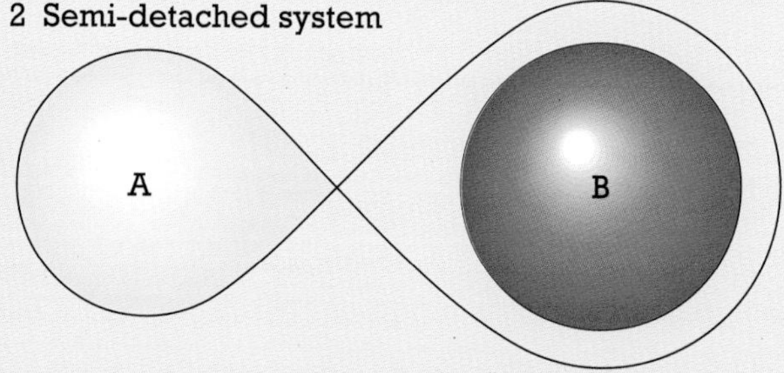

The less massive component (A) fills its Roche lobe. This short-lived stage is rarely observed.

3 Semi-detached system

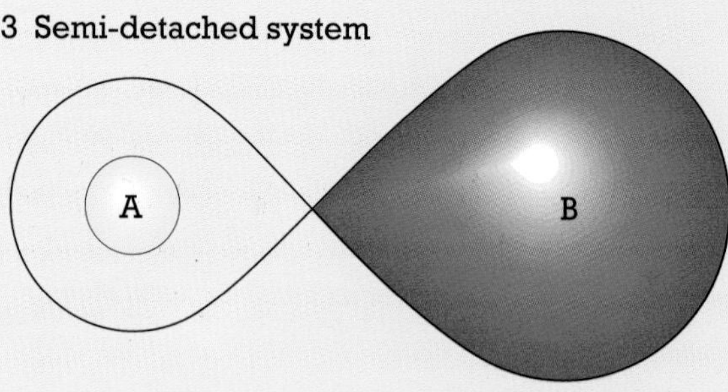

The more massive component (B) fills its Roche lobe.

4 Contact binary

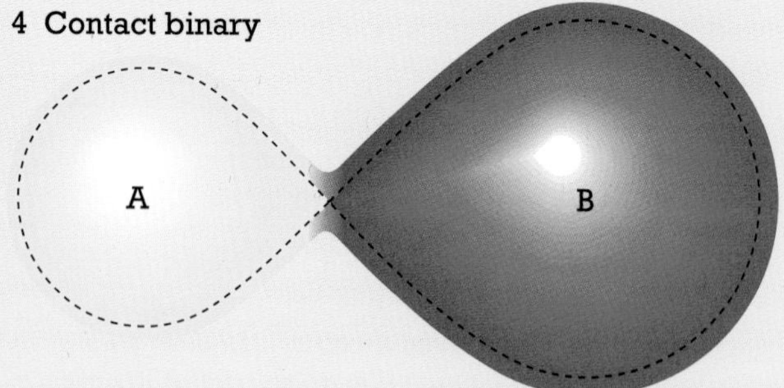

Both components overfill their Roche lobes and share a common envelope.

The huge cloud of glowing hydrogen gas in the **Rosette Nebula** contains large areas of dark obscuring dust which appear to be condensing into round blobs of material known as Bok globules. These globules are eventually expected to form new stars. The nebula glows by the radiation from stars associated with it.

revolution, in astronomy, usually the orbital revolution of one body about another, as for example the Moon about the Earth or the Earth about the Sun. The time taken for one revolution with respect to the stellar background is termed the *sidereal period of revolution. The *synodic period of a planet is defined in terms of its revolution in its orbit. The month is defined by the revolution of the Moon about the Earth.

Rhea, a satellite of Saturn, discovered by Giovanni Domenico *Cassini in 1672. It has a nearly circular equatorial orbit at 527,000 km from the planet's centre. Its diameter is 1,530 km and its density is 1,300 kg/m^3. The latter value implies a predominantly water ice composition. *Voyager images have shown a bright, icy, and heavily cratered surface, resembling the Moon's highlands. As Rhea moves in its orbit the leading and trailing hemispheres are of different brightnesses.

richest field telescope, a *telescope of moderate aperture and low power, giving a wide field of view, usually of some degrees in width and designed to reveal a large number of stars in its field. In many respects a good pair of binoculars fits the specifications for a richest field telescope.

Rigel, a star in the constellation of Orion, and also known as Beta Orionis, with an apparent magnitude of 0.12. It is a hot *supergiant multiple star at an estimated distance of 400 parsecs, with about 140,000 times the luminosity of the Sun, and is thus one of the most luminous stars known. It has a seventh-magnitude hot companion, and the pair probably form a *binary star with an orbit of very long period; the companion is a spectroscopic binary with a period of ten days. Rigel is the brightest member of the Orion stellar association.

right ascension *celestial sphere.

Rigil Kentaurus *Alpha Centauri.

rille, a narrow V-shaped valley on the surface of a lunar *mare. Rilles are typically a kilometre or so across and a few hundred metres deep. There are three major varieties: straight, curved, and sinuous. Straight and curved rilles cut across *craters and maria with complete indifference to the surface topography and are reminiscent of terrestrial fault troughs. Sinuous rilles resemble terrestrial river channels, but lunar water was not responsible. They are thought to be collapsed channels through which mare basaltic lava has flowed; some of these channels were roofed and the roofs as well as the walls have collapsed. Sometimes pools of lava have formed at the downstream end. Many rilles are associated with the edges of mare basins. Unlike water-river beds, the rille is deepest where it is widest.

ring, planetary, a thin disc of particles orbiting a planet. Rings are found around all the *giant planets, but by far the most impressive are those of *Saturn. These were first seen by *Galileo in 1610, but he thought that the planet had 'bumps'. In 1659 the Dutch astronomer Christian Huygens recognized that Saturn was surrounded by an equatorial, thin, flat ring that nowhere touched the planet. In 1675 Giovanni Domenico *Cassini discovered a dark division now named in his honour, separating the A ring from the B ring. By the late 19th century the US astronomer James Keeler used spectroscopic measurements of the wavelength of the light scattered by the rings to deduce that the outer edge was travelling more slowly than the inner edge. He thus showed that the rings were not solid but were composed of a host of individual particles each with a unique circular orbit. In the 1970s infra-red spectroscopy and radar indicated that the ring particles were mainly composed of water ice and their sizes were typically between a millimetre and 10 m; there are many more small ones than large ones. Saturn's rings extend from 7,000 km above the planetary surface to a distance of more than 70,000 km, and the thickness of the system is about 20 m. The total mass is about 3.4×10^{21} kg and if the particles in Saturn's ring could be combined into an icy satellite it would be about 1,800 km across. In fact the material is organized into tens of thousands of ringlets. The vast majority of the particles lie within the *Roche limit, this being the distance at which the tidally disruptive force exerted by the planet is equal to the gravitational force holding a satellite together. Material within this limit cannot combine to form a large *planetesimal and slowly grinds itself up by interparticle collisions.

The occultation of a star by Uranus in March 1977 indicated that it has a system of thin rings, and the *Voyager images of Jupiter taken in 1979 showed a very tenuous primary ring 54,000 km from the planet and 5,000 km wide. Voyager 2 images of Neptune taken in August 1989 showed two main rings, 53,000 and 63,000 km from the planet. The very thin rings have been found to be influenced by small *shepherding satellites in nearby orbits. For table of planetary ring systems see page 197.

ring mountains, characteristic features on the Moon and a number of other planetary satellites, being the result of the impact of meteoroids. The impact raises the crust, sometimes in huge multiple rings as in the lunar Apennines and the Mare Centralis. Such features indicate a period of gigantic collisions with debris experienced by the major bodies in the solar system in its first few hundred million years.

Ring Nebula, a *planetary nebula, also known as M57, in the constellation of Lyra. Of ninth magnitude, this beautiful object appears like an elliptical smoke ring in a telescope of even moderate size. The central star ionizes the surrounding gas so that the greenish hue in the inner shell is produced from the light emitted by doubly ionized oxygen. The reddish outer region is due to emission from hydrogen. For photograph see planetary nebula.

ring plains, very old and often large features of the Moon, created some 3.5 billion years ago when lava flooded on to its surface. Darker and relatively smoother than the rest of the surface, they indicate a turbulent episode in the Moon's early life when impacts with cosmic debris were still frequent. These **basins** or lunar *maria are for the most part found on the Moon's near side. The number of impact craters on them, fewer than elsewhere, indicates that they are younger than the lunar highlands. Similar features are found on Mercury and on some of the satellites of the other planets.

rising, the appearance of a celestial object as it rises above the observer's horizon. If the place where the observer stands has an atmosphere, *refraction will cause the object to be seen to rise shortly before it actually does. If the object has a visible disc, the time of rising has to be defined by choosing which point of the object is referred to.

Ritchey–Chrétien optics, a variation of the optics used in the Cassegrain *reflecting telescope which gives a better, less distorted image over a wider *field of view. Developed by the astronomers George Ritchey in the US and Henri Chrétien in France, it corrects spherical aberration and coma, two defects that limit the clarity of an image at distances away from the theoretical focus.

Roche limit, the minimum distance from the centre of a planet at which small solid particles orbiting in the neighbourhood of the planet can combine and form a sizeable satellite. Outside this distance their mutual gravitational attraction exceeds the tidal force (see *tides) of the planet, whereas inside this limit tidal forces overcome the gravitational attraction between the particles and prevent them combining. The Roche limit, named after the French mathematician Edouard Roche who defined it in 1846, is approximately 2.46 times the planetary radius provided the densities of the particles and the planet are equal. All the four *giant planets have rings inside the Roche limit, and sizeable satellites only occur further out than the Roche limit. A satellite or a stray body passing inside the Roche limit would be disrupted only if the rigidity of its constituent material were overcome by tidal stresses.

Roche lobe, a region of space around one of two massive objects moving about one another in a circular orbit under *gravitation, as in the case of a close binary star. The lobe about each component is bounded by a surface of equal gravitational *potential which rotates with the binary. The two lobes form a figure-eight shape, meeting at the inner (L_2) *Lagrangian point. In a detached system, neither component fills its Roche lobe. In a semi-detached system, one component (usually the less massive) fills its Roche lobe; the star cannot expand further as any attempt to do so results in material being lost in the direction of its companion, through the inner Lagrangian point. In a contact binary, both components fill their Roche lobes, or overfill them and share a common atmosphere; such systems form the W Ursae Majoris type of *eclipsing binaries. Roche lobes were first investigated by the 19th-century French mathematicians Edouard Roche and Joseph Louis Lagrange. Their

That the Earth **rotates** on its axis with respect to the celestial sphere can be seen from this time exposure of star trails over the dome of the 4.2 m William Herschel telescope at the La Palma Observatory. The camera is pointing to the Pole Star so that the stars trace out circular arcs as a result of the Earth's motion.

application to binary stars is a more recent development. (For an illustration of Roche lobe see page 138.)

Roemer, Ole (or **Olaus) Christensen** (1644–1710), Danish astronomer, mathematician, and designer of scientific instruments who provided the first reliable method of calculating the *velocity of light. From 1672 to 1681 he worked in Paris, making many observations of Jupiter as it eclipsed its moons. He found that the time between the eclipses depended upon the Earth's distance from Jupiter and deduced that this was due to the time that light took to travel from Jupiter to the Earth. He published the results of his observations in 1676, giving his calculated value for the speed of light as 225,000 km/s (140,000 miles per second). Although this differs from the currently accepted value, it was due to inaccuracies of measurement rather than to any defect in his method.

Roque do los Muchachos Observatory *La Palma Observatory.

Rosette Nebula, a *nebula in the constellation of Monoceros. It is important because of the presence within it of small round dark areas, taken by many astronomers to be stars condensing out of the gas and dust. The shell of gas is ionized by the two bright stars in the star *cluster within it.

rotation, in astronomy, usually the turning of a celestial object about an axis fixed in it, as for example the rotation of the Earth once in 24 hours. The time taken for one complete rotation with respect to the stellar background is the *sidereal period of rotation. Many of the solar system's satellites have periods of rotation equal to their periods of *revolution about their planets. This means that these motions are *synchronous.

Royal Greenwich Observatory (RGO), an *observatory founded by King Charles II and built by Christopher Wren in 1675 on the foundations of Greenwich Castle in a royal park east of London. Its original purpose was to produce accurate *star charts and *ephemerides of lunar and planetary motions so that they could be used at sea to enable longitude to be calculated correctly. The first director and Astronomer Royal was *Flamsteed. In 1767 the observatory began publishing the annual *Nautical Almanac*, a compendium of navigational information which is still being produced today. James Pond, Astronomer Royal from 1811 to 1835, introduced the time ball, a visual signal to indicate noon. The meridian at Greenwich of the Airy Transit Circle, an instrument for determining times of *transit of stars, has become the *prime meridian of the world, separating the eastern hemisphere from the western hemisphere. Greenwich has always been associated with the accurate measurement of time and was responsible for the testing of the marine chronometers constructed by John Harrison that eventually solved the problem of measuring longitude when at sea. The transit of the mean sun across the Greenwich Meridian is the basis for *Greenwich Mean Time (GMT) on which the international time zone system is based. The observatory also developed other aspects of observational astronomy and contained a 12.75-inch (32 cm), a 26-inch (66 cm), a 28-inch (71 cm), a 30-inch (76 cm), and a 36-inch (91 cm) telescope. In 1946 Herstmonceux Castle in Sussex was purchased and the observatory moved there over the following years. In the early 1980s the *Isaac Newton telescope was relocated to the *La Palma Observatory and by 1990 the observatory staff had moved to Cambridge (UK) from where the Isaac Newton telescope can be remotely controlled.

RR Lyrae star

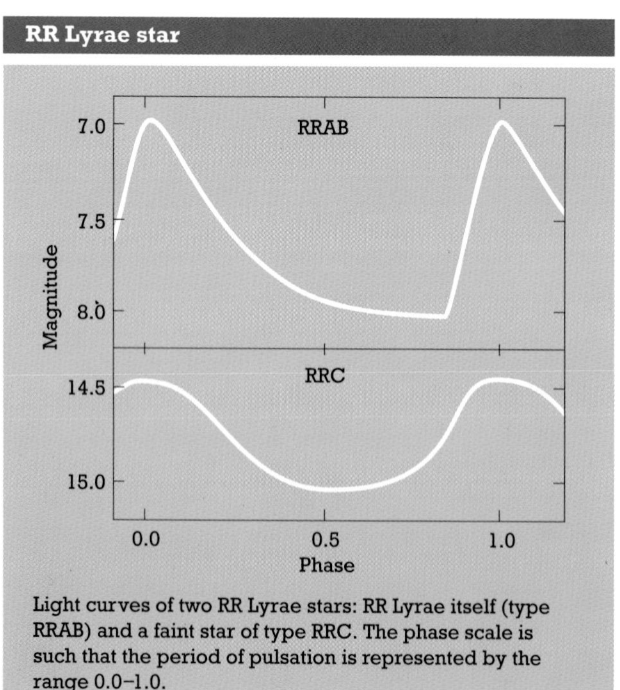

Light curves of two RR Lyrae stars: RR Lyrae itself (type RRAB) and a faint star of type RRC. The phase scale is such that the period of pulsation is represented by the range 0.0–1.0.

Royal Observatory, Edinburgh, an *observatory founded in 1818 and given its royal charter four years later. It was originally built on Calton Hill, Edinburgh, but in 1896 the observatory moved to Blackford Hill, south of the city, to obtain clearer skies. In the early 20th century the observatory became a centre for spectroscopic astrophysics and high-precision photographic *photometry. Research concentrated on the development of autoguiders and machines for automatically measuring the brightness and position of images on photographic plates. Instruments were also constructed to obtain ultraviolet observations from space, these being launched on the *European Space Agency's TD-1A spacecraft. The observatory is responsible for the UK 1.2 m (47-inch) Schmidt telescope at the *Siding Spring Observatory in Australia and the UK 3.8 m (150-inch) infra-red telescope sited at *Mauna Kea in Hawaii.

RR Lyrae star, a pulsating *variable star of short period (0.2–1.2 days), commonly found in globular *clusters and in the central spherical bulge of the Galaxy. They may vary by 0.2–2 *magnitudes. RR Lyrae stars are hot *giant stars on the horizontal branch in the *Hertzsprung–Russell diagram, with mean absolute magnitude about 0.75. They are also easily recognized in nearby galaxies and, using the *period–luminosity law, are useful distance indicators out to 200 kiloparsecs. The variations probably have a similar mechanism to the *Cepheid variable stars. RR Lyrae itself varies between magnitudes 7.1 and 8.1 in a period of 0.57 days. It has an asymmetrical light curve with a steep ascending branch (subtype RRAB). Other stars that have symmetrical curves form subtype RRC. Often the shape of the light curve and period vary periodically; this is the **Blazhko effect**.

Russell, Henry Norris (1877–1957), US astronomer and astrophysicist, who developed the work of *Hertzsprung in

RV Tauri star

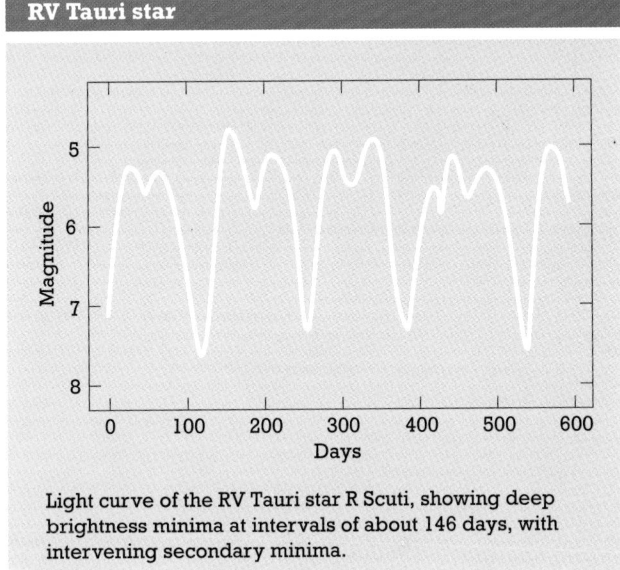

Light curve of the RV Tauri star R Scuti, showing deep brightness minima at intervals of about 146 days, with intervening secondary minima.

graphical form in what is now known as the *Hertzsprung–Russell diagram. This is a graph of the relation between the surface temperature (or colour) and absolute brightness of stars. Russell also conducted extensive work on binary stars.

RV Tauri star, a *supergiant *variable star resembling a *Cepheid variable star of W Virginis subtype, but with longer period and alternating deep and shallow minima. This behaviour may be due to the presence of pulsations in more than one harmonic mode, that is, more than one set of pulsations superimposed upon one another. The periods, measured from one deep minimum to the next, range from 30 to 150 days, and the light changes may amount to three or four *magnitudes. Some stars have constant mean magnitude (subtype RVA), but in others there is a slow periodic change in the mean magnitude (subtype RVB). There are also irregularities like those of semi-regular variables. Some 120 RV Tauri stars are known. RV Tauri itself varies between magnitudes 8.8 and 11 in a period of 79 days.

S

Sagitta, the Arrow, the third-smallest *constellation. It lies in the northern half of the sky and was one of the forty-eight figures known to the ancient Greeks; it represents an arrow that, according to various myths, was shot by Apollo, Hercules, or Cupid. It lies in a dense part of the Milky Way where novae are common, but its brightest stars are of only fourth magnitude and its arrow shape is best seen in binoculars. It contains the globular cluster M71.

Sagittarius, the Archer, a *constellation of the zodiac, one of the forty-eight constellations known to the ancient Greeks who identified it as Crotus, a son of Pan and the inventor of archery. Sagittarius contains rich Milky Way star fields towards the centre of the Galaxy; the exact centre of the Galaxy is believed to coincide with the radio source *Sagittarius A. Beta Sagittarii is a wide pair of unrelated fourth-magnitude stars, visible separately to the naked eye. W and X Sagittarii are both Cepheid variable stars that range from fourth to fifth magnitude every seven days. RY Sagittarii is the southern equivalent of R Coronae Borealis, dipping suddenly and unexpectedly from sixth to fourteenth magnitude. M22 (NGC 6656) is a superb globular *cluster for binoculars, while M23 (NGC 6494) is a large and rich *open cluster for binoculars. In all there are fifteen Messier objects in Sagittarius, more than any other constellation; they include the *Lagoon Nebula; the *Omega Nebula (known also as the Swan Nebula); and the *Trifid Nebula. The Sun passes through the constellation from mid-December to mid-January.

Sagittarius A, the unusual *radio source at the centre of the Galaxy. It lies in the constellation of *Sagittarius, at a distance of 10,000 parsecs (30,000 light-years), behind dense clouds of gas and dust which prevent optical observations of the region. Radio maps reveal a puzzling structure: a small spiral some 1.6 parsecs (5 light-years) across surrounded by many fine filaments. At the very centre is a compact radio source only 2 light-hours (2 billion km) across. Some astronomers speculate that this is the central energy source responsible for activity in the nucleus of the Galaxy. In this theory the spiral is material falling into a central *black hole. This would make Sagittarius A a scaled-down version of the powerful central energy sources of *quasars.

Sagittarius arm, a *spiral arm of the Galaxy lying nearer to the galactic centre than the *Orion arm in which the Sun is situated. It is mapped out in the vicinity of the Sun by plotting the distribution of bright stars of *spectral classes O and B and the associated emission nebulae contained in it.

Sagittarius B2, a powerful *radio source near the galactic centre, and one of the most active sites of star formation in the Galaxy. The young stars are forming from a massive cloud of gas and dust some 12 parsecs (40 light-years) across. The cloud contains over a million solar masses of material (enough for an entire globular *cluster), but only a small percentage is expected to form stars. The gas is mainly hydrogen molecules and helium atoms, with small traces of other molecules. Over fifty different molecules have been detected in the cloud by *radio astronomy, and radio *spectral lines have also been detected from as yet unidentified molecules. The known molecules include simple organic molecules such as formic acid, acetaldehyde, and ethyl alcohol.

Saha, Meghnad (1893–1956), Indian astrophysicist who, in 1920, formulated an equation, now known as *Saha's equation, describing the ionization of gases in stellar atmospheres. His work in this field has greatly clarified the relationships between stellar spectra and the temperature, pressure, and composition of stellar atmospheres and has provided a means of determining the temperature of a star and the abundance of *elements within it.

Saha's equation, an important formula derived by *Saha, relating the fraction of atoms ionized in a stellar atmosphere to the temperature of the gas and the electron density, that is, the number of free electrons per unit volume in the atmosphere. Since its development in 1920, Saha's equation has provided a remarkable insight into understanding the structure of stellar atmospheres.

Saros, a period of time between similar *eclipses of the Sun and Moon. The Babylonians of the 2nd–1st millennia BC discovered from their eclipse records that eclipses of the Sun and Moon occur in almost exactly similar detail in a period of 18 years and 10 or 11 days. This period of time, called the Saros, is almost equal to 223 *lunations (6,585.32 days) and 19 *synodic periods of the Moon's *node (6,585.78 days). In addition, 239 anomalistic *months equal 6,585.54 days. These periods ensure that at the end of a Saros the geometrical configuration of the Sun, Moon, and Earth is almost exactly the same as that occurring at the beginning of the Saros, so that an eclipse is then always repeated.

satellite, artificial, an artificial object in orbit round a celestial body such as a planet or a moon. Although the first artificial satellite, Sputnik 1, was launched on 4 October 1957, the idea of an artificial satellite goes back as far as Newton who showed in the *Principia* (1687) that if an object were projected above the Earth's atmosphere with a speed in excess of 8 km/s (5 miles per second) in a direction roughly parallel to the Earth's surface it would go into an elliptical orbit about the Earth. In the early 20th century, *astronautics pioneers such as Constantin Tsiolkovsky, Robert Goddard, Robert Esnault-Pelterie, and Hermann Oberth showed that by means of chemical rockets this ambition could be realized. Since the launching of Sputnik 1, thousands of artificial Earth satellites have been launched, the vast majority by the USA and the Soviet Union who have also put satellites into orbit about Mars, Venus, and the Moon. Other nations including China, France, India, and the UK, have also successfully launched satellites.

Once in Earth orbit a satellite remains there for ever unless any part of its orbit is within the Earth's atmosphere, or it has a motor capable of decreasing its orbital velocity sufficiently to cause it to re-enter the atmosphere. Atmospheric drag causes the satellite to spiral in, the rate of decay of its orbit depending on the density of the air it passes through and the ratio of the satellite's cross-section to its mass. Valuable information about the structure of the atmosphere with height and the daily and seasonal changes that occur in it have resulted from continuous monitoring of the changes in the orbits of artificial satellites over the past thirty years. In addition, such tracking has also yielded information about the precise shape and structure of the Earth since the nature of the Earth's gravitational field depends on the distribution of mass throughout the planet. It is only if the Earth were a point mass or a perfect sphere with its material uniformly distributed that the orbit of the satellite would be a fixed ellipse neglecting the effects of air resistance and certain other disturbing forces of small magnitude. Fortunately the orbital changes due to the Earth's departure from a perfect sphere are different from those due to atmospheric drag and so the precise shape and structure of the Earth have been deduced.

Satellites carry scientific instruments, radioing the information collected by such instruments back to the tracking stations on Earth. There is a wide variety of uses for artificial satellites, including meteorology and weather forecasting, long-range communication, military surveillance, navigation, location of Earth resources, and astronomy. Satellites in orbit round the Moon and Mars have sent back detailed photographs of the surfaces of these bodies. A particular advantage of using an artificial satellite as a platform from which to study the universe is that it is above the Earth's atmosphere. The whole range of electromagnetic radiation given out by the Sun and other celestial objects therefore impinges upon the satellite instruments. The new branches of X-ray, gamma-ray, ultraviolet, and infra-red astronomy are only fully realistic using artificial satellites and many new discoveries about the universe have been made through their pursuit.

Sputnik 2, launched in November 1957 carried the dog Laika. Since then many humans, animals, insects, and plants have been put into orbit about the Earth to study the effects of weightlessness on living organisms. Space medicine has made tremendous advances and it is now clear that the dream of permanently manned space stations in orbit about the Earth is capable of realization before the end of the 20th century. Cities in space could well be a feature of the 21st century.

satellite, natural, any natural body that orbits around a planet. The sixty known satellites of the solar system span a wide range of sizes. The seven largest are the Earth's *Moon, the four *Galilean satellites of Jupiter, *Titan, and *Triton. These each exceed 2,500 km in diameter and are really small 'worlds', with complex geological histories and sometimes even an atmosphere. All the other satellites have sizes comparable to those of the *asteroids, namely from about 10–1,500 km. Natural satellites also display a variety of other characteristics. Their compositions can be rocky or icy, their shapes can vary from nearly spherical to very irregular, and their surface ages may be very ancient and heavily cratered or may still be subjected to changes originating in their interiors. The sizes of their orbits vary from less than two planetary radii to a few hundred planetary radii from the centre of the parent planet, and they may take from several hours to more than a year to orbit their planet. It is thought that some satellites have been captured by the gravitational attraction of their planet. These satellites have irregular orbits and some orbit the planet in the opposite direction to the planet's orbital revolution about the Sun. This is known as **retrograde motion**. Their orbits may be highly inclined to their planet's orbit about the Sun and may have large eccentricities consistent with their origin by capture. By contrast the extensive systems of satellites with regular orbits around the four *giant planets were probably formed from gas and dust surrounding the parent planet, thus mimicking, on a smaller scale, the formation of the planets in the solar nebula. The smaller moons, those less

When Galileo first observed **Saturn** through his telescope in 1610 he thought it was a triple planet with huge stationary moons on either side. Since then the nature of its ring system has been continually studied through Earth-based telescopes, and the two Voyager missions to Saturn have greatly increased our knowledge of the planet, its rings, and its satellites.

than a few hundred kilometres across, have irregular shapes and probably originated as fragments from destructive collisions between larger bodies. In the outer solar system they often orbit in the proximity of planetary *rings. The evolution of the orbits of satellites is dominated by tidal effects, which in most cases has synchronized their rotational and orbital periods and caused significant changes to their semi-major axes. The elements of *orbit of outer satellites, particularly eccentricities, are subject to considerable perturbation by the Sun. Several pairs, and even triplets, of satellites have orbital periods that are related in a simple way to each other, that is, the ratios of their periods are simple fractions. For example, Jupiter's satellite *Europa has a period almost exactly one-half that of *Ganymede. Such pairs are said to be locked in resonance. For table of planetary satellites see page 198.

Saturn, the sixth *planet from the Sun and the outermost planet clearly visible with the unaided eye. Saturn is the second largest planet in the solar system. Some of its main physical characteristics like size, composition, internal structure, and atmosphere suggest an overall resemblance to *Jupiter but the results of spacecraft investigations have shown that there are also important differences between the two planets. Saturn, like Jupiter, is mostly composed of light elements such as hydrogen and helium and has an overall density less than that of water. It has an inner rocky core of about ten to twenty times the mass of the Earth, roughly the same size as the core of Jupiter. However, the percentage of helium in the envelope surrounding the core is lower than that of Jupiter. This may be due to helium sinking into the interior, which could also be the explanation of the internal energy source that Saturn, like all other giant planets, has. This energy source drives mass movements in the atmosphere that show up as parallel bands, composed of darker **belts** and lighter **zones**, in motion with respect to each other. A large white spot appears every 27–30 years. It is probably a giant storm containing frozen ammonia. The large size of Saturn's core with respect to the envelope of light elements suggests that when it formed there was less gas available than for Jupiter, thus leading to the smaller mass of Saturn. A peculiarity of the magnetic field of this planet is the very small angle between the line joining the magnetic poles and the planetary rotation axis. Saturn's equator has an inclination to its orbit slightly larger than that of the Earth, thus inducing seasonal effects in the behaviour of the atmosphere. *Saturn's rings are well known and were explored in great detail by the *Voyager missions. The visibility of the ring system changes in a 29.5-year cycle. When presented edge-on to the Earth, the rings cannot be seen at all well. The satellite system is also very interesting. It is composed of seventeen bodies, of which the innermost ones orbit at the outskirts of the ring system, having strong gravitational interactions with the rings themselves. Proceeding outwards there is a group of icy satellites of small to intermediate size, some of which strongly influence each other's orbits. *Titan is the second-largest satellite of the solar system and has a substantial atmosphere. There are three more satellites beyond Titan, one of which is on a retrograde orbit, that is, it moves about Saturn in the opposite direction to Saturn's orbital motion about the Sun.

Saturn's rings, an extended, disc-like structure, easily visible with even a small telescope, that surrounds *Saturn. *Galileo was the first to observe them in 1610, but did not understand that what he was seeing in his telescope was actually a set of planetary *rings; this was achieved almost half a century later by the Dutch astronomer Christian Huygens. Another few decades later Giovanni Domenico *Cassini discovered that the ring appeared to be separated into two parts by a dark band; when this was eventually recognized to be a gap in the distribution of material, it became known as **Cassini's division**. In the 19th century it was shown by the British physicist James Clerk Maxwell that the rings of Saturn must be composed of a very large number of small particles moving about the planet in independent orbits, and that the former conception of a solid ring could not be reconciled with its long-term stability. Recent ground-based observations, and above all the results of the *Voyager spacecraft missions that have explored the Saturnian system, have shown that most of the particles composing the ring system of the planet are made out of water ice, and that the dynamical structure of the system is extremely complex, with the main classical rings finely subdivided into hundreds of ringlets, some of which are elliptical. The quantitative explanation of all the details of the Saturnian ring system is currently among the most challenging problems of celestial mechanics.

scattering, the random changes in direction of waves when they encounter a material, solid, liquid, or gaseous, which is made of particles of various sizes. Both sound waves and *electromagnetic radiation can be scattered if they meet appropriately sized particles. Stars **twinkle** because the rays of light from such tiny angular sources are scattered by turbulent small eddies in the Earth's atmosphere, whereas planets, being bigger in angular size, do not twinkle unless they are near the horizon. The reddening of objects near the horizon is due to the fact that atmospheric particles scatter blue light better than red, which is also the reason that our sky is blue. Even the interplanetary and interstellar media, though of very low densities, scatter electromagnetic radiation at radio wavelengths giving rise to so-called interplanetary and interstellar **scintillation**. Measurement of such scintillations of radio waves from some radio sources has led to a determination of their angular sizes.

Schmidt telescope *reflecting telescope.

Schönberg–Chandrasekhar limit (Chandrasekhar–Schönberg limit), the maximum mass of helium that can be produced at the core of a *dwarf star by *nuclear fusion. When this limit, about 12 per cent of the star's total mass, is exceeded, the temperature can no longer remain uniform throughout the core, and the high temperature needed just outside the core in order to maintain hydrogen burning can no longer be maintained by small changes in the star's structure. Consequently the star begins to evolve rapidly to become a *giant star. The Schönberg–Chandrasekhar limit is named after the German astronomer Erich M. Schönberg and the Indian astrophysicist *Chandrasekhar, whose calculations identified it; it is distinct from the *Chandrasekhar limit.

Schwarzschild radius, the distance from the centre of a *black hole within which the gravitational field is so strong that not even *electromagnetic radiation can escape from it. It was first calculated by the German astronomer Karl Schwarzschild in 1916. The escape velocity V from the gravitational pull of a mass M at a distance R from the centre of the mass is given by

$$V^2 = 2GM/R$$

where G is the gravitational constant. If V is equal to the velocity of light, c, then R is the Schwarzschild radius for the mass M and is

$$R = 2GM/c^2.$$

scintillation *scattering.

Scorpius, the Scorpion, a *constellation of the zodiac, representing the scorpion that stung Orion to death. Its brightest star is *Antares. Beta Scorpii, called Graffias from the Latin meaning 'claws', is divisible into a double of third and fifth magnitudes through small telescopes. Near this star lies the bright X-ray source *Scorpius X-1. There are three naked-eye doubles in Scorpius: Omega, Mu, and Zeta Scorpii. Nu Scorpii is a quadruple star similar to the famous 'Double Double' in *Lyra. M4 (NGC 6121) is a large globular cluster, one of the closest to us, nearly 2,200 parsecs (7,000 light-years) away. The curving tail of Scorpius extends into rich Milky Way star fields, where lie the outstanding naked-eye *open cluster M7 (NGC 6475) and the binocular cluster M6 (NGC 6405). NGC 6231 is a cluster like a mini-Pleiades. The Sun passes through the constellation at the end of November.

Scorpius X-1, a powerful X-ray source discovered in 1962 when long-wavelength X-rays were detected using a large-area Geiger counter flown on a rocket. Subsequent research has enabled the source to be identified with a short-period *X-ray binary system with an orbital period of about 0.8 days.

Sculptor, the Sculptor, a *constellation of the southern sky introduced by the 18th-century French astronomer Nicolas Louis de Lacaille to represent a sculptor's workshop. Its brightest stars are of only fourth magnitude. Sculptor contains NGC 253, a ninth-magnitude galaxy seen nearly edge-on in the *Sculptor Group of galaxies, and also the Sculptor System, a dwarf member of our *Local Group of galaxies.

Sculptor Group, a group of galaxies of which the largest, NGC 253, is visible in binoculars. It is as massive as our own galaxy but intrinsically about four times brighter. About 2 million parsecs (7 million light-years) away, the Sculptor Group contains a number of galaxies visible in quite modestly sized telescopes.

Scutum, the Shield, the fifth-smallest *constellation. It lies just south of the celestial equator. Scutum was introduced by Johannes Hevelius (1611–87) in honour of the King of Poland. Its brightest stars are of only fourth magnitude but it lies in a rich area of the Milky Way and contains M11 (NGC 6705), the Wild Duck Cluster, so called because its shape resembles a flight of ducks. Delta Scuti, of fifth magnitude, is the prototype of a class of variable stars that pulsate in size every few hours giving rise to small changes in brightness.

seasons, periods of the year determined by the tilt of the Earth's rotational axis to the orbital plane. The Earth's axis is inclined to the normal to the *ecliptic (orbital) plane at an angle of 23 degrees 26 minutes, which is called the obliquity of the ecliptic. In the course of the year the Earth's axis maintains a nearly fixed *sidereal direction and so its orientation to the Sun varies, as shown in the figure. This causes seasonal variations in the altitude of the Sun in the sky and in the duration of the daylight hours. These variations are more pronounced at high latitudes. At the *equinoxes the whole Earth has exactly 12 hours of daylight, while at a *solstice the hours of daylight have their maximum extent in one hemisphere, their minimum in the other, and at this time one polar region has continuous daylight while the other has continuous darkness. The Earth's orbit is not circular but is an ellipse of small eccentricity with perihelion on about 3 January. The Earth is then about 3 per cent closer to the Sun than in July, but since the Earth moves most rapidly at perihelion, the northern winter is about three days shorter than summer. In the northern hemisphere the seasons spring, summer, autumn, and winter commence when the Sun reaches the First Point of Aries, Cancer, the First Point of Libra, and Capricorn, respectively. In the southern hemisphere the arrival of the Sun at these same four points of the ecliptic mark the beginnings of autumn, winter, spring, and summer.

Seasonal variations also take place on other planets, their extent depending on the inclination of the planet's rotational axis. They are most extreme for Uranus but most clearly observed for Mars where the polar caps vary conspicuously in size throughout the Martian year.

Secchi, (Pietro) Angelo (1818–78), Italian Jesuit priest and astronomer who pioneered the recording, measurement, and classification of stellar spectra. He found that he could classify the vast majority of stars into a small number of *spectral classes and these were later refined into the currently used *Draper classification.

Secchi classification *Draper classification.

secular, in astronomy, a term referring to a change in an element of an *orbit which increases indefinitely, that is, from age to age. Thus a secular velocity is proportional to the time passed while a secular acceleration is proportional to the square of the time passed. In the elements of the planetary orbits such quantities exist, though in the case of the *semi-major axes, the *eccentricities, and the inclinations,

Seasons

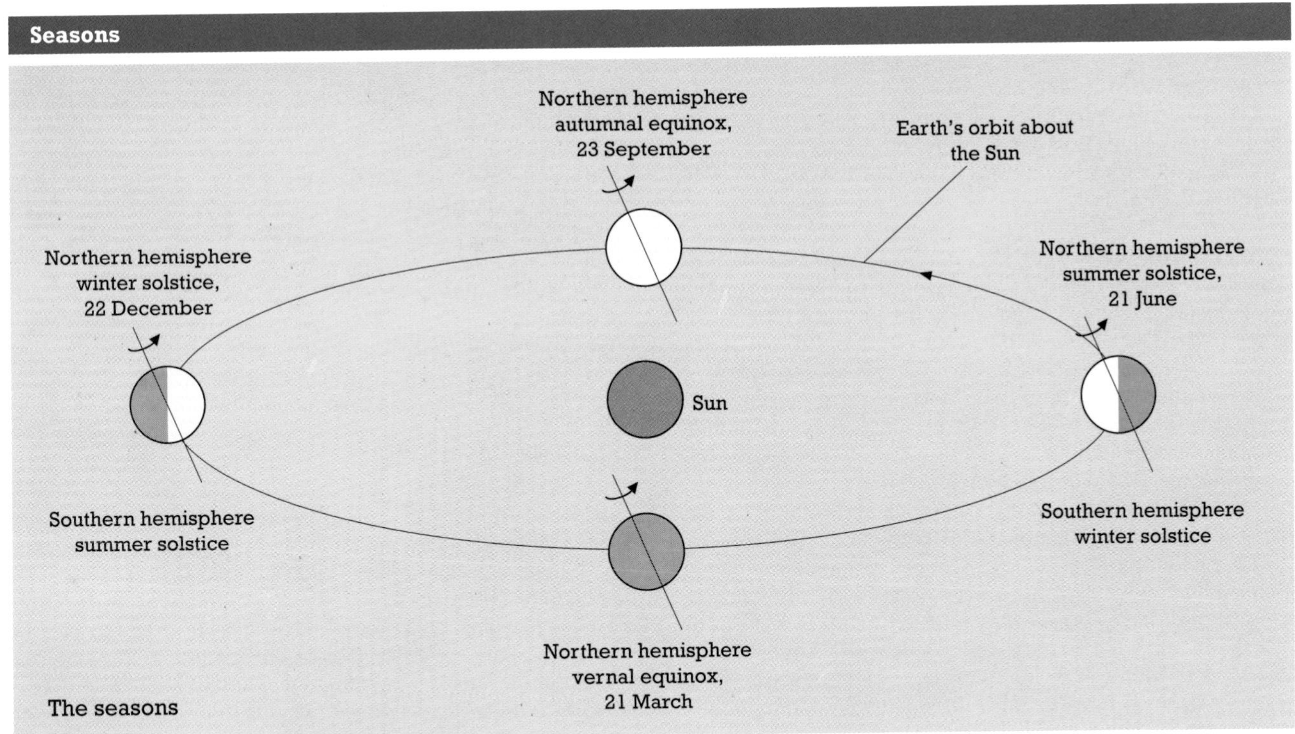

The seasons

they may be merely approximations to the actual elements which may oscillate over long periods of time.

seeing, the quality of an Earth-based telescopic image of a celestial object. The light from a star at a great distance can be considered to arrive in parallel rays and the image of the star observed through a telescope forms a *diffraction pattern with concentric rings. However, turbulence in the Earth's atmosphere scatters the light rays and blurs out the sharpness of the diffraction pattern into an amorphous blob known as the seeing disc. The quality of an observing site is often described in terms of the sizes of seeing disc which are encountered; a disc of the order of 1 arc second may be considered as good seeing. In observational terms, seeing conditions are usually described on the five-point 'Antoniadi' scale, varying from 1 (perfect, steady seeing) to 5 (very poor).

segmented mirror telescope *multiple-mirror telescope.

seismology, the study of the propagation of sound waves through the interior of a planet or satellite. These waves are generated by earthquakes or man-made explosions and a collection of well-distributed seismographs, which are instruments for measuring the horizontal and vertical movement of the planetary surface, enable the nature and time of arrival of the waves to be recorded. Three types of waves occur, these being the compressional waves (P-waves, involving a push–pull movement along the direction of propagation), the shear waves (S-waves, involving perpendicular movement), and the longitudinal waves (L-waves) that travel along the surface. All travel at different speeds and suffer different degrees of absorption. The wave velocity increases with rock density and the waves are refracted, that is, bent when the density changes. Wave characteristics can be interpreted in terms of the subsurface structure of the crust and mantle and also the physical conditions acting on these rocks. The outer boundary of the Earth's core provides a major seismic discontinuity because, being liquid, the S-waves cannot travel through it. Seismic waves were detected by the instrument packages that were deployed on the Moon's surface by *Project Apollo. These waves were produced by moonquakes and also by the impact of used pieces of space hardware.

selenography, the mapping of the Moon. This subject started in earnest with the invention of the telescope, the first telescopic maps being produced by *Galileo and by the British astronomer Thomas Harriot in 1611. The Belgian cartographer Michael van Langren (1645) drew a map with surface shading, topographical features, and a viable nomenclature. This was improved by Francesco Grimaldi, and in 1650 Giambattista Riccioli introduced a new nomenclature that has stood the test of time. The large light and dark areas were named after watery expanses, such as oceans, seas, and lakes, and the various states of the climate and weather, for example, Mare Frigoris. Features such as craters and mountains were associated with famous philosophers, scientists, mathematicians, and astronomers. About 100 years later the development of photography in the mid-19th century eventually led to the production of new atlases by the Dutch–American astronomer Gerard Kuiper, the US physicist David Alter, and the Japanese astronomer Shotaro Miyamoto. The first maps of the far side of the Moon were made after the Soviet Luna 3 spacecraft produced photographs in 1959. Detailed lunar mapping has continued with images from Orbiter and *Project Apollo missions.

selenology, the study of the Moon. The subject was essentially founded by *Galileo who first attempted to chart its topographical irregularities. The nature and origin of the surface features, particularly the *craters, was a topic of considerable debate. The 17th-century British scientist Robert Hooke thought they were bursting bubbles, William *Herschel favoured volcanic activity and as the space age approached, the impact cratering theory came to the fore.

Even today the origin of the Moon and its evolution with time are still being studied diligently.

semi-diameter, half the angular diameter of any celestial object displaying a finite disc such as the Sun, the Moon, a planet, or satellite. It is the angle subtended at the observer by a radius of the object at right angles to the line joining the object to the Earth.

semi-major axis, half the major axis of the *ellipse described by the orbit of a planet around the Sun according to the first of *Kepler's Laws, when the perturbations of the other planets are neglected. For a circular orbit it is the radius of the circle; for an orbit of low eccentricity it is approximately the average distance from the Sun. The *period of the orbit depends only upon the semi-major axis. Similar considerations apply to satellites in elliptical orbits with a focus at the centre of mass of the planet. When the planetary perturbations are taken into account, the semi-major axes of the orbits of the planets change slowly with time, oscillating around an average value; the minute changes to this average have been computed only recently. On the other hand, for strongly unstable orbits (described by *chaos theory) such as those of *comets and *Apollo–Amor objects, the semi-major axes can change significantly as a result of close approaches with the major planets.

Serenitatis Basin, a typical circular *mare on the Moon, 680 km in diameter. It was visited by Apollo 17 and Lunokhod 2. The Lunokhod was an unmanned vehicle that roamed around Serenitatis for four months, producing photographic panoramas, magnetometry readings, and X-ray analyses of the soil. The basin is noted for the striking colour contrasts of its lava and the curved *rilles around its edge. The outer edges are also highlighted by a dark outer ring of basalt that extends south-eastwards into the adjacent *Tranquillitatis Mare. The centre of the mare is crossed by a light-coloured ray that seems to originate from the crater *Tycho, 2,000 km to the south-west. The mare surface is definitely spherical but the centre of the sphere does not quite coincide with the present centre of mass of the Moon.

Serpens, the Serpent, a *constellation straddling the celestial equator, one of the forty-eight constellations known to the ancient Greeks. It represents a huge snake held in the hands of Ophiuchus who was a healer credited with the ability to bring people back to life, and the fact that snakes shed their skin was a symbol of rebirth. Serpens is a unique constellation because it is split into two halves, the head and the tail, either side of *Ophiuchus; however, it is regarded as a single constellation. Its brightest star, third-magnitude Alpha Serpentis, is known as Unukalhai, from the Arabic meaning 'serpent's neck'. M5 (NGC 5904) is a globular *cluster for binoculars. The most celebrated object in Serpens is M16 (NGC 6611), an *open cluster embedded in the Eagle Nebula, although the nebula shows up well only on long-exposure photographs.

SETI (search for extraterrestrial intelligence), a series of projects to detect signs of intelligent *extraterrestrial life. So confident are many scientists that life must exist in parts of the universe other than Earth that efforts have been and are being made to detect signs of the existence of intelligent species and to bring our existence to their attention. The *Ozma Project, in 1960, was the first attempt to use radio astronomy to search for galactic civilizations transmitting radio signals. Because the *twenty-one centimetre line of radio emission from interstellar hydrogen would be known to any civilization at or beyond our stage of scientific knowledge, scientists confined their attention to wavelengths near that value. Two nearby stars Tau Ceti and Epsilon Eridani were 'listened to' without success. In 1971 the US astronomer Bernard Oliver and the British physicist John Billingham devised the *Cyclops Project, a plan to build 1,000 steerable radio dishes, each of 100 m (330-foot) diameter, the combined array being potentially capable of radio communication with an intelligent species anywhere in the Galaxy. The project has never been funded. Other programmes of radio listening have been carried out not only in the USA but also in the Soviet Union and other countries. More than fifty have been undertaken in the past twenty-five years logging more than 120,000 hours of observations and using increasingly sophisticated apparatus and methods. None so far has obtained positive results. *NASA is currently preparing a comprehensive SETI programme for the 1990s and the *IAU in 1982 created 'Commission 51—Search for Extraterrestrial Life' to co-ordinate efforts.

It is not widely appreciated that deliberately transmitted signals are not the only radio waves to be looked for. In fact if there are any other technological species at our stage of development within 12 parsecs (40 light-years) of the Earth, they probably already know of our planet's existence, its distance from the Sun, its period of revolution about the Sun and period of rotation, as well as the existence of the human race. During that time scientists have been producing high-intensity radar bursts and television and radio programmes which would have, to any distant observer, made the Sun appear to be an unusual star in the radio part of its electromagnetic spectrum. One attempt to tell other intelligent species of the existence of the human race was the placing of engraved plaques in the *Pioneer 10 and Pioneer 11 spacecraft, and video discs in *Voyager 1 and Voyager 2, since, on completion of their missions, all four spacecraft were destined to leave the solar system and travel between the stars. The recorded information should enable any intelligent civilization to identify the solar system and the planet in it from which the spacecraft came.

SETI poses a difficult dilemma for the human race. If it turns out we are alone, then are we merely a highly improbable accident? On the other hand, if other intelligent species exist, we must be prepared to accept that we may be the most primitive civilization for we have been in existence for merely a moment in the life of the universe.

setting, the disappearance of a celestial object as it descends below the observer's horizon. If the place where the observer stands has an atmosphere, *refraction will cause the object to be seen to set shortly after it actually does. For an object with a visible disc, the time of setting depends upon which point of the disc is chosen.

Seven Sisters *Pleiades.

Sextans, the Sextant, a *constellation on the celestial equator, introduced by the Polish astronomer Johannes Hevelius (1611–87) to represent the sextant with which he measured star positions. It is exceedingly faint (its brightest star is only of magnitude 4.5) and it contains no objects of particular interest.

Seyfert galaxy, a type of *galaxy distinguished by the violent activity at its centre. First discovered by the US

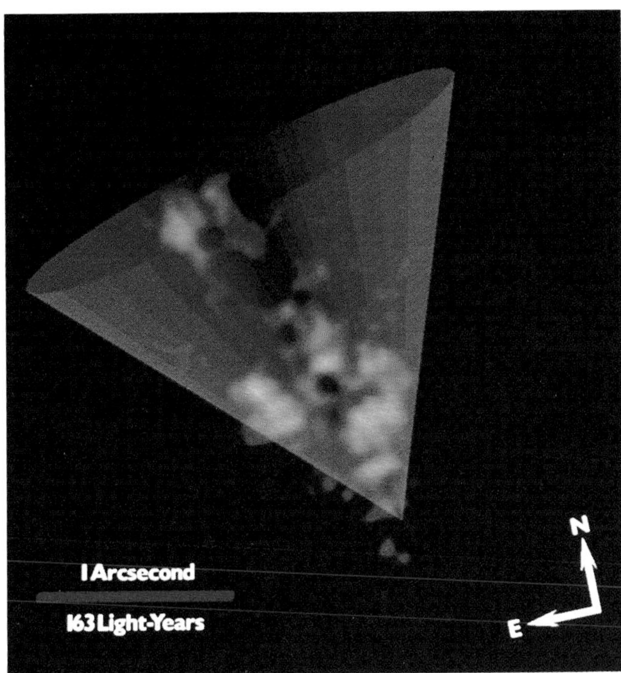

The central part of the **Seyfert galaxy** NGC 1068 contains bright clouds of ionized gas, photographed here by the Hubble space telescope. The clouds are glowing because they are caught in the 'searchlight' of ionizing radiation, represented schematically here by the green cone, beamed out of the galaxy's energetic nucleus.

astronomer Carl Seyfert in 1943, over a hundred examples have since been found. It is possible that Seyfert galaxies, with their bright nuclei and ejection of hot gas from their centres, are temporary stages in the lives of normal galaxies.

shadow bands, parallel bands of shadow seen crossing any light-coloured surface during a total solar *eclipse, shortly before and after the Sun becomes completely hidden by the Moon. They may be caused by the irregular *refraction of sunlight from the thin crescent Sun.

Shapley, Harlow (1885–1972), US astronomer who proposed that the Sun was not at the centre of the universe but at 15,000 parsecs (50,000 light-years) from the galactic centre. (The current estimate is approximately 10,000 parsecs.) In 1911, based on the work of *Russell, he developed a method of determining the size of *eclipsing binary stars by measuring the variation in their magnitude as the components orbit each other. Shapley was the first to propose that *Cepheid variable stars are pulsating stars.

shell star, a star that continuously or at irregular intervals sheds a ring of gas at its equator or a roughly spherical shell of gas. A shell star is a *giant star or *dwarf star of *spectral class B that is rapidly spinning and close to the point of breaking up. Ejection of a shell is usually accompanied by a fade in magnitude, as in Pleione. However, it may give rise to a brightening as in Gamma Cassiopeiae in the 1930s; this star lends its name to a class of *variable stars. In other objects there may be little change in magnitude, the shell being detected only from the star's *spectrum, which is dominated by sharp absorption lines arising in an extensive envelope between the core and the atmosphere of the star. Similar stars whose spectra are dominated by emission lines are sometimes classed separately as **Be stars**, but a given star may show emission or absorption spectra at different times.

shepherding satellites, small natural satellites thought to confine planetary *rings within sharply defined narrow bands. The existence around the giant planets of rings of very small radial width and with sharp edges has suggested that the ring material might be confined against spreading by the gravitational action of these neighbouring shepherding satellites. Such small moons orbiting in the proximity of rings have been observed by the *Voyager probes. In Saturn's system, *Atlas orbits near the outer edge of the A ring, while Prometheus and Pandora appear to confine the F ring. Other examples were found in the systems of Uranus and Neptune. It is possible that more shepherding satellites exist, but are too small to be detected.

shooting star *meteor.

sidereal, the term normally used to describe time periods measured with reference to the fixed stars. For example, the Earth's period of axial rotation relative to the stars is called the sidereal day. The sidereal month and year are defined similarly. The sidereal day is about four minutes shorter than the ordinary solar day, the difference being caused by the Earth's orbital motion. In fact, the number of sidereal days in one year exceeds the number of solar days by exactly one. For many astronomical purposes it is convenient to measure time by the stars. The sidereal time is defined geometrically as the *hour angle of the vernal *equinox. In this way the sidereal day is divided into 24 hours, and clocks

When Voyager 2 encountered Uranus in January 1986 it took these images of two tiny **shepherding satellites**. These are thought to prevent the nine-ring system from spreading out. The rings range in width from 100 km at the widest part of the outermost ring, to only a few kilometres for most of the others.

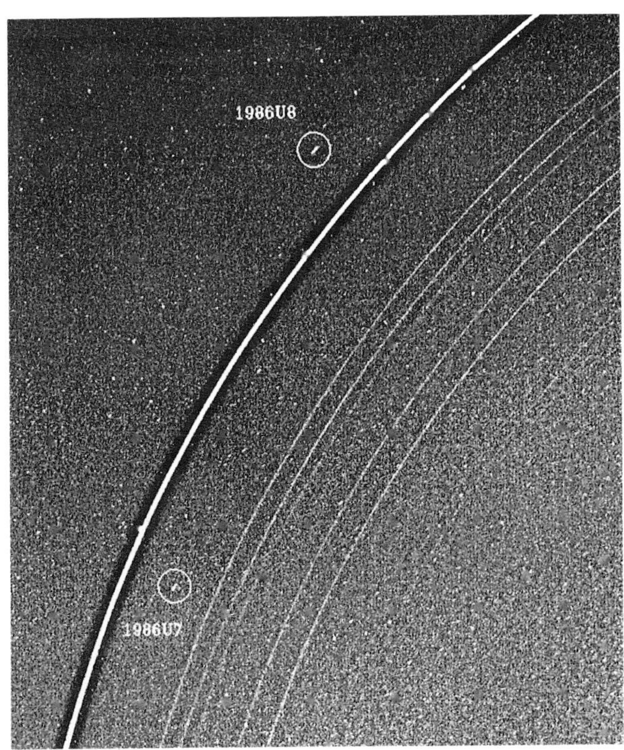

keeping sidereal time will gain four minutes each day on solar time clocks. The sidereal time of *transit of any celestial object is simply its right ascension so that the setting and operating of telescopes is simplified.

siderolite *meteorite.

Siding Spring Observatory, an *observatory located on a mountain top near Coonabarabran, some 300 km (200 miles) due north of Canberra, Australia. Several organizations' telescopes are sited there, including: the 3.9 m (154-inch) **Anglo-Australian telescope** opened in 1975; a 1.2 m (47-inch) Schmidt *reflecting telescope, originally used for a southern sky survey by the *Royal Observatory, Edinburgh, but now under joint British–Australian control; and several telescopes of the Australian National University, including the 2.3 m (90-inch) Advanced Technology telescope, which, at its opening in 1985, was the first telescope to incorporate a thin primary mirror, an alt-azimuth *mounting, computerized control, and a rotating, domeless building in a single, relatively low-cost instrument. In recent years spectacular infra-red images of star-forming regions have been obtained with the Anglo-Australian telescope, and advances in photographic techniques associated with the Schmidt telescope have improved knowledge of the fine and faint structure in extended objects such as the faint shells surrounding elliptical galaxies.

silvering, the process, now obsolete, of applying a silver coating to the mirror of a *reflecting telescope. Modern telescopes use vacuum-deposited aluminium (US aluminum) as the reflective coatings so that the term 'silvering' has now been superseded by the term 'aluminizing'.

singularity, a point in space where the curvature of space is effectively infinite, a situation said to occur in a *black hole. In such circumstances the laws of physics are no longer applicable. In fact an *event horizon surrounding the black hole or singularity prevents observation of events within that horizon.

Sinope, a small satellite of Jupiter. It has a diameter of 36 km and moves in a highly eccentric orbit in the opposite direction to Jupiter's orbital motion, at a mean distance of 23.7 million km in a period of 758 days. Discovered in 1914, it is one of four small moons orbiting Jupiter at much the same distance.

Sirius, the brightest star visible to the naked eye (after the Sun) and one of our nearest stellar neighbours. Located in the constellation of Canis Major, and also known as Alpha Canis Majoris or the **Dog Star**, its magnitude is −1.46. It is a hot *dwarf star at a distance of 2.6 parsecs, with twenty-three times the luminosity of the Sun. It has an eighth-magnitude *white dwarf companion in a 50-year orbit.

Small Magellanic Cloud *Magellanic Clouds.

solar activity *Sun.

solar apex *apex.

solar constant, a measure of the amount of radiant energy emitted by the Sun. Any surface in space exposed perpendicularly to the Sun's direction will intercept a fraction of the Sun's radiant energy. The solar constant is defined to be the energy reaching a square metre of such a surface in one second at a distance from the Sun's centre equal to the mean distance to the Earth. Its value is about 1.36 kilowatts per square metre.

solar cycle *sunspots.

solar eclipse *eclipse.

solar energy, the energy received from the Sun at the Earth's surface. The total **solar radiation** arriving at the Earth's orbit amounts to 1.36 kilowatts per square metre or roughly 10^{14} kilowatts over the Earth's hemisphere. Though this energy governs our climate, winds, and sea-waves the great bulk of it is ultimately re-emitted into space as radiant heat, apart from a tiny fraction which is absorbed by plants and stored chemically in their living tissue and also their fossil remains. Thus, apart from the latent nuclear energy of matter which may be extracted from heavy elements by *nuclear fission or from light elements by *nuclear fusion, the Sun is the ultimate source of all mankind's fuels. Direct utilization of solar energy will be of vital importance to our future, if our energy needs continue to grow, for two reasons. Firstly, resources of fossil and nuclear fission fuels are limited and use of the latter potentially dangerous. Secondly, use of fuel reserves results in pollution of the environment either through waste matter or through waste heat, the latter causing dangerous climatological changes such as smog around cities. Immediate utilization of solar energy either by direct absorption in solar panels or via wave, wind, or hydro-electric power obviates both of these problems by using energy which would otherwise be re-radiated into space. Total harnessing of the solar radiation incident on 0.01 per cent of the Earth's surface would provide 10 kilowatts for each of the world's current population, roughly the current US consumption per person.

solar flare *flare.

solar motion, the speed and direction of the Sun's movement with respect to the local group of some 30,000 stars. The velocity of about 19.5 km/s is in the direction of a point in the constellation of Hercules, and is near the bright star Vega (which is in the constellation Lyra). The direction of the solar motion is called the solar *apex and has coordinates on the *celestial sphere of right ascension 270 degrees, declination +30 degrees. The solar motion is a combination of the Sun's orbital motion about the centre of the Galaxy and the gravitational influence of the local group of stars.

solar parallax, the angle subtended by the Earth's equatorial radius at the centre of the Sun, at the time when the Earth is at its mean distance from the Sun. The value of this *parallax angle is about 8.8 arc seconds. The radius of the Earth (6,378.164 km) divided by the solar parallax in radians (4.26×10^{-5} radians) gives the *astronomical unit (1.50×10^{8} km), that is, the average distance from the Earth to the Sun.

*Aristarchus used the lunar phase to estimate (incorrectly) that the Sun was twenty times further away from Earth than the Moon and had a parallax of 180 arc seconds. *Tycho Brahe assumed the same value but *Kepler concluded, from Tycho's observations of Mars, that the solar parallax was certainly below 60 arc seconds. Vendelinus (1630) used a telescope to repeat Aristarchus' observation and obtained a value of 15 arc seconds. The parallax of Mars was measured in 1672 and this direct measurement led (by *Kepler's Laws) to a value for the parallax of the Sun of 9.5 arc seconds. *Halley noted that the timing of the *transits of the Sun by the inner planets would lead to even better values and subsequently a transit by Venus led to a value of 8.6 arc seconds. These methods, including the use of the orbits of asteroids, are now of purely historical interest. The solar parallax (and therefore the Sun's distance from the Earth) is now accurately found from observations of the distances of spacecraft that have become artificial planets, an accuracy limited only by the accuracy of our knowledge of the velocity of light.

solar spectrum, the continuous *spectrum of the Sun. The spectrum is crossed by dark *Fraunhofer lines indicating that the *photosphere of the Sun acts like an opaque body surrounded by a cooler gas which emits little radiation. By measuring the wavelengths and structures of the absorption lines it is possible to determine the abundances of the elements present in the surface regions of the Sun, and also the temperature, pressure, magnetic field, and winds in the region being observed. The opaque (somewhat lower) regions of the solar photosphere emit a continuous spectrum that is characterized by its temperature. The radiation intensity in the solar spectrum in the visible and infra-red wavelength regions is similar to that of a *black body, that is, a perfect radiator, with a temperature of 6,000 K. The maximum intensity occurs at around 500 nm. The visible and infra-red emission varies by only about 1 per cent with the eleven- and twenty-two-year solar cycles. The same cannot be said of the long-wave (millimetre and radio) and short-wave (ultraviolet and X-ray) spectra. Much of the energy here is emitted by the solar *corona, a highly variable tenuous *plasma that has a temperature of just over a million degrees. The visible spectrum of the corona is dominated by the emission lines of multiply ionized atoms of iron, magnesium, silicon, and oxygen.

solar system, the collection of planets, satellites, asteroids, comets, and cosmic dust particles that are mostly in elliptical, coplanar orbits around the Sun. The total mass of the solar system is around 450 times the mass of the Earth and about 1/750 of the mass of the Sun. About three-quarters of that mass is concentrated in the gaseous giant planet Jupiter. *Cosmogony, the study of the origin of planets, indicates that the cloud of gas and dust that condensed to form the solar system must have had a mass a few hundred times greater than that of the present system. The planets formed about 4.6 billion years ago and in the early days suffered a great deal from bombardment by asteroids. The rate of these collisions has decreased by about 2,000 times since then. The surface temperature of the planets decreases roughly as the inverse square root of their distance from the Sun. The inner planets, Mercury, Venus, Earth, and Mars have a rock and metal composition, low relative mass, and few satellites. The Moon is unusual in as much as it has a mass of 1/81 of that of Earth. Only Charon, Pluto's satellite, compared to Pluto's mass, has a larger ratio. Phobos and Deimos, the two small satellites of Mars, are thought to be captured asteroids. The outer planets, Jupiter, Saturn, Uranus, and Neptune, have rock–metal cores of ten to twenty or so times the mass of the Earth and are surrounded by huge atmospheres of hydrogen and helium gas. These planets have large collections of orbiting satellites and all have ring systems, although the rings of Saturn are by far the most elaborate. Pluto, the outermost planet, has a mass of only one-quarter that of the Moon and is probably an escaped satellite. There was relatively little matter in the pre-planetary nebula in the region of Pluto and it would have taken a long time for a planet to form at that distance from the Sun, so it is probable that there are no major planets beyond Neptune. However, many astronomers believe that a large spherical shell of comets, the *Oort–Öpik Cloud, surrounds the planetary system and extends almost one-third of the way to the nearest stars. Therefore the planetary system probably has a radius of only about forty times that of the Earth's orbit, whereas the distance to the nearest stars is some 270,000 times the radius of the Earth's orbit. See the figure on pages 152–3.

solar wind, a stream of charged particles, mainly protons and electrons, that are ejected from the Sun into the interplanetary medium. The solar wind escapes from holes in the hot *corona and forms the *heliosphere, which can reach distances of about 100 astronomical units from the Sun. The effect of the solar wind is vividly seen in the tail of a comet. The tail, which is always directed away from the Sun, is formed by the solar wind sweeping material out from the nucleus of the comet.

solid state camera, in astronomy, an instrument that creates an electronic image of a celestial object using any one of a variety of types of light-sensitive microelectronic chips. These convert light into electrical signals and include linear diode arrays and charge injection devices. However, the *CCD camera has gained greatest popularity in astronomy after being developed for use in television cameras. Compared with the best photographic emulsions, it reduces exposure times from hours to minutes.

solstice, either of the two points on the *ecliptic that are farthest away from the *celestial equator. They correspond to Cancer and Capricorn, two of the constellations of the zodiac. The two solstices therefore correspond approximately to the Sun's position in the celestial sphere at the beginning of summer and winter and are related to the *seasons. The northern summer solstice occurs at about 21 June when the north pole is at maximum tilt towards the Sun and is in continuous daylight; the Sun never sets there. The south polar region is then in continuous darkness. At the winter solstice (about 22 December) the northern pole is at maximum tilt away from the Sun and is in continuous darkness while the south pole has continuous daylight.

Sombrero Galaxy, a *spiral galaxy in the constellation of Virgo seen edge-on and resembling a sombrero hat. Though clearly visible in a telescope of 20 cm (8 inches) aperture, its details are best seen when long-exposure photographs are made. Among these details is a prominent dark lane of obscuring material as well as a large number of globular clusters.

south celestial pole *celestial sphere.

Southern Coalsack, a light-absorbing dark dust cloud in the constellation of Crux. The cloud is about 170 parsecs distant and at its northern edge the beautiful *Jewel Box open cluster is located. The cluster, containing over a hundred stars, is young, being only some 10 million years old.

Southern Cross *Crux.

Southern Crown *Corona Australis.

Solar system

The solar system consists of the Sun, the planets and their satellites, comets, meteorites and asteroids, and interplanetary gas and dust. The positions and orbits of the planets in the solar system are illustrated here.

Earth

This is the only planet in the solar system known to support life. It is the largest planet of the inner solar system. The individual plates that make up the Earth's surface are constantly moving. Almost three-quarters of the Earth's surface is covered with water and at any one time about 50 per cent of the Earth's surface is covered with cloud.

Mars

Often known as the red planet, Mars is covered with iron-rich dust. It has north and south polar caps that consist of frozen water and frozen carbon dioxide.

Mercury

This is the innermost and most dense planet in the solar system. It has a very thin atmosphere and like Venus does not have a natural satellite. Its cratered surface looks very similar to the Earth's Moon.

Venus

Apart from the Sun and the Moon Venus is the brightest object in th sky and the nearest planet to Ea It has a cloud covering of 100 per cent and an atmosphere compose primarily of carbon dioxide.

The outer solar system

Orbits of planets

Saturn

Consisting mainly of hydrogen and helium, Saturn is the second-largest planet in the solar system. Before the development of the telescope, Saturn was the most distant planet known to man. Saturn's rings have a diameter of more than 170,000 km which is equivalent to almost three-quarters of the distance from the Earth to the Moon.

Uranus

This is the third-largest plane the solar system but it has on the fourth-greatest mass, tho of Jupiter, Saturn, and Neptur being greater. Uranus's atmosphere, similar to that of Neptune, contains hydrogen with smaller amounts of heliu and methane.

Jupiter

Being one-tenth the size of the Sun, Jupiter is the largest planet in the solar system. The surface of Jupiter is subjected to windstorms, the most noticeable of which is the Great Red Spot which is three times the size of Earth and has been raging ever since it was first noticed over three hundred years ago.

	Period of rotation	Rotation around Sun
Mercury	59 days	88 days
Venus	243 days	225 days
Earth	23.9 hours	365.25 days
Mars	24.6 hours	687 days
Jupiter	9.9 hours	11.86 years
Saturn	10.2 hours	29.46 years
Uranus	17.2 hours	84 years
Neptune	18.4 hours	165 years
Pluto	6.4 days	248 years

The table *(left)* details the period of time it takes for each planet to rotate once on its axis and also to revolve once around the Sun. The comparative sizes of the planets and Sun are also shown below.

Sun Mercury Venus Earth Mars

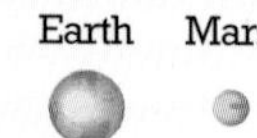

Jupiter

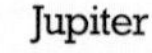

The scale (below) indicates the relative distances of each planet away from the Sun in millions of kilometres.

Sun

Mercury (57.9)

Venus (108.2)

Earth (149.6)

Mars (227.9)

Jupiter (778.3)

Saturn (1427)

Uranus

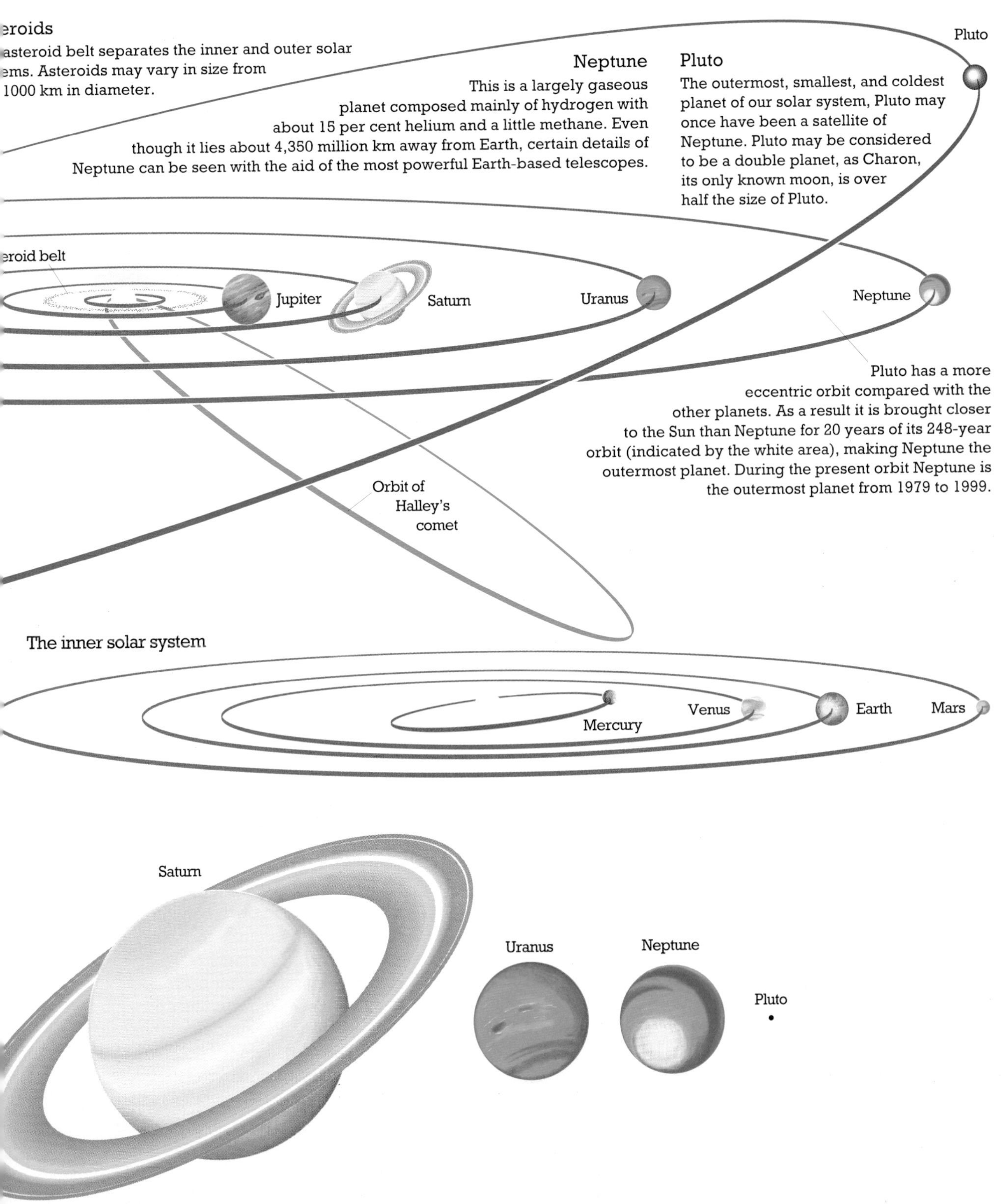
eroids
asteroid belt separates the inner and outer solar
ems. Asteroids may vary in size from
1000 km in diameter.
Neptune
This is a largely gaseous planet composed mainly of hydrogen with about 15 per cent helium and a little methane. Even though it lies about 4,350 million km away from Earth, certain details of Neptune can be seen with the aid of the most powerful Earth-based telescopes.
Pluto
The outermost, smallest, and coldest planet of our solar system, Pluto may once have been a satellite of Neptune. Pluto may be considered to be a double planet, as Charon, its only known moon, is over half the size of Pluto.
Pluto
eroid belt
Jupiter
Saturn
Uranus
Neptune
Pluto has a more eccentric orbit compared with the other planets. As a result it is brought closer to the Sun than Neptune for 20 years of its 248-year orbit (indicated by the white area), making Neptune the outermost planet. During the present orbit Neptune is the outermost planet from 1979 to 1999.
Orbit of Halley's comet
The inner solar system
Mercury
Venus
Earth
Mars
Saturn
Uranus
Neptune
Pluto
Neptune (4496.6)
Pluto (5900)

Southern Fish *Piscis Austrinus.

Southern Triangle *Triangulum Australe.

south galactic pole *galactic pole.

space, a region of the universe empty of matter. The term is often used loosely to refer to interplanetary space, interstellar space, or intergalactic space, being the regions surrounding planets, stars, or galaxies. These regions, however, contain dust and gas, though in very small densities. In *cosmology, space, though empty, has properties according to which cosmological model is being used. Such properties are described by the general theory of *relativity and determine the evolution of the universe.

spacecraft, a vehicle designed to travel in space. Since the dawn of the space age on 4 October 1957 with the launching of Sputnik 1 by the Soviet Union, thousands of spacecraft have left the Earth's surface to go into Earth orbit, or to the Moon, the planets, or to Halley's Comet. The first spacecraft were unmanned artificial satellites in Earth orbit. Sputnik 2, carrying the dog Laika, was launched on 3 November 1957. The first US satellites were the Explorer series launched in 1958. Explorer 1 discovered the *Van Allen radiation belts. Over a period of many years the Soviet Union launched the large series of Cosmos spacecraft, and the Molniya series of communications satellites. Syncom, Telstar, and Early-Bird (Intelsat 1) were US communications satellites. Many series of satellites, such as Landsat, Navstar, and Metsat, were launched for navigation, meteorology, and surveillance purposes.

Scores of satellites devoted to various fields of scientific research have been launched, each containing a set of scientific experiments to be carried out in orbit. Astronomical satellites have contained equipment to accept radiation from various parts of the *electromagnetic spectrum, including the infra-red, ultraviolet, radio, X-ray, and gamma-ray regions. The *Hipparcos satellite and the *Hubble space telescope are examples of how observation from above the Earth's atmosphere has enabled spectacular advances in particular scientific fields to be made.

Soviet manned missions have included several types of spacecraft. Vostok was the first manned spacecraft. This was a series of six one-manned vehicles the first of which put Yuri Gagarin into orbit on 12 April 1961. The sixth, on 16 June 1963, orbited Valentina Tereshkova, the first woman in space. The Voskhod series, also launched in the 1960s, consisted of two spacecraft each capable of carrying three cosmonauts. They were followed by the Soyuz series which is still being used today; the first was launched on 23 April 1967. This series has been used to ferry cosmonauts to the Soviet Salyut space station. A similar space station, *Mir (meaning 'peace'), was put into orbit on 20 February 1986. The Soviet Union has also built and flown the Buran shuttle, similar to the US space shuttle. The US manned space flight programme began with the Mercury series of six flights, of which the first two were suborbital. The first, launched on 5 May 1961, took Alan Shephard into space for only 15 minutes. The second series, the Gemini Project of the mid-1960s, made ten flights, each carrying two astronauts into orbit. On 11 October 1968, Apollo 7 was tested in Earth orbit prior to the dispatch to the Moon of later Apollo spacecraft. The late 1970s saw the approach and landing tests of the space shuttle before the first space flight by a shuttle began on 12 April 1981. Since then regular flights into Earth orbit have been made for a wide variety of space missions, apart from the long gap after the Challenger disaster on 28 January 1986, when the vehicle exploded, killing its crew of seven. The first US space station, Skylab, was put into orbit on 14 May 1973 by a Saturn V rocket. Three successive crews inhabited it until it was finally abandoned in February 1974. It burned up in the Earth's atmosphere in July 1979. At the time of writing plans are well in hand for the construction of a new space station to be in use by the middle of the 1990s.

In the Luna series of twenty-four Soviet spacecraft, Luna 9 made the first soft landing on the Moon on 3 February 1966. Luna 3 took the first pictures of the far side of the Moon in October 1959; Luna 10 was put into orbit about the Moon on 3 April 1966. Luna 17 and Luna 21 each landed an eight-wheeled vehicle on the Moon's surface. The first, Lunokhod 1, operated under remote control for ten months in the Mare Imbrium, conducting a series of measurements of the magnetic field and making chemical analyses. The second, Lunokhod 2, landed near the Mare Serenitatis and operated for four months. Lunas 16, 20, and 24 sent back samples of lunar soil to Earth from three different locations. The series was terminated with Luna 24, launched on 15 August 1976. The US Ranger series of nine spacecraft for lunar exploration were launched between August 1961 and March 1965. The last three in the series obtained the first close-up pictures of the lunar surface. Between May 1966 and January 1967, the Surveyor series of seven spacecraft were landed on the Moon to investigate the Moon's surface properties. Each had a television camera and over 80,000 pictures were transmitted. The Lunar Orbiter series of five space probes, put into lunar orbit between August 1966 and August 1967, made photographic examinations of the proposed Apollo landing sites. Irregularities in their orbits led to the discovery of *mascons, large regions of higher than usual density within the Moon.

The US *Project Apollo essentially consisted of Apollo 7 to Apollo 17, each flight made with a crew of three, though only Apollos 11, 12, 14, 15, 16, and 17 made lunar landings. The last three were each equipped with a lunar roving vehicle for transport. Missions 7 to 10 tested various stages of the operation; Apollo 13 suffered an explosion which rendered the mission inoperable though the three astronauts survived. More than 380 kg (837 pounds) of samples were returned from the Moon by the Apollos to be subsequently analysed. Special packages of instruments left at the landing sites operated for several years, transmitting their scientific data back to Earth.

Soviet missions for planetary exploration have included the Zond, Mars, and *Venera series, though only Venera has been completely successful. The Zond series consisted of a mixture of lunar and planetary probes. Zonds 1 and 2 flew past Venus and Mars in 1964 and 1965 but returned no information. Zond 3 in 1965 returned good-quality pictures of the far side of the Moon while Zonds 5 to 8 circumnavigated the Moon in the years 1968–70. Zond 4 was probably a failure. The Mars series of seven Martian probes, launched at intervals between 1962 and 1973, were all only partially successful. Mars 1 was lost en route; Mars 2 and 3 each consisted of an orbiter and a landing capsule, but only the Mars

The Magellan **spacecraft** was deployed from the cargo bay of the space shuttle *Atlantis* on 4 May 1989 to explore Venus. Once in orbit about Venus, its novel radar system penetrated the dense atmosphere to provide photographic images of the surface.

USA

3 capsule landed safely and ceased transmission almost immediately. The two orbiters, however, sent back information about the Martian surface and atmosphere. Mars 4 was unsuccessful, passing Mars in February 1974. Mars 5 did enter Martian orbit but failed after a few days' operation. Mars 6 was also a failure while the landing capsule of Mars 7 did not reach the planet's surface. Two additional probes, the Phobos series, launched in the late 1980s to explore the Martian moon Phobos, failed to complete their missions.

In contrast to these misfortunes, the successful Soviet Venera series to the planet Venus has greatly enlarged our knowledge of that planet's atmosphere and surface features. Though the first three Venera probes failed to send any data from Venus, Veneras 4 to 8 all ejected capsules which descended by parachute through Venus's dense atmosphere. All transmitted data on temperature, pressure, and composition during the descent, though the first three failed before reaching the surface; Veneras 7 and 8 lasted on the surface for 23 and 50 minutes respectively, sending further data. Veneras 9 to 14 followed similar programmes, their capsules also withstanding the harsh conditions both during and after the parachute descent. As well as measurements, panoramic photographs from the surface were relayed back to Earth. Later Veneras in the series have continued to gather information about Venus.

The USA, following their initial failures, launched two series of important spacecraft, *Pioneer and *Mariner. Pioneers 1, 2, and 3 were failed lunar probes; Pioneers 4–9, launched between 1960 and 1968, entered solar orbit to measure the Sun's activity. They gave warning of solar flares during the Apollo programme. Pioneer 10, launched in March 1972, visited Jupiter. In its fly-past of that planet it transmitted photographs not only of Jupiter but also of three of the Galilean satellites: Europa, Ganymede, and Callisto. Pioneer 11, launched in April 1973, visited Jupiter and subsequently, in September 1979, made a fly-past of Saturn. Both spacecraft have now left the solar system. The Mariner programme comprised a series of interplanetary vehicles that explored Venus, Mercury, and Mars. Mariners 2 and 5 made fly-pasts of Venus in 1962 and 1967 while Mariner 10 flew past Venus once on 5 February 1974 and Mercury three times (on 29 March and 21 September 1974, and 9 March 1975). Much of the surface of Mercury was photographed in detail. Mariners 4, 6, and 7 flew past Mars, Mariner 4 in July 1965, Mariner 6 on 3 July 1969, and Mariner 7 on 5 August 1969. All three sent back photographs of the surface as well as making a wide variety of measurements of temperature, pressure, and composition of the Martian atmosphere. Mariner 9 was put into orbit about Mars on 13 November 1971. Thousands of pictures of the surface of Mars were obtained together with much other information about the atmosphere and surface. Phobos and Deimos, the two small satellites of Mars, were also photographed. In the series, only Mariners 1, 3, and 8 failed. Mariners 11 and 12 were renamed *Voyager 1 and 2.

In 1975, the USA launched two spacecraft, *Viking 1 and Viking 2, to Mars. Both were spectacularly successful. Each consisted of an orbiter and a lander. The orbiters became artificial satellites of Mars, photographing the surface, taking measurements of atmospheric water vapour and surface temperature, and relaying data from the landers back to Earth. The landers studied surface conditions, made meteorological measurements, and tested for signs of life. As far as the presence of life was concerned, the results were inconclusive. Perhaps the most successful US planetary mission so far has been the *Voyager programme. Voyager 1, launched in 1977, reached Jupiter in March 1979 and Saturn in November 1980. It then left the solar system. Voyager 2, also launched in 1977, flew past Jupiter in July 1979, Saturn in August 1981, Uranus in January 1986, and Neptune in August 1989. It also, on completing its planetary mission, has left the solar system. The Voyager mission's treasure of data about the giant outer planets, their ring systems, and their moons, is still being analysed. More recently launched missions include Galileo, to explore the Jupiter system; Magellan, to map the surface of Venus in greater detail; and Ulysses, to explore the solar environment above and below the north and south poles of the Sun well away from the ecliptic.

On the return of Halley's Comet to the vicinity of the Sun in 1986, a number of spacecraft were sent by Japan, the Soviet Union, the United States, and Europe to rendezvous with the comet. The European Space Agency's *Giotto spacecraft, launched in 1985, came within 605 km of the nucleus in March 1986, collecting valuable data concerning the comet's composition, activity, and tail.

space exploration, the exploration of the solar system using artificial satellites and spacecraft. Our knowledge of the Sun, planets, natural satellites, and comets in the solar system, as well as the interplanetary medium, has been greatly advanced in recent years by the *Mariner, *Venera, *Pioneer, and *Voyager series of spacecraft. More recently, space exploration is being extended by the launching of artificial satellites into Earth orbit carrying advanced technological packages to further our astronomical knowledge of the wider universe.

Spacelab, a manned laboratory designed by the *European Space Agency to be put into Earth orbit and returned to Earth by a space shuttle. It is kept in the shuttle's cargo bay for the entire flight, which can last up to ten days. Within it, scientists carry out experiments or make observations with the shuttle's cargo bay doors open. The laboratory consists essentially of two parts, a pressurized module in which the scientists can work without spacesuits, and a segmented pallet in which instruments can be placed. The pallet can be modified between missions to suit the needs of the planned scientific programme. The first mission (Spacelab 1) was launched on 23 November 1983.

space telescope *Hubble space telescope.

space–time, a four-dimensional model of the universe in which the position of a particle is specified by four numbers, three of them being space-like coordinates, the fourth being time. The trajectory of this particle drawn in this so-called block universe is its *world line, each point of which has a space–time position. The Lithuanian mathematician Hermann Minkowski (1864–1909) said that the concept of space–time would henceforth take the place of space and time as separate entities. Space and time are linked in a precise way by the special and general theories of *relativity, with the result that measurements of distance and time are no longer absolute as in classical Newtonian mechanics but depend upon the velocity of the observer. Furthermore, the presence of matter causes curvature of various kinds in space–time and the density of such matter, if high enough, will even produce a finite, closed space–time universe.

speckle interferometry, a technique for restoring distorted images of stars. Turbulence in the Earth's atmosphere can distort light from a star so that, at any instant, the image when seen through a large telescope looks like an irregular group of dots (speckles). Several photographs of very short duration are taken and combined using a computer to produce a single high-quality image. The technique is frequently used to separate close binary and multiple stars into their components.

spectral classes, a system of classifying stars by their *spectrum, first introduced by *Secchi. His original classification as a series of letters, A, B, C, and so on has been rearranged as knowledge of stellar structure has improved. The colour of a star indicates its surface temperature and stars are now arranged in the *Draper classification in order of decreasing temperature, the classes in that order being O, B, A, F, G, K, M, R, N, and S. Each class is divided into ten steps, as for example A0 to A9. The hottest stars of class O are bluish in colour with surface temperatures up to 100,000 K; B-class stars also have a bluish colour. A-class stars are blue-white, F are white, G are yellow-white (the Sun belongs to this class with a surface temperature of about 5,000 K), while K-class and M-class stars are respectively orange and red. The classes R, N, and S contain only a small percentage of stars with surface temperatures so low that molecules can exist in their atmospheres. The classes R and N are sometimes united in a single class designated by the letter C. An echo of the original Secchi classification is heard when O-, B-, and A-class stars are referred to as *early-type stars and K and M stars are termed *late-type stars.

spectral line, a line appearing in a *spectrum. Spectral lines result from the interaction of *electromagnetic radiation with matter and occur at discrete *wavelengths characteristic of the material producing them. An **absorption line** shows up as a missing wavelength or dark line in a spectrum, and results when electromagnetic radiation is absorbed by a gas or plasma. When radiation is emitted by a gas or plasma, a bright **emission line** results. Spectral lines can occur in the non-visible parts of the spectrum, such as at radio, infra-red, and X-ray wavelengths.

spectroheliograph, an instrument used to produce a photographic image of the Sun in light of a particular wavelength. The sunlight passing through the telescope is intercepted by a spectroscope to produce a *spectrum. A slit just in front of the photographic plate selects light of a particular wavelength by its position in the spectrum. Passing through the slit, this light falls on a narrow band of the plate. A narrow strip of the Sun's image is thus captured. By moving the equipment sideways the whole image of the Sun in light of that wavelength is progressively registered on the plate, producing a spectroheliogram. For example, if a spectral line due to hydrogen is selected, subsequent study of the picture will give a measure of the way hydrogen is distributed at that time across the solar surface.

spectrometer, in astronomy, an instrument attached to an optical telescope in order to obtain the *spectrum of a celestial object. It normally uses a diffraction grating to disperse or spread the light into its component colours. Older instruments employ glass prisms and some special designs have elements such as Fabry–Perot interferometers. The spectrum is recorded on a photographic plate or is stored electronically by a solid state device such as a *CCD camera. Depending on the brightness of the object and the astrophysical problem being investigated, different amounts of spectral spread or dispersion may be applied. The scale of the spectrum is expressed in angstroms per millimetre. Spectrometers facilitate the study of stellar atmospheres, radial velocity measurements via the *Doppler effect, and stellar magnetic fields.

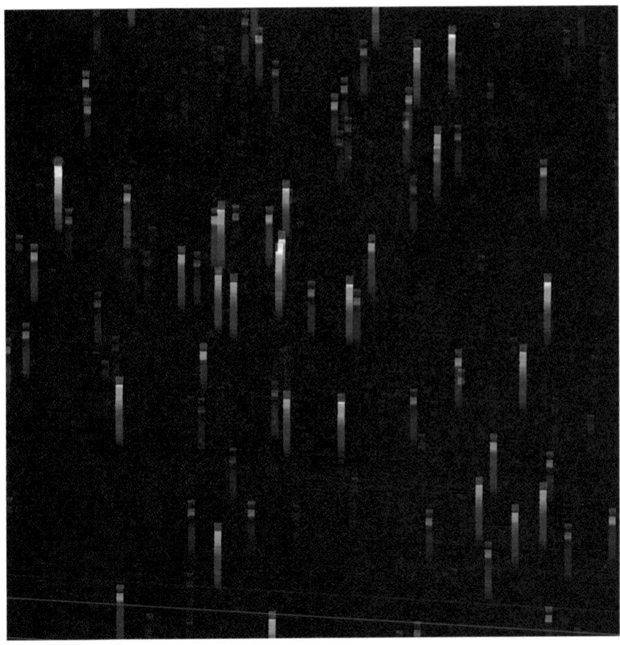

The **spectra** of the Hyades open star cluster reveal many things about the member stars. The Doppler effect gives information about the motion of the stars; analysis of the bright emission lines enables the stars' constituent elements to be deduced; and the dark absorption lines indicate the nature of the gas and dust which lies between the stars and the Earth.

spectrophotometer, a laboratory instrument for analysing a *spectrum that has been recorded on a photographic plate. By passing a narrow beam of light through the recorded image the original intensity of the light from the source at the various wavelengths is recovered. This information can then be displayed as a graph or stored as a digitized spectrum for further processing.

spectroscopic binary *binary star.

spectrum (pl. spectra), the image formed by rays of light or other *electromagnetic radiation in which the parts are arranged in a progressive series according to wavelength. At visible wavelengths it is the band of colours produced when light emerges from a prism. If the light first passes through a narrow slit, images of the slit appear as dark parallel *spectral lines in the spectrum. The particular wavelengths at which these lines appear are governed by the make-up of the transparent interstellar gas the original beam passed through. In contrast, the light making up the bright part of the spectrum, known as the **emission spectrum**, originates from the energy released by excited atoms, molecules, and ions. A study of the spectrum therefore enables much information to be obtained regarding the make-up of the interstellar gas and also of the hot dense gas of a star or nebula producing the beam. Such studies of the spectra of celestial objects, pioneered by Gustav Kirchhoff, Robert Bunsen, William Huggins, and others, laid the foundations of the science of *astrophysics.

speculum metal, a brittle white alloy consisting of two parts of copper to one of tin. It takes a very high polish and in former times *reflecting telescope mirrors were made from it. A disadvantage of the speculum mirror was its tendency to tarnish rapidly. The development of silvered and aluminized mirrors has now rendered the use of speculum obsolete.

speed of light *velocity of light.

spherical astronomy, that branch of astronomy which deals with the directions of celestial objects, the relations between those directions, and the transformation of *coordinate systems from one form to another by means of spherical trigonometry. One of the oldest branches of astronomy, it is still one of the most useful but now includes such topics as spherical trigonometry, the *celestial sphere and coordinate systems, and the correction of observational data to remove the effects of atmospheric *refraction, stellar *aberration, *parallax, *precession, and *nutation. It is often sufficient to know the directions of celestial objects and for this reason spherical astronomy is concerned with the projections of these objects on to the celestial sphere. Nevertheless spherical astronomy, with its emphasis on spherical trigonometry, is also concerned with the tracking of artificial satellites and interplanetary probes, and is therefore an essential part of the modern topics of astrodynamics and space research.

Spica, a first-magnitude *variable star in the constellation of Virgo, and also known as Alpha Virginis. It is a spectroscopic *binary star with a period of four days, comprising two hot *dwarf stars. Its distance is 70 parsecs, and the combined luminosity is about 1,600 times that of the Sun.

spicules, jets of hot gas ejected from the solar *chromosphere into the *corona to heights of a few thousand kilometres. They are associated with the strong magnetic fields near the edges of large cells in the chromospheric *granulation and each spicule lasts several minutes before being absorbed into the corona.

spiral arms, the regions of dust, gas, and stars that spring from opposite sides of the nucleus of a *spiral galaxy and coil round it in its equatorial plane. In the case of a *barred spiral galaxy the arms spring from each end of the bar. Spiral and barred spiral galaxies are labelled by the prefix S in the *Hubble classification.

spiral galaxy, a galaxy exhibiting spirally coiled arms of dust, gas, and stars springing from a central nucleus. It is one of the most common types of galaxy and is labelled So to SC in the *Hubble classification. The majority of spiral galaxies possess two arms arising from opposite sides of the nucleus. In general, the stars within the arms are young *Population I stars.

sporadic meteor *meteor.

Spörer's Law, the continual change in the distribution of *sunspots in solar latitude as a sunspot cycle progresses. The number and location of sunspots vary in a regular manner. Every eleven years there is a lull in sunspot activity, known as a sunspot minimum. The first spots after a sunspot minimum appear in latitudes 30–35 degrees north and south. Thereafter, succeeding spots appear nearer and nearer to the solar equator with the last spots of the cycle appearing near the equator as the first spots of a new cycle appear again in latitudes 30–35 degrees north and south. This effect can be illustrated by the *butterfly diagram.

spring equinox *equinox.

spring tide *tides.

star, a self-luminous gaseous body, such as our own Sun. A star is therefore unlike a *planet, which can be seen only by light reflected from it. Stars are formed directly out of the interstellar medium (the gas and dust that exists in the vast spaces between stars) whereas planets form from material in orbit around a young star. The *Galaxy contains some 100 billion stars. The stars are traditionally grouped into *constellations, but for the most part they are unconnected and are at greatly differing distances. The brightest stars in each constellation are usually designated by letters of the Greek alphabet, and many have proper names, most of which were given by Islamic astronomers. Fainter stars are designated by their number in a catalogue.

Apart from the Sun, stars are at great distances from the Earth; the nearest, Proxima Centauri, is 1.3 parsecs or 40 million million kilometres away. Consequently most stars appear as fixed points of light that cannot be seen as discs by even the largest telescopes. However, the discs of a few stars are large enough to be measured by interferometry, or by observing the time they take to disappear behind the Moon during an occultation. The distances of nearby stars can be determined by observing their apparent change in position (*parallax) relative to the background of more distant stars as the Earth orbits the Sun, or by indirect methods such as the *distance modulus. The velocity of a star will in general be a combination of motion towards or away from the Earth (*radial velocity) and motion across the sky (*proper motion). The radial velocity of a star can be determined from the *Doppler effect by examining its spectrum, and many stars are near enough for us to measure their proper motion directly. These two velocities can be combined to give the stellar motion with respect to the Sun.

Much can be learnt about stars by studying their spectra. They can be classified according to their temperature and luminosity. Most stars derive their energy from nuclear fusion, though in some the *Kelvin–Helmholtz contraction releases gravitational energy. *White dwarfs are powered by thermal energy, and *neutron stars by rotational energy. Many stars are members of *binary star or *multiple star systems. In such cases their stellar mass can be calculated from their motion under the gravitational influence of their companions. It is also common to find *variable stars in which some characteristic, usually magnitude, changes periodically. Theories of *stellar structure and *stellar evolution are tested by seeing whether they can explain the characteristics and behaviour of the stars we see, particularly in *clusters which are groups of stars that were formed at about the same time. Stellar evolution can be conveniently summarized on the *Hertzsprung–Russell diagram.

The twenty nearest and twenty brightest stars are listed in the tables on page 199, with the Sun for comparison.

star chart, a representation of the sky on a flat surface. The first printed star chart, as distinct from hand-drawn manuscripts, was engraved by Albrecht Dürer in 1515, depicting the forty-eight *constellations listed by Ptolemy in the *Almagest*. The constellation figures appear in reverse, as on a celestial globe, a convention that was followed in 1603 by Johann Bayer's *Uranometria*, the first great star atlas. Bayer devoted a chart to each of the Ptolemaic constellations, with an additional chart showing the twelve new southern constellations invented by Dutch navigators. Bayer plotted over 2,000 stars in the *Uranometria*, twice as many as Dürer, and introduced *Bayer letters, the system of labelling stars with Greek letters.

The invention of the telescope greatly improved the precision of celestial mapping, leading to the in fluential *Atlas Coelestis* by *Flamsteed, the first Astronomer Royal, which was published posthumously in 1729. This work inspired Johann Bode's *Uranographia* of 1801, the first atlas to show all naked-eye stars (plus many fainter ones, giving a total of 17,000 stars). The *Uranographia* also established the concept of constellation boundaries. Like earlier great atlases, Bode's work depicted the Greek constellation figures as well as the stars. However, during the 19th century the pictorial aspect of the constellations was abandoned in favour of increasingly precise plotting of the positions and brightnesses of stars, as exemplified by the monumental 1863 *Bonner Durchmusterung* atlas by the Prussian astronomer Friedrich Wilhelm Argelander (1799–1875).

By the 20th century, photography had revolutionized celestial charting, culminating in the 1950s with the National Geographic Society–Palomar Sky Survey made with the 48-inch (1.2 m) Schmidt telescope at the Mount Palomar Observatory. A new and improved Palomar survey is under way, as is a southern survey by the European Southern Observatory and the Anglo-Australian Observatory. However, the traditional star chart has not been superseded. Perhaps the most popular star chart of the 20th century has been *Norton's Star Atlas*, first published in 1910 by the British amateur Arthur P. Norton (1876–1955) and regularly revised. It was joined in 1981 by *Sky Atlas* 2000.0 by the Dutch amateur Wil Tirion (1943–)

static universe *cosmology.

stationary point, a point in space such that a particle placed there could stay in relative *equilibrium, keeping the same distances with respect to the other bodies in some *N-body problem. When $N = 3$ and the two major masses are in a circular orbit around the common centre of mass, there are five such relative equilibria, known as *Lagrangian points, with the three bodies turning with the same angular velocity around the centre of mass: the three collinear configurations, with the small particle either in between or outside the other two but on the same line, and the two triangular configurations such that the three bodies always form an equilateral triangle. The *Trojan group of asteroids oscillate around either one or the other of the triangular stationary points of the Sun–Jupiter–asteroid system; Saturn and its satellites Dione and Helene are also close to a triangular configuration.

Stationary point also refers to the two directions in which the apparent motion of a planet with respect to the stellar background is neither direct nor retrograde.

Steady State theory, any theory of *cosmology which postulates that the universe is the same in all places and at all times. The Steady State theory requires that matter is continually being created. Although it is one of the most appealing theories it cannot satisfactorily explain the radio source counts and the *microwave background radiation and so is not now generally accepted.

Stefan's Law, the relation between the total power emitted as *electromagnetic radiation by a *black body and the temperature of the black body. The total power P emitted per unit surface area, in watts per square metre, is given by $P = \sigma T^4$, where $\sigma = 5.67 \times 10^{-8}$ W m^{-2} K^{-4} is the Stefan–Boltzmann constant, and T is the absolute, or Kelvin, temperature of the black body. Although astronomical sources rarely radiate precisely like black bodies, their emission often approximates to it. The fourth-power temperature dependence is very strong, and the total power radiated per square metre can vary from 3 microwatts for the *microwave background radiation ($T = 2.7$ K), through 500 watts of mostly infra-red radiation from the dark side of the Moon ($T = 300$ K), and 75 megawatts of mostly visible light from the surface of the Sun ($T = 6{,}000$ K), to 14,000 gigawatts of ultraviolet light from the surface of a superhot, proto-*white dwarf star like PG1159-035 ($T = 125{,}000$ K).

stellar association, a group of young stars born at the same time in a particular place in the Galaxy's spiral arms. The stars, usually numbering from ten to a hundred, may exist in a region several hundred parsecs across and may comprise *early-type stars of spectral classes O and B. Such an OB association is often associated with an *open cluster. The *Great Nebula in Orion is the centre of an OB association while the Perseus OB1 association is centred on the *Double Cluster in Perseus. Other associations include T associations which consist of faint, low-mass *T Tauri stars.

stellar dynamics and kinematics, the branch of astronomy dealing with the movements of stars within galaxies or within globular clusters or open clusters. In the 18th century it became clear that the stars were not fixed relative to the Sun but had their own intrinsic velocities through space. From then on information concerning the *proper motions of the stars was collected and by the late 19th century it was clear that, under *Newton's Law of Gravitation, stars moved in *trajectories through space. Stellar dynamics and kinematics attempts to understand the movements of stars, and in the 20th century studies by the Dutch astronomer Jan H. Oort, *Eddington, the British astronomer Henry Plummer, and many other astronomers have shown that in globular and open clusters, stars pursue orbits dictated by the general gravitational field of the cluster. Within the Galaxy, it is clear that stars, including the Sun, follow orbits about the galactic centre; in the case of the Sun the orbital period is in the vicinity of 200 million years. In recent years computer simulated studies of stellar systems have produced results aiding astronomers in their understanding of the dynamics of stellar systems such as our own galaxy.

stellar evolution, the study of the life history of stars. Stars are believed to be formed by the collapse under *gravitation of material, mainly hydrogen, in *nebulae. They heat up through the *Kelvin–Helmholtz contraction, and when their central temperature reaches about 10 million K they begin to draw their energy from *nuclear fusion, converting hydrogen to helium in the *carbon–nitrogen cycle. Such stars are called *dwarf stars, and are found on the main sequence in the *Hertzsprung–Russell diagram. This stage lasts a million years or less in the most massive stars, or about 10^{10} years for stars like the Sun. It continues until the mass of helium at the star's core reaches the *Schönberg–Chandrasekhar limit. Hydrogen burning then continues in a shell surrounding the core, and the star begins to evolve

Stellar evolution

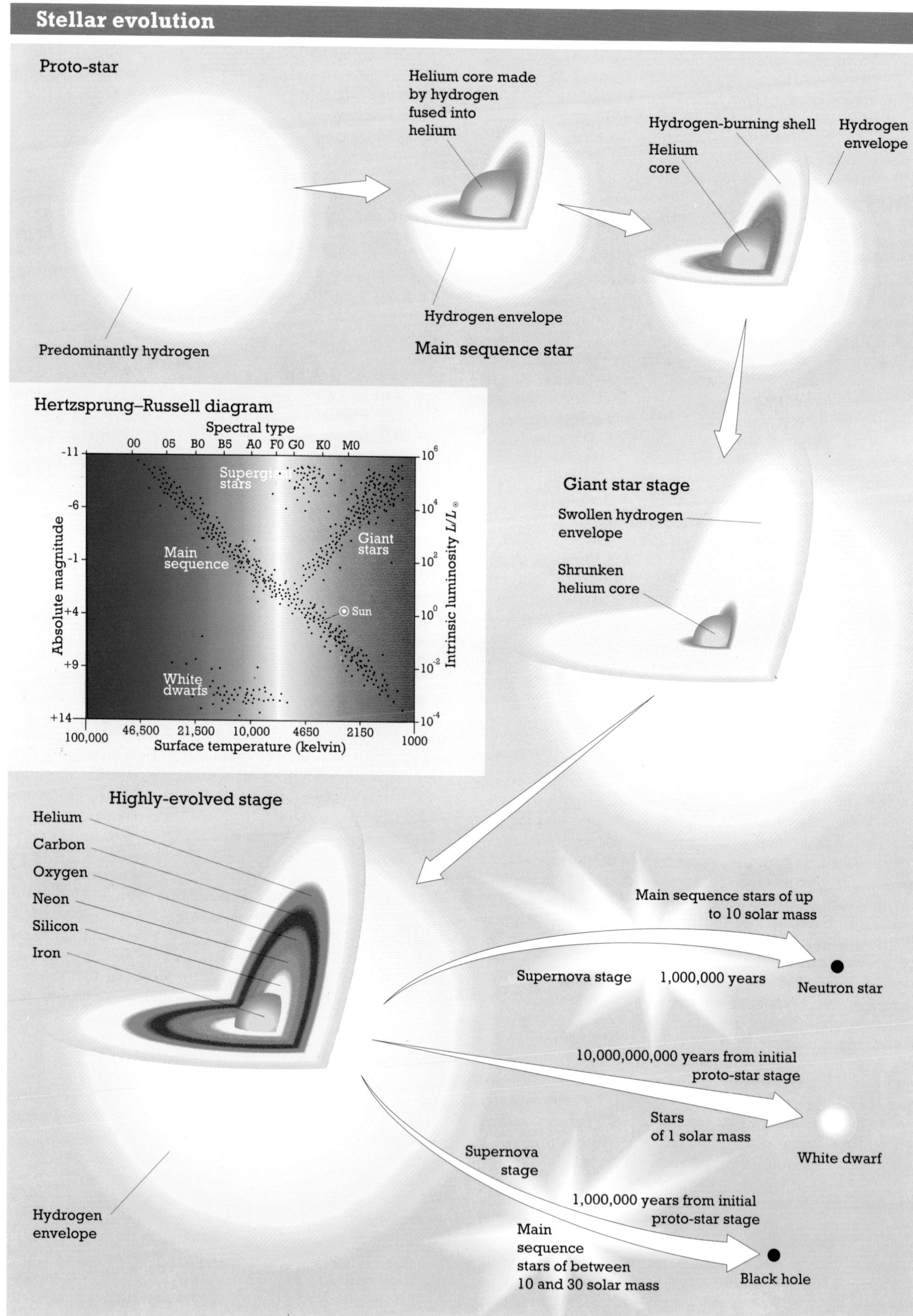

towards the *giant star stage, with increasing radius and decreasing surface temperature. Stars with a mass greater than 0.4 times that of the Sun can go on to convert helium to carbon, and carbon to heavier elements by *nuclear synthesis. Eventually the star exhausts its nuclear fuel. Most stars shed their outer layers, exposing the helium-rich core which becomes a *white dwarf. One such star is *FG Sagittae, which is at present evolving very rapidly. In massive stars, the core may collapse in a *supernova explosion, forming a *neutron star or even a *black hole. Such a collapse is thought to occur if the star's mass remains greater than the *Chandrasekhar limit. The evolution of close *binary stars is more complicated, as the components may exchange material with one another.

stellar mass, the mass of a star. It can be directly measured only for the components of a *binary star whose orbits have been determined and whose distance is known, using the law of gravitation. The masses of only a few dozen stars have been found in this way, but many more can be estimated using the *mass–luminosity relation. The stars of greatest known mass, forty or more times that of the Sun, are very rare. Most stars are less massive than the Sun, but stars with less than 0.1 times the Sun's mass are also rarely found. Their central temperature cannot reach the level at which *nuclear fusion can start, and they radiate energy gained solely from the *Kelvin–Helmholtz contraction. Such objects are called **brown dwarfs**.

stellar motion, the movement of a star relative to the Sun. All stars, including the Sun, have their own velocities through space but are so distant from the Sun, and therefore the Earth, that the first *proper motion of a star was not detected until 1718 when *Halley showed that the positions of Sirius, Aldebaran, and Arcturus had altered from the positions given by *Hipparchus almost two millennia before. Study of the movements of stars is the province of *stellar dynamics and kinematics and reveals much about the dynamical structure of the Galaxy. Photography at different epochs is used to detect changes in the positions of the nearer stars against the background of fainter and presumably more distant ones. The artificial satellite Hipparcos was specially designed to provide very highly accurate measurements of stellar motions.

stellar parallax *parallax.

stellar populations, the classification, introduced by *Baade, of stars into two main classes. **Population I** contains young blue *supergiants while **Population II** contains old *red giants. Such populations show a marked preference for various parts of the Galaxy, with Population I stars being found in open, galactic *clusters and the *spiral arms while Population II stars make up globular clusters and are also found in the region of the galactic centre. Population II stars seem to have only traces of the heavy elements in their make-up in contrast to Population I stars which may have up to ten times more. This is consistent with the belief that the *interstellar medium is being continually enriched by the ejection of heavy elements created in *supernova explosions.

stellar structure, the internal organization of a star. For some stars, information may be obtained more or less directly about the *stellar mass, size, *luminosity, *surface temperature, and chemical composition. To deduce the internal structure of stars, the laws of physics governing gravitation, thermodynamics, and atomic and nuclear processes must be used. The course of *stellar evolution for stars with various initial masses and chemical compositions can then be modelled by computer. However, the results are rather uncertain for the earliest and final stages. *Dwarf stars on the main sequence of the *Hertzsprung–Russell diagram are thought to remain approximately chemically homogeneous until they have exhausted the available hydrogen for *nuclear fusion as determined by the *Schönberg–Chandrasekhar limit. *Red giant stars are not homogeneous: they can have a core in which helium undergoes fusion to form carbon, surrounded by a shell in which the 'burning' of hydrogen to helium continues. At a later stage, the helium burning is confined to a shell around the core. Very massive stars can undergo further stages of fusion to arrive at a structure in which there is an iron-rich core surrounded by successive shells composed mainly of (1) silicon, (2) carbon and oxygen, (3) helium, and (4) hydrogen with helium. Such objects are the precursors of Type II *supernovae. *White dwarfs and *neutron stars represent the end of the evolutionary process and are composed of *degenerate matter.

stellar temperature, the temperature of a region in a star. No single temperature characterizes a star completely. The temperature of the Sun, for example, is 15 million K in the thermonuclear core, falling to a minimum of 4,000 K at the top of the *photosphere, and rising to some 2 million K in the tenuous *corona. Most of a star's light is radiated from its photosphere, and the photospheric temperature is for many purposes most characteristic of the outer layers of a star. Even here, however, the situation is not simple because there is a range of temperatures through the photosphere (4,000–8,000 K for the Sun). In addition, temperature is a thermodynamic quantity which may depend upon the *electromagnetic radiation and the various atoms, ions, and nuclear particles present at any place in a stellar atmosphere. These various temperatures are often condensed into a single *effective temperature. The effective temperature actually relates to the radiation of a star, but is representative of particle properties in the photosphere. Effective temperatures for normal stars range from 3,000 to 40,000 K; the Sun's is 5,780 K.

Stéphan's Quintet, a grouping of five galaxies in the constellation of Pegasus. The galaxies appear to be close together and their brightness and apparent tidal interaction arising from their mutual gravitational fields suggest that they are all at the same distance and are part of the same group. However, one galaxy in the Quintet (NGC 7320) has a much lower red shift than the other four, in conflict with the Hubble Law which predicts that galaxies at similar distances will have similar red shifts and therefore similar velocities away from the Earth. This has led to the controversial suggestion that galaxy red shifts might be due to causes other than the *Doppler effect. However, most astronomers believe that NGC 7320 is a nearby galaxy that very unusually happens to lie in the same direction as the other four galaxies. The Quintet was discovered in the 1870s by the French astronomer Edouard Stéphan at the Marseilles Observatory.

Steward Observatory, an *observatory belonging to the University of Arizona at Tucson which operates a number of telescopes at the *Kitt Peak National Observatory. The Steward Observatory collaborates with the Smithsonian

Stéphan's Quintet is a cluster of strongly interacting galaxies. The mutual tidal effects between two of the galaxies in particular have swept material out from the galaxies and across intergalactic space.

Institution in operating a *multiple-mirror telescope on Mount Hopkins near Tucson.

Stickney, a large crater on the surface of the Martian moon *Phobos. With a diameter of about 10 km, it reaches half the size of the whole satellite. The surrounding system of linear grooves may be a result of global fractures provoked by the impact that formed Stickney.

stratopause, the boundary between the *stratosphere and the *mesosphere at a height of about 60 km (38 miles). It is the layer at which the air temperature stops increasing with height. Its temperature is about 10 °C and is caused in the main by the absorption of solar ultraviolet radiation by *ozone. Similar hot layers do not occur in the atmospheres of the other planets because these atmospheres lack significant amounts of oxygen. The ozone at the stratopause shields the Earth's inhabitants from harmful ultraviolet radiation. Ozone also acts as an effective greenhouse gas by absorbing infra-red radiation that is attempting to leave the Earth's surface and in doing so contributes to the *greenhouse effect. The main region of ozone formation is above 25 km (16 miles) above the Earth's surface (the photochemical region of the atmosphere) where the oxygen atoms produced by the break-up of oxygen molecules still find plenty of molecules of oxygen with which to combine.

stratosphere, a region in the Earth's atmosphere from 18 to 60 km (11–38 miles) above the Earth's surface. It lies between the *troposphere and the *mesosphere. Together these three layers form the **homosphere**, the principal constituents being molecular nitrogen (78 per cent), molecular oxygen (21 per cent), with argon, water vapour, and carbon dioxide making up most of the remaining 1 per cent. The *ozone layer exists at a level of the stratosphere some 25 km (16 miles) high as a result of the break-up of molecular oxygen by ultraviolet radiation, with subsequent combination of oxygen atoms with oxygen molecules. Various human activities of monumental foolishness are not only increasing the carbon dioxide content of the homosphere, thus producing the *greenhouse effect, but also damaging the ozone layer. The human race's attitude to these matters may well be the deciding factor on whether a tolerable future will exist for it during the 21st century.

Strömgren sphere, the volume around luminous blue *supergiant stars of *spectral class O in which the interstellar hydrogen and helium are completely ionized by the ultraviolet radiation from the star. The sphere shows up as an emission *nebula with fairly sharp boundaries. Typical dimensions of the sphere are 25–100 parsecs depending on the spectral classification of the embedded star. The ionization of the gas is balanced by the recapture of electrons by the lone nuclei with the emission of visible light. These balanced processes were first investigated theoretically by the Danish astronomer Bengt Strömgren in 1939.

Struve family, several generations of Struve who were all eminent astronomers. **Friedrich Georg Wilhelm von Struve** (1793–1864) was one of the greatest 19th-century astronomers, whose skilful and exhaustive observations of binary stars led to the publication in 1837 of his *Stellarum Duplicium Mensurae Micrometricae*, a catalogue listing over 3,000 such stars. Born in Germany, as a young man he emigrated to Russia where he studied at Dorpat (Tallinn) University in Estonia and later became director of the observatory there. In 1835 he was commissioned by Czar Nicholas to build and direct the new observatory at Pulkovo near St Petersburg where he remained for the rest of his life. His Russian-born son, **Otto Wilhelm Struve** (1819–1905), spent his entire working life at the Pulkovo Observatory, succeeding his father as director. His work covered many areas of observational astronomy, including binary stars and a new determination of the Earth's *precession. He was one of the first to establish that the Sun is moving relative to the nearby stars. **Otto Struve** (1897–1963), a great-grandson of Friedrich von Struve, was born in Russia but emigrated to the US in 1921, where he worked for the rest of his life. In 1938, he discovered the presence of hydrogen in interstellar space by observing the spectra of distant stars. He also used spectroscopy to investigate stellar rotation, binary stars, turbulence in stellar atmospheres, and galactic nebulae, and discovered widespread interstellar calcium in the plane of the Galaxy. Otto Struve was one of the greatest of 20th-century observational astronomers and certainly the most prolific. During his career, he was director of four different observatories and was president of the *IAU (International Astronomical Union) for three years from 1952. In this capacity he had considerable influence in re-establishing world-wide astronomical co-operation after World War II. Three other members of the family, **Hermann Struve** (1854–1920), **Ludwig Struve** (1858–1920), and **Georg Struve** (1886–1933) were also distinguished astronomers.

sub-, in connection with planets or satellites, a prefix used to indicate a place on the surface of a planet or satellite immediately below another celestial object. Thus:

sublunar point is that point on the Earth's surface where the Moon is in the *zenith at that moment;

subsolar point is that point on the surface of a planet or satellite where the Sun is in the zenith at that moment;

substellar point is that point on the surface of a planet or satellite where a particular star is in the zenith at that moment.

subdwarf, a member of a small class of stars that are 1.5–2 magnitudes fainter than a normal star on the main sequence of the *Hertzsprung–Russell diagram of the same *spectral class. They include some stars with individual peculiarities, but most subdwarfs are old *Population II stars that are members of the galactic halo. Compared with Population I stars in the disc of the Galaxy, their atmospheres contain a lower proportion of elements heavier than hydrogen and helium. This accounts in part for their lower luminosity. No subdwarfs are found among the naked-eye stars.

subgiant *giant star.

sublunar point *sub-.

subsolar point *sub-.

substellar point *sub-.

summer solstice, that time, about 21 June, when the Sun, moving along the *ecliptic, is at its maximum northerly distance from the *celestial equator. The Sun's declination is then 23 degrees 26 minutes and lies in the zodiacal constellation of Cancer. The Sun passes directly through the *zenith at noon on the Tropic of Cancer. The Earth's northern hemisphere then has its longest period of daylight and shortest period of darkness; the southern hemisphere has its shortest period of daylight and longest period of darkness. The winter *solstice occurs on about 22 December when the Sun is at its maximum southerly distance from the celestial equator. The Sun passes directly through the zenith at noon on the Tropic of Capricorn. The Earth's southern hemisphere then has its longest period of daylight and shortest period of darkness; the northern hemisphere has its shortest period of daylight. It should be noted that when summer is beginning in the northern hemisphere, winter is beginning in the southern hemisphere.

The **Sun's** disc, with a large prominence at the top, is photographed here in ultraviolet light by the Skylab 2 space station. When prominences erupt they follow the lines of the solar magnetic field. 'Ghost images' of the Sun can be seen at other wavelengths to the left of the main image.

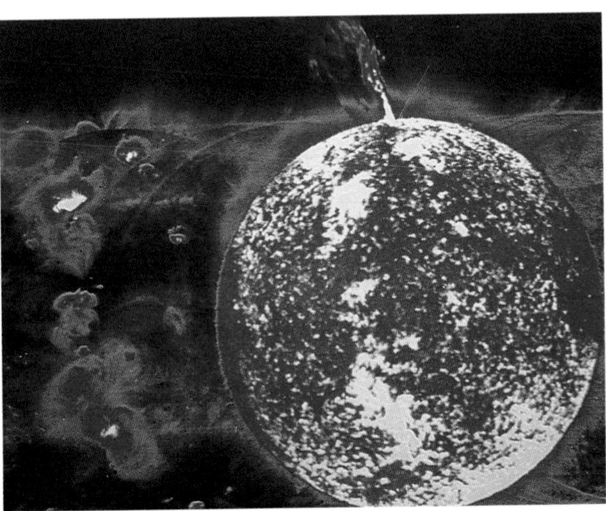

summer triangle, a large figure formed by the stars *Altair, *Deneb, and *Vega, in the constellations *Aquila, *Cygnus, and *Lyra, respectively. The summer triangle is near the *meridian in the evening sky during the months of July to October.

Sun, the star around which the Earth orbits. An ordinary *main-sequence star of *spectral class G2, the Sun is a self-luminous gaseous mass, comprising some 71 per cent hydrogen and 26 per cent helium by mass, having an absolute visual *magnitude of +4.83 and an effective surface temperature of 5,770 K. At the solar centre, the temperature rises to 1.5×10^7 K to provide the pressure necessary to support the overlying mass against gravity which, at the Sun's radiating surface (the *photosphere), is twenty-seven times that of the Earth. This high temperature also sustains, by the *nuclear fusion of hydrogen into helium, a solar radiation output which totals 3.8×10^{26} watts emanating from the photosphere.

For descriptive purposes the Sun may be subdivided into a series of roughly spherical shells. The **core**, in which the energy for the solar radiation is released by nuclear fusion, occupies less than one-thousandth of the Sun's volume, but is 160 times more dense than water, while the photosphere is 10^{-7} times less dense than water. Because of the enormous mass of the Sun and the opacity of its material, radiation leaks from the core to the photosphere extremely slowly, taking longer than about 10 million years to do so. During this leakage, X-rays are weakened and emerge as visible light. However, *neutrinos generated in the fusion process can escape freely through the Sun and provide, in principle, a means of direct observation of the core. The failure to observe the predicted number of solar neutrinos has generated much debate about the solar structure. In the final 15 per cent of the way to the solar surface from the centre, convection plays a major role in transporting energy outwards and the circulating motions from this convection zone appear in the fine structure of the visible solar surface, in particular its *granulation. Convective motions also play a part in the transport of magnetic fields, generated by currents in the rotating solar interior, to the surface where they appear in complex forms, known as **solar activity**, on both large and small scales, the most intense fields appearing in *sunspots. Outward from the photosphere lies the solar atmosphere in which the temperature declines to a minimum of 4,200 K and thereafter rises, owing to the dissipation of shock waves coming from subphotospheric convection, through the *chromosphere and the transition layer, where it jumps to a value of 2×10^6 K characterizing the *corona. The high coronal temperature results in a continuous escape of matter into interplanetary space as the *solar wind. Both the chromosphere and the corona exhibit a wide range of variable or energetic phenomena, which are further examples of solar activity. These are associated with magnetically active regions, of which some of the most important are solar *flares, *prominences, and holes in the corona.

The mass of the Sun is 1.99×10^{30} kg and the mean radius of the approximately spherical photosphere is 7.0×10^5 km. These figures are respectively 3.3×10^5 times the Earth's mass and 110 times the Earth's radius; the solar volume

Sun, structure of

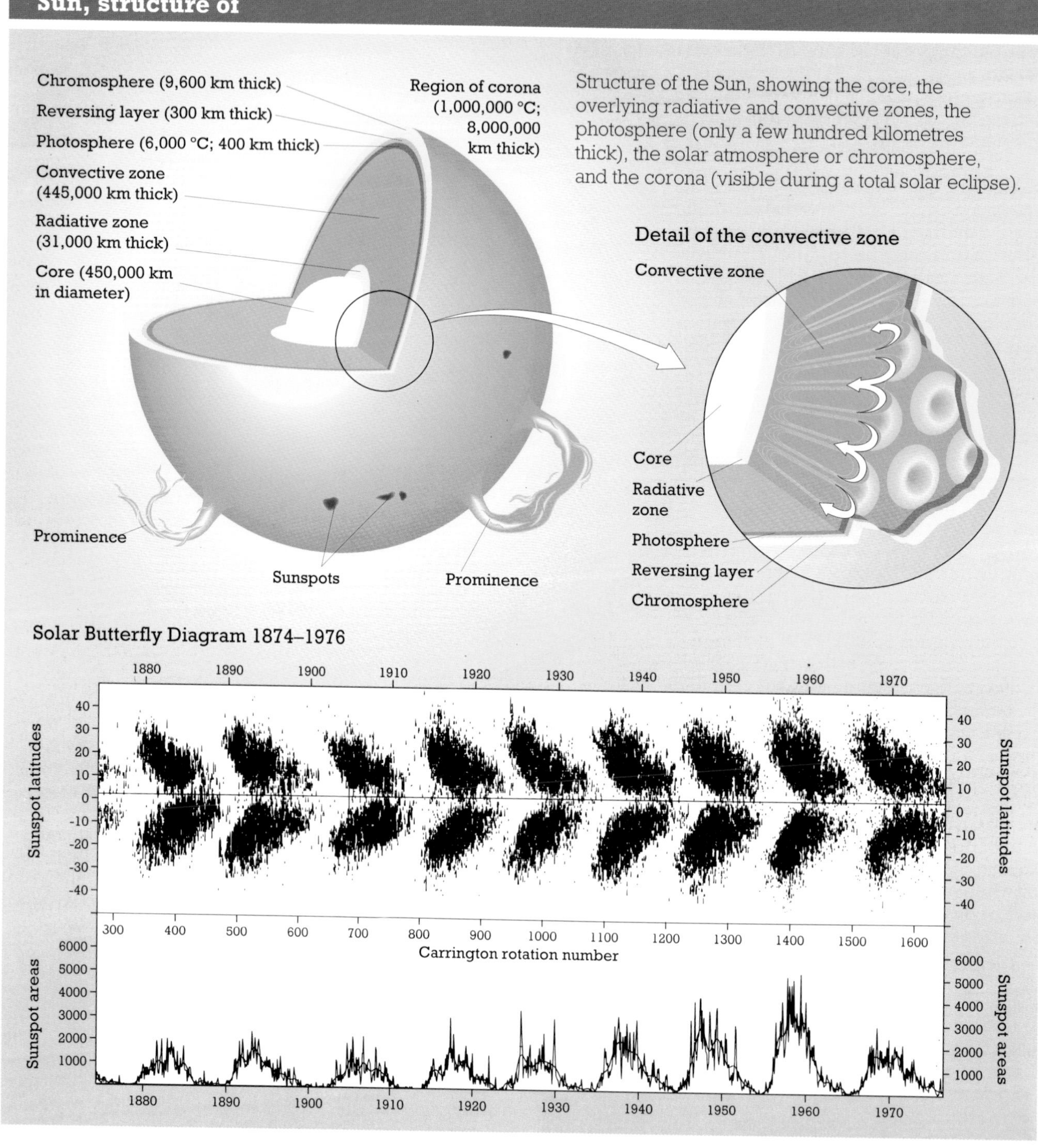

Structure of the Sun, showing the core, the overlying radiative and convective zones, the photosphere (only a few hundred kilometres thick), the solar atmosphere or chromosphere, and the corona (visible during a total solar eclipse).

would thus contain 1.3 million Earths. Solar rotation is observed in the motions of surface features, such as sunspots, in the photosphere and overlying layers. The mean rotation period is 25.4 days about an axis inclined at 7.25 degrees to the *ecliptic but ranges from 25 days at the solar equator to 41 days near the poles. This rotation is also related to the *oblateness of the Sun's disc, amounting to a 0.005 per cent deviation from a sphere. The effect of this flattening on the Sun's gravitational field contributes to the *precession of Mercury's perihelion and so complicates the use of this as a test of the general theory of *relativity.

The importance of the Sun, both to astronomers and to mankind, lies in its proximity to the Earth, the mean distance being about 150 million kilometres, or 1 astronomical unit. In consequence, the solar radiation maintains terrestrial conditions, particularly the Earth's surface temperature, within habitable limits, as well as providing daytime illumination directly and nocturnal illumination by reflection from the Moon. Solar energy is the ultimate source of all the energy we utilize, except radioactivity. To the astronomer, the Sun is unique for three reasons. Firstly, it is the only star close enough to the Earth to be readily observable as an extended object, that is, as other than a point of light: the photosphere has an apparent angular diameter of 0.53 degrees. This enables astronomers to observe the structure of the Sun's atmosphere in great detail. By contrast, the angular diameter of a similar star at the distance of *Proxima Centauri (1.7 parsecs) would be only 0.007 arc seconds.

Secondly, it is the only normal star close enough to detect its very weak radiation in the radio and X-ray bands, although other stars are visible at these wavelengths during periods of strong activity such as flares or *nova explosions. Thirdly, the Earth orbits within the solar wind, which forms the outer layers of the solar atmosphere, permitting direct sampling of solar material by spacecraft in the vicinity of the Earth and direct observation of the three-dimensional structure of the solar atmosphere by spacecraft in solar orbit. Furthermore, the comparatively short distance to the Sun means that the paths of charged particles (the solar *cosmic rays) may be traced back to their sources on the Sun, whilst neutrons travelling at nearly the speed of light from the Sun do not have time to decay before reaching the Earth.

sundial, a device for measuring the apparent solar *time by the shadow of a rod or triangular edge cast by the Sun on a scaled dial. It consists of two parts, the **gnomon** and the **plate**. The gnomon is the rod or triangular edge and acts as the indicator. It is set parallel to the spin axis of the Earth. The geometry is such that the angle that the gnomon makes with the horizontal is equal to the latitude of the sundial. The gnomon is always in the north–south plane. The markings forming the scaled dial on the plate depend on its orientation. The simplest case is when the plate is perpendicular to the gnomon. The plate is then in the plane of the *celestial equator and, as the Sun apparently moves around the Earth's axis at the uniform rate of 15 degrees each hour, the hour lines on the plate are uniformly placed and are also 15 degrees apart. Other plate orientations require trigonometrical calculations to position the lines. The sundial suffers from two problems. If its position is, say, x degrees of longitude west of the time meridian it reads $4x$ minutes slow; if it is east it reads fast. Secondly, the *equation of time has to be taken into account.

sunspot cycle *sunspots.

sunspots, dark spots visible on the surface of the Sun. Sunspots have been recorded since ancient times and are now recognized as regions of the *photosphere cooled some 2,000 K below their surroundings through insulation by strong magnetic fields of about 2,000 gauss. Sunspots have a relatively dark central area (the umbra) surrounded by a lighter outer region (the penumbra). The flow of gas from the umbra to the penumbra is described by the *Evershed effect. The numbers and latitudes of sunspots vary in an eleven-year period, known as the **solar cycle** or **sunspot cycle**. This cycle is described by *Spörer's Law and is illustrated graphically in the *butterfly diagram. At any instant the **Zurich relative sunspot number** indicates the total area covered by the sunspots. The basic eleven-year cycle has longer-term variations superimposed upon it. For example, sunspots reverse their magnetic polarity about every twenty-two years, a cycle discovered by *Hale, but the most extreme example of long-term fluctuations was the *Maunder minimum of 1645–1715 when virtually no sunspots were observed. Although it is recognized that the sunspot variations are governed by the diffusion of magnetic fields upwards from the rotating solar interior, the process is still far from understood. Numerous other manifestations of solar magnetic activity are known, including *flares and *prominences, the latter being giant loops of relatively cool gas suspended by magnetic fields in the hot *corona. In recent years there has been a resurgence of interest in the possible correlation between sunspot activity and diverse terrestrial phenomena ranging from the weather to the profits of the Hudson Bay Company! However, while there seem to be a number of irrefutable short-term effects on the weather, the existence of long-term relationships between climate and sunspot activity, particularly the eleven-year cycle, remains highly debatable because of the difficulties in analysing the data in a statistically rigorous fashion.

supergiant, a star of very high *luminosity (6,000 to 250,000 or more times that of the Sun), large diameter (20 to 500 or more times that of the Sun), and low mean density (from one hundredth to one ten-millionth that of water). Supergiants are rare and no supergiant is near enough to determine its distance directly by measuring its trigonometric *parallax. Owing to their high luminosity, however, there are several supergiants among the bright naked-eye stars, including Antares, Betelgeuse, Deneb, and Rigel.

superior planet, a planet revolving about the Sun in an orbit further from the Sun than the Earth's orbit. Mars, Jupiter, Saturn, Uranus, Neptune, and Pluto are superior planets. Unlike the *inferior planets, Venus and Mercury, superior planets can be visible throughout the night and are therefore easier to observe.

supernova, a star that temporarily brightens by twenty *magnitudes or more. A supernova is a more destructive event than a *nova and can be classified in two ways. Type I supernovae are old, small stars that have few heavy elements. The explosion usually destroys the star completely leaving only a supernova remnant of expanding gas. They are probably *white dwarf components of *binary stars that undergo detonation by *nuclear fusion. Type II supernovae are massive young stars that have exhausted the available elements for energy generation and whose cores collapse to form a *neutron star or *black hole, ejecting a shell of gas as they do so. The material ejected from a Type II supernova eventually enriches the interstellar medium with heavy elements. The last supernova to be seen in the Galaxy was in 1604, and the one before that was *Tycho's Star in 1572. However, several are detected each year in other galaxies,

Supernova 1987A

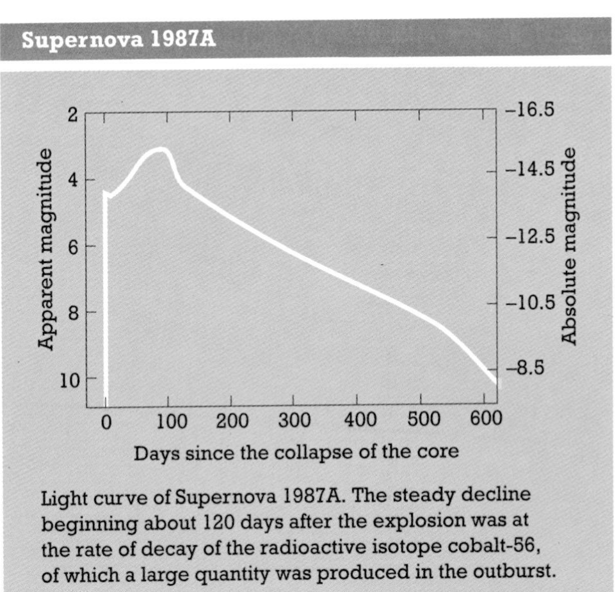

Light curve of Supernova 1987A. The steady decline beginning about 120 days after the explosion was at the rate of decay of the radioactive isotope cobalt-56, of which a large quantity was produced in the outburst.

Supernova

Light curves of typical supernovae.

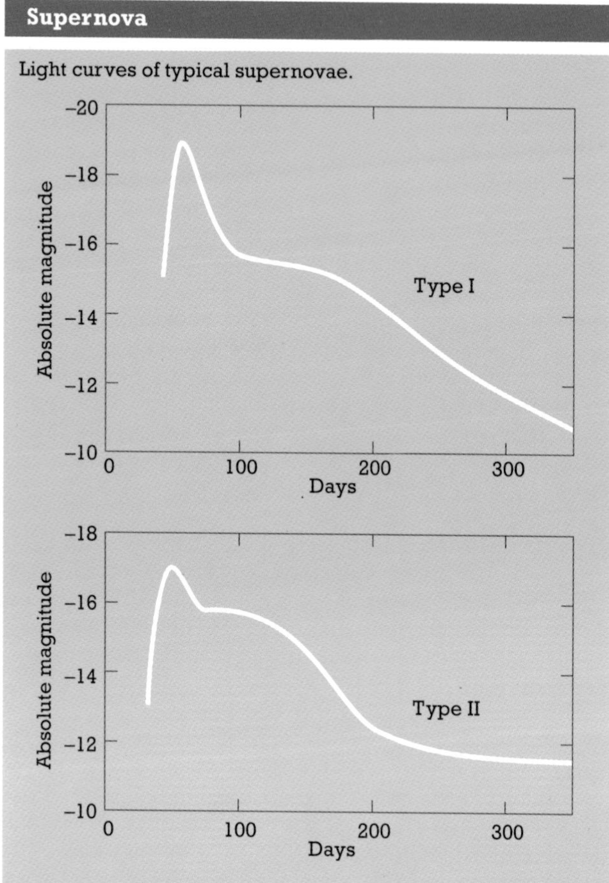

such as *Supernova 1987A in the Large Magellanic Cloud. The shock wave from a supernova about 4.6 billion years ago in the vicinity of the Sun is thought by some astronomers to have triggered the collapse of a gas cloud to form the solar system and to have provided the chemical elements necessary for life to evolve.

Supernova 1987A, a *supernova in the Large Magellanic Cloud, the brightest to be seen since 1604. The progenitor star was already catalogued as the blue *supergiant Sanduleak −69° 202, of *magnitude 12. The supernova was of Type II, although not typical. A burst of *neutrinos emitted when the star's core collapsed was recorded by two detectors on 24 February 1987. On the same day, the outburst was detected optically by the Canadian astronomer Ian Shelton and others, and the star rose to magnitude 2.9 in late May before beginning its slow fade. Supernova 1987A rose more slowly to its maximum brightness than most other Type II supernovae, which have *red giant progenitors. This is accounted for by the star's different initial structure. An *optical pulsar, which may be a remnant of the supernova, may possibly have been detected in January 1989 with the exceptionally short period of 0.0005 seconds, but attempts to confirm it in subsequent months were unsuccessful. Supernova 1987A is now surrounded by a bright ring of expanding gas. See also the figure on page 165.

supernova remnant, the gaseous remnant of a star that exploded in a *supernova outburst. An exploding supernova ejects most of its material to create an expanding, very roughly spherical *nebula. Examples of such supernova remnants are the *Crab Nebula, observed by Chinese astronomers in 1054, and the *Veil Nebula in the constellation of Cygnus.

surface gravity, the acceleration experienced by an object when it is in free fall under *gravity close to the surface of a celestial body. The value of this acceleration, at a specific point, is equal to the product of the mass of the celestial body and the constant of gravitation, divided by the square of the radius of the body. An object on the surface of a celestial body has a weight equal to the product of its mass and the gravitational acceleration. The mean surface gravity on the Earth and Moon are 9.82 and 1.62 m s^{-2}, respectively. On a spinning body the effective surface gravity varies as a function of distance from the spin axis because of centrifugal force. On Earth it is 0.034 m s^{-2} less at the equator than it is at the poles.

surface temperature, in astronomy, the temperature of the surface of a celestial body. The surface temperature of a star is the temperature of the *photosphere from where most of the star's light is radiated. As stars are gaseous, the

Supernova 1987A, the brightest supernova seen since 1607, resulted from the explosion of a blue supergiant star in the Large Magellanic Cloud. The upper photograph was taken in 1969 and the lower one on 26 February 1987, two days after the supernova exploded. Supernova 1987A is the bright star on the right of the lower image. The aftermath of the explosion is still being detected.

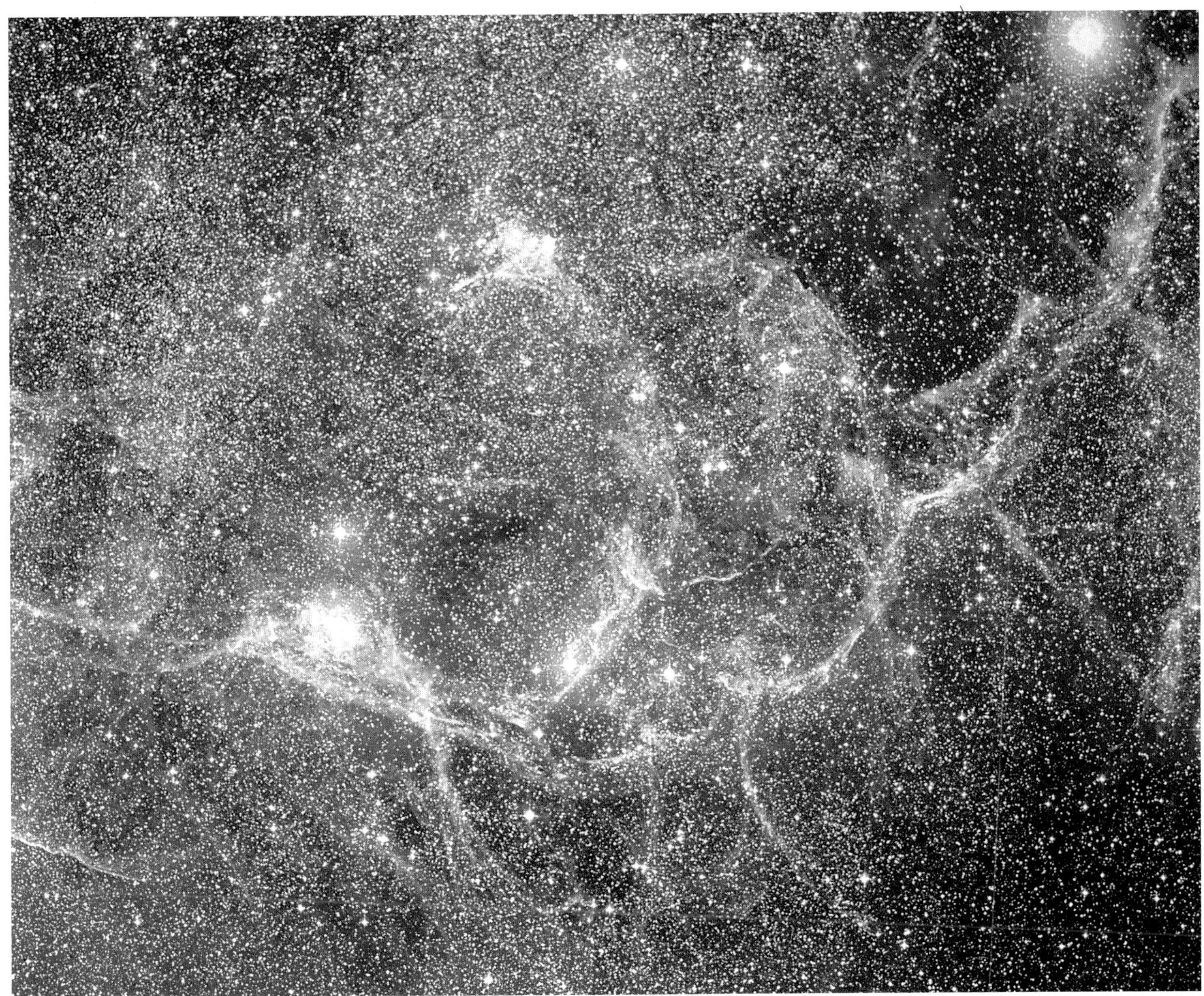

The Vela **supernova remnant** is an extensive network of fine luminous filaments. It is the debris from a supernova explosion some ten thousand years ago. The shock wave from the explosion is still travelling through the tenuous interstellar gas, heating it, and causing it to emit visible light. Embedded in the gas is the Vela pulsar, a rapidly spinning neutron star believed to be the remains of the exploded star itself.

observer sees light coming from a range of heights in the star's atmosphere each having different temperatures, so the surface temperature is really an average of the stellar temperatures of a series of atmospheric layers. The surface temperature of a planet depends on the relative amounts of solar radiation absorbed by the planet, and the radiation emitted from the planet's surface. The mean planetary temperature varies as the square root of the planet's distance from the Sun. The temperature also varies from day to night and from equator to pole; these variations in temperature are much smaller on a planet with an atmosphere, like the Earth, than they are on a slowly rotating airless satellite like the Moon.

surges, sudden and short-lived events on the Sun in the vicinity of the penumbra of *sunspots or from active regions such as *flares. In a surge, gas from the *chromosphere is ejected at a velocity of some 100–200 km/s into the solar *corona, sometimes to a height of 150,000 km before falling back along the curved path by which it ascended. The event is usually completed in a few minutes.

Swan Nebula *Omega Nebula.

symbiotic star, a stellar object whose *spectrum contains emission lines and combines the characteristic features of a *red giant star and a hot object. The hot object is generally thought to be a *white dwarf or *dwarf star, or an *accretion disc around such a star, and the hot star excites the spectral emission in the cool red giant's extended envelope between the core and the atmosphere. However, single-star models have been proposed for some symbiotic stars. Symbiotic stars include *variable stars, very slow *novae, and some recurrent novae. In addition they include stars that do not vary in magnitude.

synchronous, taking the same period of time. For example, a planetary satellite in a synchronous orbit has an orbital period equal to the period of *rotation of the planet so that the satellite, if in an equatorial orbit, remains above the same region of the planet's surface. When the planet is the Earth this is called a *geosynchronous orbit. In a synchronous rotation, a satellite rotates in a period equal to its period of *revolution about its planet. Most of the solar system's satellites are in synchronous rotation; for example the Moon always keeps the same face turned towards the Earth.

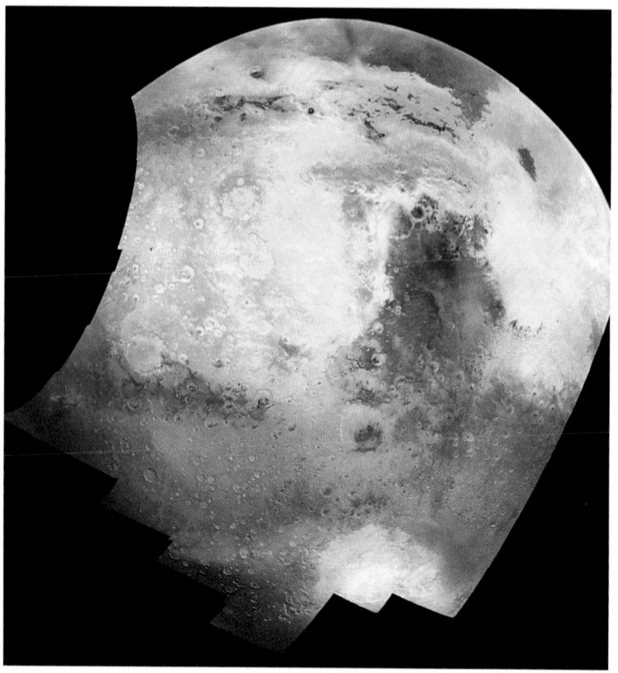

This mosaic of Mars was made from about a hundred images obtained by the Viking 1 orbiter in 1980. The dark blue area at the right of centre is **Syrtis Major**, probably a basaltic volcanic region. The heavily cratered yellow-brown region to its left is called Arabia, and the bright area at the far right is Isidis Planitia, an ancient impact basin. The white area in the far south is covered by carbon dioxide frost.

synchrotron radiation, *electromagnetic radiation emitted by an electron moving at a high speed in a magnetic field. The name arises from the fact that in the nuclear accelerators called synchrotrons, electrons moving at speeds near the velocity of light emit this light. Some of the light coming from the Crab Nebula is generated by this synchrotron mechanism from electrons spiralling in the mesh of magnetic fields within the nebula.

synodic month, the time between successive new or full moons. It is one of several definitions for the month. Alternatively called a **lunation**, the synodic month, associated with the *phase of the Moon, was an obvious method of measuring the passage of time to early man. *Lunar calendars were in use in various parts of the world several thousand years ago.

synodic period, the time interval between successive similar positions of a body with respect to the Sun and Earth. The synodic period of a planet, for example, is the time between two successive identical *planetary configurations of opposition or conjunction. The synodic period for the Moon it is the time between successive *phases of full moon or new moon.

Syrtis Major, a dark triangular plateau close to the equator of Mars, lying between longitudes 283 degrees and 298 degrees and latitudes 20 degrees north and 1 degree south. Syrtis Major is easily visible through telescopes from the Earth and the seasonal winds blow around the lighter coloured dust periodically exposing the underlying darker rocks. In the early 20th century this colour variation was thought to be caused by the annual change in the form of vegetation. Wind-blown dust can reflect light with varying efficiencies. The dune-like hills are made of coarser material whereas the white streaks that extend tangentially around crater rims are produced by the deposition of finer dust. The cameras on board *Mariner 9 recorded notable boundary changes in the Syrtis Major markings in periods of less than three weeks.

syzygy, the geometrical configuration in which a celestial body is nearly in alignment with the Sun and the Earth. It is applicable to the Moon when it is new, that is, between the Sun and the Earth, or full, that is, on the opposite side of the Earth to the Sun. At syzygy, sea *tides on Earth of maximal range (spring tides) are caused. Syzygy is also applicable to Mercury and Venus when they are at the *planetary configurations of superior and inferior conjunction, and to conjunctions and oppositions of the outer planets.

tail, the streamer of gas and dust that extends from the *coma of a comet. Most comets develop tails when they are closer than about 2 astronomical units from the Sun. The tail, which consists of two parts, one of gas and one of dust, always points away from the Sun. The gas tail consists of ionized molecules emitted by the nucleus. It has a bluish colour, distinct boundaries, and is typically a million kilometres wide and tens of millions of kilometres long. The structure can change, break up, and re-form within hours. The individual molecules have velocities away from the comet of between 10 and 100 km/s and are being accelerated by a combination of *radiation pressure and the *solar wind. The dust tail is more bland and curved, the curvature depending on the mass of the dust particles. This dust has been ejected by the cometary nucleus and is being pushed away from the comet by gases released as the comet absorbs solar radiation.

Tarantula Nebula, a luminous emission *nebula in the Large *Magellanic Cloud, visible to the naked eye. It extends half a degree in the night sky and has bright stars within it. At the centre is a highly luminous object which may be a dense star *cluster or one extremely massive object. The name 'Tarantula' is a reference to the pattern, seen in telescopes, of tendrils of gas and dust, reminiscent of the spider of that name.

Tau Ceti, a nearby third-magnitude *dwarf star in the constellation of Cetus. Slightly cooler than the Sun, its luminosity is half that of the Sun and its distance is 3.6 parsecs. It is the nearest Sun-like star that is not a binary star, and so may have a solar system like ours. The star is therefore regarded as a promising target for SETI, the search for extraterrestrial intelligence, and was one of the stars included in the *Ozma Project.

Taurids, a *meteor shower associated with *Encke's Comet. This is a very old shower and varies little from year to year. It lasts about two months as the Earth passes through the *meteor stream, which is approximately 30 million kilometres wide. The Earth intersects the stream twice, once in early November and again in early July. The July shower, which is known as the Beta Taurids, is visible in daytime. The orbit of the meteor stream is significantly perturbed by Jupiter.

Taurus, the Bull, a *constellation of the zodiac. In mythology it represents the bull into which Zeus turned himself to abduct Princess Europa. Only the front half of the bull is depicted in the sky. Its eye is marked by the red giant *Aldebaran, and its horns are tipped by second-magnitude Beta Tauri (also known as Elnath, Arabic for 'the butting one') and third-magnitude Zeta Tauri. Near Zeta Tauri lies the *Crab Nebula. The face of the bull is represented by a V-shaped star cluster, the *Hyades. On the bull's back lies another famous cluster, the *Pleiades. The Sun passes through the constellation from mid-May to late June. The *Taurid meteors radiate from the constellation every autumn.

Taurus A, a bright *radio source associated with the *Crab Nebula. The nebula is the remnant of a powerful *supernova explosion which was observed by oriental astronomers in 1054. The radio and optical emissions from the nebula are strongly polarized, showing that they are *synchrotron radiation. At the centre lies an unusual compact radio source, the Crab *pulsar, a neutron star spinning thirty times a second. The pulsar radiates in a narrow beam, and is observed from the Earth as a series of regular pulses. The pulses have been detected across the whole electromagnetic spectrum, from radio waves through visible light to X-rays and gamma rays.

Taurus–Littrow, a valley on the Moon produced during the formation of the *Serenitatis Basin. It was the landing site of Apollo 17, the last manned mission to the Moon in the *Project Apollo programme. The valley is totally enclosed by mountains some of which tower more than 2,000 m above the valley floor. Much soil has slipped down the sides of these mountains forming layers in the base of the valley.

Taurus X-1, a strong X-ray source whose position coincides with the *Crab Nebula supernova remnant. Discovered in 1963, it was the second X-ray source to be recorded, but the first to be identified with an optical object. It is possible that the X-rays come from a shell of gas heated by the *pulsar at the centre of the Crab Nebula.

tectonics, the deformation of the crust of a planetary surface. In the case of the Earth the tectonically significant layer extends into the mantle to a depth of about 400 km. The surface of the Earth can be seen in places to be creased into folds, often tens of kilometres across. The form which the deformation takes depends on the chemistry, water content, and porosity of the rock, as well as its pressure, strength, and temperature and the rate at which the compression takes place. Fast compression can lead to brittle fractures, whereas slow compression simply results in viscous flow. Sedimentary rocks, formed at surface temperatures and pressures, can find themselves pushed down to lower regions, where the

The **Tarantula Nebula**, in the Large Magellanic Cloud, is an active region of star formation; one-third of all the young stars and most of the Wolf–Rayet stars in the Cloud are to be found within it. The nature of the extremely bright central object is, however, unknown. The bright star some distance from the Tarantula Nebula is Supernova 1987A shortly after it exploded.

temperature and pressure is much higher. Tectonic processes in these regions can lead to a rearrangement of the original constituents of the rock and a recrystallization, these changes being known as metamorphism. Tectonic processes can occur after the crust readjusts itself to the release of the pressure that was exerted by overlying glaciation. Sideways pressure exerted by continental drifts also leads to tectonically produced mountains and faults.

tektite, a small glassy object found in certain localized areas of the Earth's surface. These areas include the southern half of the Australian continent, Indo-Malaysia, the Ivory Coast, Czechoslovakia, and Texas. Tektites have many different ages. For example, the Australites are 610,000 years old and the Moldavites are 14.7 million years old. About a million tektites have been found so far. They weigh up to 300 grams and are spheroidal, pear-, lens-, or disc-shaped. Their surfaces show evidence of having melted and re-solidified, and the glass is brittle. Many scientists think that tektites were formed by the fusion of terrestrial material during the impact of giant *meteorites, but some believe they were created by explosive volcanoes on the Moon.

telescope, an instrument for observing objects at a distance. Most wavelengths of *electromagnetic radiation can be collected by appropriately designed telescopes. In astronomy, an optical telescope is used to produce magnified images and to collect light from faint sources, particularly those invisible to the naked eye. Compared with the unaided eye, a telescope has more light-gathering power and provides better angular resolution, so that objects can be observed with more detail in the magnified image. A *refracting telescope uses a large lens (the objective) to collect the light and bring it to a *focus and the image is observed through one or more smaller lenses (the eyepiece). However, almost every modern form of collector of electromagnetic radiation employs a reflector. For example, the most common optical telescope used by professional astronomers is the *reflecting telescope. In addition to observing the image with the eye, an optical analysing instrument, such as a *photometer or *spectrometer, can be placed at the secondary focus of a Cassegrain or coudé telescope. For most types of telescope, if celestial objects are to be examined for any length of time, the *mounting supporting the telescope must be moved continuously to follow the apparent movements of the objects caused by the Earth's rotation.

As well as light-collecting telescopes there are instruments that are designed to collect electromagnetic radiation of other wavelengths. For example, the Lovell *radio telescope at Jodrell Bank has a large steerable metal dish that reflects and focuses incoming radio waves on to an aerial (US antenna). Other radio telescopes are based on the principle of *interferometry and can measure the position of radio sources with great accuracy. Telescopes have also been designed to detect wavelengths shorter than visible light, particularly X-rays and gamma rays. Because X-rays are absorbed by the Earth's atmosphere, an *X-ray telescope is usually carried on an *X-ray satellite or balloon. *Gamma-ray astronomy also makes use of telescopes carried on satellites. In addition, gamma rays can be detected indirectly at ground level by observing the flashes of light (known as *Cerenkov radiation) which are produced when energetic gamma rays enter the upper atmosphere producing a *cosmic ray event.

Telescopium, the Telescope, a *constellation of the southern sky introduced by the 18th-century French astronomer Nicolas Louis de Lacaille. Its brightest stars are of only fourth magnitude and it is devoid of interesting objects apart from a few faint galaxies, an unremarkable globular *cluster, and a small *planetary nebula.

Telesto, a small satellite of Saturn. Telesto moves in an almost circular orbit of radius 294,660 km in a period of 1.888 days. Two other Saturnian satellites, *Tethys and Calypso, have very similar orbits. Telesto is not spherical but rather ellipsoidal, with a mean diameter of 30 km.

Tempel's Comet, the first comet to have its *spectrum analysed. Emission bands were seen in the spectrum of the comet when it was discovered by the German astronomer William Tempel in 1864 and these were later recognized as being due to carbon, CN, and CH. Coupled with these was an underlying solar spectrum produced by sunlight scattered from the cometary dust. Many other comets are associated with Tempel. For example, Comet Tempel–Tuttle, discovered in 1866, is the parent of the *Leonid meteor shower. It has a period of about 33 years and the meteoroid dust that has been emitted from the nucleus is still close to the comet and has not yet spread out around the orbit. Therefore, although the Earth intersects the orbit every year, a major meteor swarm is seen only every 33 years.

temperature, the degree of hotness of an object. It describes the mean *kinetic energy of the individual molecules in the object. The mean kinetic energy of a molecule is equal to the mass of the molecule multiplied by the mean of the square of its velocity, and this is equal to the product of the temperature, measured on the Kelvin scale, and 1.5 times Boltzmann's constant (1.380622×10^{-23} J K^{-1}). Temperature is measured in terms of degrees and fixed points. The Celsius scale of temperature uses two fixed points. The melting-point of pure water under normal Earth-surface conditions is taken to be 0 °C, and the boiling-point of water is taken to be 100 °C. Stars have surface temperatures that vary between around 3,000 °C and 80,000 °C. The centres of stars have temperatures of many millions of degrees. This temperature is associated with high densities and pressures and the combination of these conditions with the high temperature enables the material in the star's centre to be transmuted by *nuclear fusion into elements of higher atomic mass, together with release of energy. Planets are much less massive than stars and the central temperatures are too small for this nuclear processing to take place.

tenth planet, a hypothetical major planet yet to be discovered. Since the mass of the outermost planet Pluto is too small to be the cause of the differences between theory and observation in the orbits of Uranus and Neptune, another planet might be responsible. Ever since the discovery of Pluto in 1930 the search for the next planet has been under way. The planet, if it exists, must be either fainter than Pluto, or located in a region of the sky overcrowded with stars, or have an orbit of very high inclination to the *ecliptic. Its influence on comets entering the inner solar system would be considerable; but there are insufficient cometary data to draw any firm conclusions. Sceptics claim that the observations of Uranus and Neptune which appear to support the tenth-planet hypothesis are old and were made when observational techniques and data analysis methods were less accurate, and that without a complete review reliable predictions cannot be made. The problem remains to be solved, either by discovering a new planet or by improving the *ephemerides of the outer planets.

terminator, the boundary separating the illuminated part of a non-self-luminous celestial object from the part in darkness. Planets and satellites, illuminated by the Sun, have terminators which move over these bodies' surfaces according to the changing position of the Sun with respect to these surfaces.

terrestrial planet *inner planet.

Tethys, a satellite of Saturn, discovered by Giovanni Domenico *Cassini in 1684. It has a nearly circular equatorial orbit at 294,660 km from the planet's centre. Its diameter is 1,060 km and its density is 1,200 kg/m^3, indicating a predominantly icy composition. All parts of its surface are heavily cratered. Two outstanding topographic features are the giant **Odysseus** crater, 400 km in diameter, and a trench or large valley, **Ithaca Chasma**, about 100 km in width and several kilometres deep.

Tharsis Ridge, the main area of volcanic activity on Mars. There are four main volcanoes on the ridge: *Nix Olympica (Olympus Mons), *Pavonis Mons, Arsia Mons, and Ascraeus Mons. The ridge has become elevated above the general level of the Martian surface by about 7 km. Volcanic activity is also exhibited by the fractures and faults, some up to several thousand kilometres in length, that are found over the ridge region. The average slope on the flanks of the Tharsis volcanoes is only about 3 degrees.

Thebe, a small satellite of Jupiter. It was discovered by the US astronomer S. P. Synnott when the *Voyager images were examined and was initially called 1979J2. With *Metis, *Adrastea, and *Amalthea it forms a group of four minor satellites that have orbits very close to the Jovian equatorial plane and which are within the orbit of *Io. Thebe has an average diameter of 100 km and its orbit has a radius of 221,900 km.

third quarter *phase.

three-body problem *N-body problem.

Thuban, a fourth-magnitude *binary star in the constellation of Draco, and also known as Alpha Draconis. One of its components is a hot *giant star at a distance of 71 parsecs, with about 150 times the luminosity of the Sun. Thuban is a spectroscopic *binary star with a period of 51 days. In about 2830 BC Thuban was within an angular distance of 10 arc minutes of the north celestial pole; owing to *precession, it will again be the pole star around AD 23,000.

tidal bulge *tides.

tidal friction *tides.

tides A massive body, by its gravitational attraction on a finite-sized distant body, will raise tides on the second body. If the second body has an atmosphere and water or other fluid on its surface, the tides produced by the first body in the atmosphere and fluid will be much higher than those produced in the solid body. Both the Sun and Moon produce tides in the Earth's atmosphere, oceans, and crust, the Sun's tide-raising force being about one-third that of the Moon. *Newton was the first to give an adequate explanation of ocean tides although the connection between the tides and the phase of the Moon had been noted from antiquity. Suppose the Moon were to act alone. Then at the Earth's centre, centrifugal force, due to the Earth being in orbit about the centre of mass of the Earth–Moon system, is balanced by these bodies' mutual gravitational attraction. On the side of the Earth nearer the Moon, however, the Moon's gravitational attraction is greater than centrifugal force while on the opposite side the centrifugal force is greater. In both cases, the consequence is that a **tidal bulge** is produced. Both tidal bulges follow the Moon in its orbit about the Earth. If the Earth were not rotating, therefore, an island on the Earth's surface would experience a high tide once every 13.65 days, or half the Moon's *sidereal period of revolution. Because the Earth is rotating once every 24 hours in the direction of the Moon's rotation the combination of these two periods produces at a given place on Earth two high tides every 24 hours 50 minutes.

The Sun's effect, being only one-third that of the Moon's, is evident in the relationship between the phase of the Moon and the phenomena of **spring** and **neap** tides. When the Moon is full or new (see *syzygy), the tide-raising forces of Sun and Moon reinforce each other so that the particularly high tides known as spring tides occur (see *syzygy). At lunar first quarter or third quarter, however, the Sun's effect opposes that of the Moon so that the high tides then, known as neap tides, are not nearly as high as the spring tides. The Moon's and the Sun's orbital planes do not lie in the Earth's equatorial plane; in addition the Moon's and Sun's geocentric orbits are elliptical. These factors, together with the fact that the distribution of land to sea is irregular, cause the simple picture outlined here to be modified to some extent.

The long-term effect of the tides on the Earth–Moon system, known as tidal evolution, is important. Transfer of angular momentum from the spinning Earth to the Moon in its orbit takes place via **tidal friction** between the waters of the Earth held back by the Moon and the ocean floors. The Moon slowly spirals outwards from the Earth, a process that will continue until the day (the Earth's rotation period) and the month (the Moon's sidereal period of revolution about the Earth) are equal and both are of length about forty of the present day's length. Thereafter, the braking effect of the Sun's tidal force will reverse the process bringing the Moon spiralling back towards the Earth to be finally broken up into fragments when it is close enough. This evolution due to tidal friction has been estimated to take something like 10 billion years.

time, the indefinite continued existence of the universe in the past, present, and future regarded as a whole. Time may be measured astronomically either with reference to the stars or to the Sun. The former, called *sidereal time, is conceptually simpler. Local sidereal time is defined with respect to the stars. For example, the sidereal *day is essentially the time of one apparent revolution of the *celestial sphere. More precisely, sidereal time is the *hour angle of the vernal *equinox. The *precession of the equinoxes, and sometimes *nutation, are allowed for but these effects are very small, so that sidereal time is almost entirely dependent on the Earth's rotation.

Civil timekeeping is based on solar time which is defined in terms of the position of the Sun in the sky. This depends both on the Earth's rotation and on its orbital motion. While the Earth's axial rotation appears uniform, its orbital motion is not, because its orbit is elliptical. A further complication arises because of the inclination of the Earth's equator to its orbit. For these reasons, the Sun itself is not a suitable point of reference for solar time. Instead, a fictitious body called the **mean sun** is adopted, which moves above the equator at a uniform rate. The **mean solar time** is then defined with respect to the position of the mean sun. It is the hour angle of the mean sun plus 12 hours, so that 12.00 corresponds to noon. The mean solar time on the Greenwich Meridian is chosen as a world standard and is called **Universal Time** (UT) or *Greenwich Mean Time. Civil time used in different parts of the world generally differs from UT only by an exact number of hours or half-hours.

Both sidereal and Universal Time depend on the Earth's rotation for their apparent uniformity. In the past, the Earth could be regarded as the standard timekeeper, as there was no clock capable of detecting irregularities or changes in the Earth's rotation rate. Even so, such irregularities were indicated by observations of the planets and the Moon, particularly *occultations which occurred systematically early in UT, indicating that the Earth's rotation period was gradually increasing, but in an unpredictable manner. Therefore in 1952 a new time-scale was introduced called **Ephemeris Time** (ET). This is a modified solar time-scale correcting variations in the Earth's rotation rate. It is defined to correspond to the uniform time that occurs in the motion of the solar system, and it is independent of the Earth's rotation. ET is used to predict the positions of the Sun, Moon, and planets, but it can be determined only with difficulty after observations of these bodies have been analysed. By contrast, UT is readily accessible, as it is directly related to sidereal time obtained from *transit observations of the stars. At present ET is about 55 seconds ahead of UT.

Irregularities in the Earth's rotation rate can now be measured directly by timing transit observations with atomic clocks. Atomic time-scales have been available since 1956 and **International Atomic Time** (TAI) was introduced in 1972. This is the most precisely determined time-scale available today. It is not an astronomical time-scale, as its basic unit, the SI second, is defined from atomic processes but in such a way that it does not differ significantly from the ephemeris second. It is conceivable that an astronomical time-scale like ET might not be uniform compared with an atomic time-scale. This would involve a variation in the gravitational constant and, as yet, no such variation has been established. For practical purposes, the two time-scales are in step and

$$\text{ET} = \text{TAI} + 32.18 \text{ seconds.}$$

The special theory of *relativity shows that different observers measuring the same interval of time may deduce different values. This has almost no effect on daily time-keeping but can become significant when speeds near the *velocity of light are involved.

Titan, the largest satellite of Saturn, discovered by the Dutch astronomer Christian Huygens in 1665. The diameter of its solid body is 5,150 km, slightly smaller than that of *Ganymede, but Titan also has an impenetrable atmosphere 200 km thick. After the *Voyager encounters, Titan's atmosphere (first detected by the Dutch-American astronomer Gerard Kuiper in 1944) has revealed a fascinating variety of chemical and physical phenomena. The surface pressure is about 1.5 times that of the Earth's and the temperature is about 94 K. Methane, ethane, and argon are the main atmospheric constituents. The atmospheric chemistry, under the influence of the solar ultraviolet radiation and energetic electrons in Saturn's *magnetosphere, produces complex molecules including hydrogen cyanide, an important compound in simulations of pre-biological organic chemistry on the early Earth. These trace constituents comprise the ubiquitous haze that totally hides Titan's surface from view. Conditions near the surface, however, have been inferred from models of the atmosphere, and it seems likely that they permit the formation of a global ocean, or at least lakes and seas, of methane, ethane, and other hydrocarbons. Deposits of complex organic polymers 'snowing' down from the atmosphere are also likely. On the other hand, photochemically produced hydrogen continually escapes the satellite's gravity, forming a detectable hydrogen ring around Saturn. A global wind circulation exists in the high atmosphere, as well as seasonal effects. The mean density of Titan is 1,880 kg/m^3 indicating the presence of a rocky core surrounded by an icy mantle.

Titania, a satellite of Uranus, discovered by William *Herschel in 1787. It has a nearly circular equatorial orbit about Uranus at 435,910 km from the planet's centre. It is about 1,600 km in diameter and its density of 1,600 kg/m^3 suggests that the interior contains a mixture of ice and rock. Photographed by *Voyager 2, its icy surface shows many small craters, a few large impact basins, and an extensive network of faults up to 5 km deep. Evidence was also found of an early period when the surface was re-formed.

total eclipse *eclipse.

totality *path of totality.

trajectory, in astronomy, the path in space followed by a natural or artificial object under gravitational or other forces. For all massive natural objects, the dominating force is *gravity and the trajectory, frequently an *orbit, is accurately described by *Newton's Laws of Motion. However, for smaller particles, especially ionized particles, electromagnetic forces can dictate the form of the trajectory.

Tranquillitatis Mare, an extremely level basin close to the equator of the Moon. It is the site of the first manned landing on the Moon and was chosen because of its flatness and its location. The spacecrafts Ranger 8 and Surveyor 5 had already visited the region to investigate its suitability. The Apollo 11 mission reached the Moon in mid-July 1969 and on 20 July, Neil Armstrong and Edwin Aldrin walked on the lunar surface. Rock and soil samples were collected and a seismometer, a laser reflector, and an aluminium (US aluminum) foil for collecting *solar wind particles were deployed. Tranquillitatis is an old lunar basaltic lava flow, its rocks being older than the oldest rocks known on Earth. They are bluish in colour, a consequence of their high titanium content. The surface of the mare is covered with the *craters of small asteroids which have collided with the Moon's surface.

transfer orbit *Hohmann transfer orbit.

transient lunar phenomena, minor coloured glows and localized regions of haze that occur at the surface of the Moon. They are typically seen around the rims of circular *maria and in the vicinity of *craters such as Aristarchus and Alphonsus. The frequency of occurrence maximizes when the Moon is near the perigee of its orbit. At this time the

Trapezium

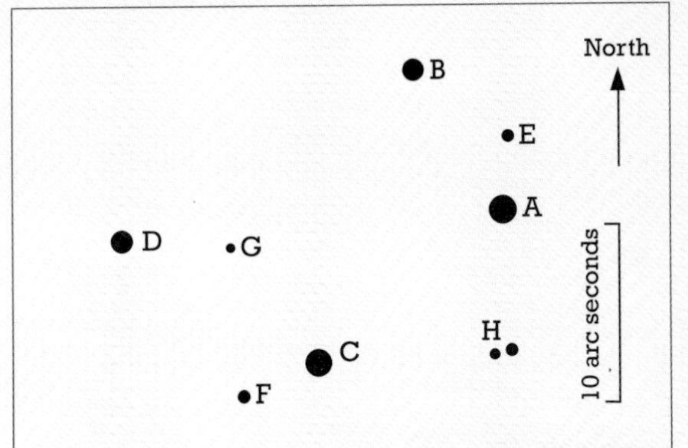

Telescopic view of the Trapezium, the multiple star Theta[1] Orionis. Stars A and B are eclipsing binaries. Of the fainter components, E and F can be seen in a telescope of 15 cm (6 inches) aperture.

lunar crust is under the maximum gravitational strain, moonquakes are most frequent, and gas and dust are being released from certain regions.

transit, the passage of a celestial body across a reference line. For example, it can refer to the time when a star or an artificial satellite crosses the *meridian of an observatory, or an inner planet is seen to cross the face of the Sun, or a Jovian satellite is seen to cross the disc of the planet. Through the history of observational astronomy transit phenomena have played an important role. For example, it was from variations in the timings of predicted transits and *occultations of satellites according to the distance of Jupiter from the Earth that *Roemer measured the velocity of light in 1675. Solar transits of the inner planets, particularly those of Venus, were used in the early determinations of the scale of the solar system. James Cook, the British explorer, recorded the transit of Venus of 1769 for this purpose at Tahiti in the Pacific Ocean as one of his scientific studies.

The recording of nightly passages of stars across an observatory meridian provides perhaps the best-known of transit phenomena, originally providing an accurate measure of the passage of *time. In the 19th century, time, particularly important for navigation, was obtained from transit circles with the *Royal Greenwich Observatory being chosen as the *prime meridian and providing the basis of *Greenwich Mean Time. With the advent of atomic clocks with their high precision and stability, transit observations are now used to monitor the variations of the rotation of the Earth caused, for example, by the melting of the polar ice caps.

Trapezium, a *multiple star at the centre of the *Great Nebula in Orion. Small telescopes show four components, apparently in the shape of a trapezium. Two of these are *eclipsing binaries, so there are at least six stars present, while several fainter nearby stars may be connected. The Trapezium probably forms part of a *cluster of hot, young stars, perhaps no more than 20,000 years old, located in and illuminating the nebula. Most of the cluster members are obscured behind the dust clouds. Widely spaced multiple stars elsewhere in the sky, which seem unlikely to have stable orbits, are called **Trapezium-like objects**; they may be considered as loose star clusters.

Triangulum, the Triangle, a *constellation of the northern sky, one of the forty-eight figures known to the ancient Greeks who called it Deltoton. Its brightest stars are of third magnitude. Triangulum contains M33 (NGC 598), the third-largest *spiral galaxy in our *Local Group of galaxies, visible as a large smudgy patch through binoculars in a dark sky. M33, sometimes known as the Triangulum Galaxy, is an Sc spiral, 900 kiloparsecs distant, with an absolute magnitude of −18.9, and an estimated true diameter of 16 kiloparsecs.

Triangulum Australe, the Southern Triangle, a *constellation of the southern sky introduced by the Dutch navigators Pieter Dirkszoon Keyser and Frederick de Houtman in the late 16th century. Its brightest star, Alpha Trianguli Australis, named Atria, is of second magnitude. The scattered *open cluster NGC 6025 is just visible to the naked eye.

Trifid Nebula, a beautiful example of an emission *nebula powered by the ultraviolet emission of young, hot stars. Its glowing red colour is produced when ionized hydrogen recombines to form neutral gas. The dark lanes within it give it its name since they divide the nebula into three parts. The Trifid Nebula is in the constellation of Sagittarius.

triple-alpha process, a *nuclear fusion reaction occurring in the later stages of a star's life when the central temperature has risen to 10^8 K. Three alpha particles or helium nuclei fuse to form a carbon-12 atom in a process where energy is released.

triple star, a *multiple star containing three components, normally bound together by gravitation. Two of the stars form a close *binary star, with the third star moving in a larger orbit about the system's centre of mass. In some

It is believed that the **Trifid Nebula** is an active region of star formation. The reddish part of the nebula is characteristic of glowing hydrogen gas, whereas the blue region is cooler gas that is scattering the light emitted by the star at the centre of the blue area.

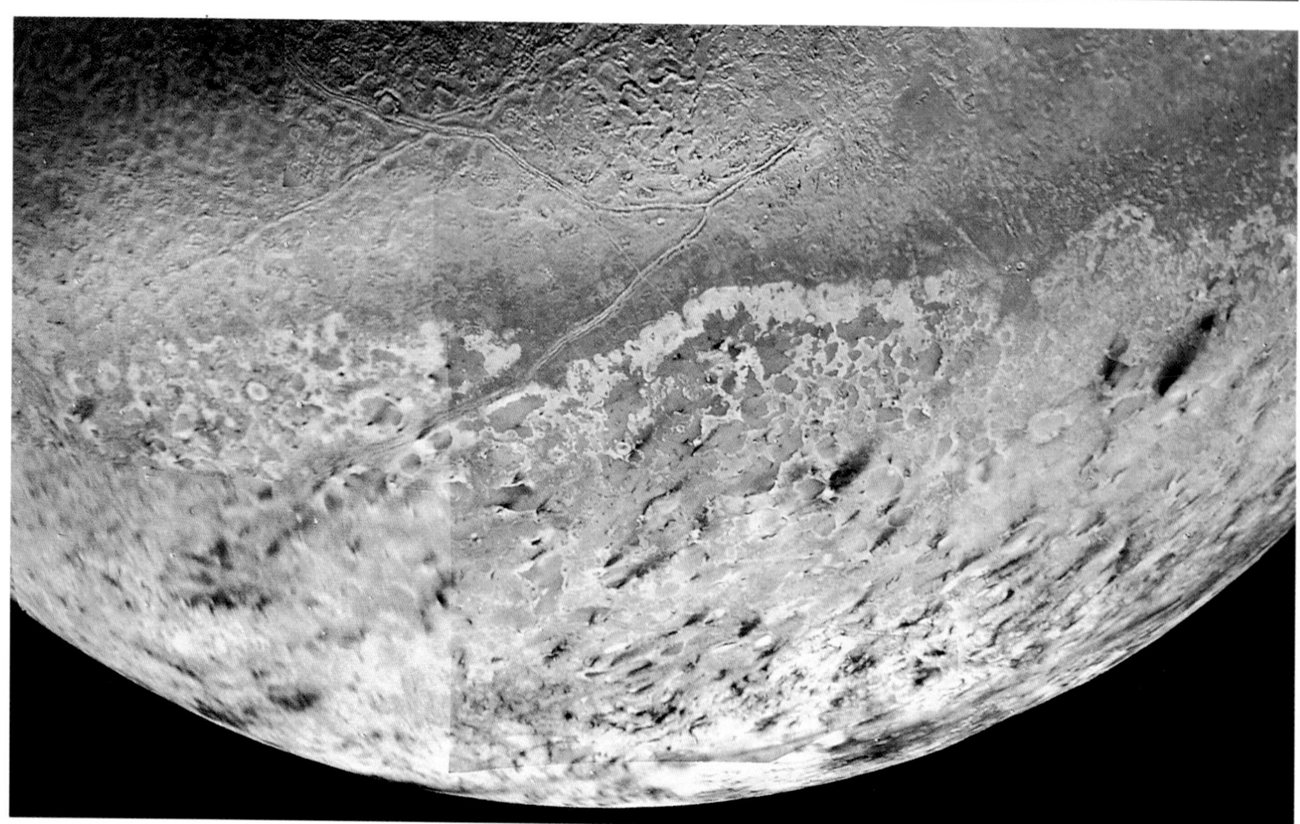

The great variety of surface features of Neptune's largest moon, **Triton**, is visible in this mosaic compiled from images taken by Voyager 2. The large south polar cap (at the bottom) is thought to be a slowly evaporating layer of nitrogen ice deposited during the previous winter. North of the cap, on the western (left) side of the disc, is the 'cantaloupe' region, characterized by small dimples with upraised rims and shallow central depressions. Long fractures cross the region and extend into the polar cap; towards the south this terrain has a light covering of frost. The north-east of the image shows smooth plains which may indicate recent outflows on to the surface.

apparently triple stars, however, one component is an unconnected star lying nearly in the same line of sight, as in the case of an *optical double star. The nearest stars to our solar system, *Proxima Centauri and the two components of *Alpha Centauri, form a triple system. The components in a triple may be too close all to be resolved in a telescope, but they may be detected as spectroscopic binaries or as *eclipsing binaries. Owing to the finite speed of light, the presence of a third component in an eclipsing binary such as *Algol can also be inferred if the observed times of eclipses show a periodic displacement consistent with orbital motion of the centre of mass of the eclipsing pair; this is the **light–time effect**.

Triton, the largest satellite of Neptune, discovered by the British amateur astronomer William Lassell in 1846. Somewhat smaller than the Earth's Moon, Triton orbits Neptune in the opposite direction to Neptune's orbital motion and with a significant inclination to Neptune's equator. This suggests that Triton may be a captured satellite, whose initially highly elliptical orbit was subsequently shrunk and made circular by the effects of *tides. The average density of about 2,000 kg/m^3 is very close to that of Pluto, implying that a mixture of ice and rock forms the interior below the icy crust. In general Triton and Pluto are probably very similar objects. *Voyager 2's close approach revealed that the surface of this satellite has extremely complex features and bright markings. These include a large white frozen methane and nitrogen polar cap, marked by dark spots and streaks. These are probably sites where liquid nitrogen below ground suddenly boils and jets out in a geyser-like manner, releasing plumes of darkened carbon-rich dust which is then deposited down wind. Much of Triton's equatorial zone consists of geologically young terrain crisscrossed by a network of low ridges and valleys which are probably faults in the crust from which viscous slush oozed out. Frozen lakes with terraced edges are also present in this zone, showing that water-based slurries erupted from the interior, flooded local lowlands, then receded before freezing. Large-scale lava flooding is common on the broad smooth plains, ranging in age from very ancient regions with closely packed superimposed craters, to comparatively young areas free of craters. This lava was probably a mixture of water, ammonia, and methane ices, which can flow much better than pure water ice does at Triton's surface temperature of 38 K. This volcanic activity was probably driven by tidal effects which heated Triton's interior. Triton's atmosphere is 60,000 times more rarefied than the Earth's and consists of nitrogen with a very small fraction of methane. However, it supports a haze of very fine particles from about 3 to 6 km above the surface.

Trojan group of asteroids, a group of *asteroids that travel on the same orbit as Jupiter but either 60 degrees in front of the planet or 60 degrees behind, that is, the angle between the asteroids and the planet is 60 degrees as seen from the Sun. One would expect that any slight difference in velocity would cause the asteroids to catch up with the planet or vice versa. The French mathematician Joseph Louis Lagrange solved the problem in 1772 when he discovered that an asteroid's position is stable if the Sun, Jupiter, and the asteroid form an equilateral triangle. The two places where this occurs are known as the L4 and L5 *Lagrangian points. Asteroids do not remain precisely at these points, but oscillate by around 20 degrees about them, these oscillations taking up to 150 years. There is no evidence that any of the known Trojans can escape from these points. *Achilles was the first leading Trojan to be discovered, in 1906. Collisions between Trojans are very gentle and *Hektor, which is 150 km wide by 300 km long, might even be two asteroids stuck together.

Trojan satellites, small bodies orbiting in the vicinity of the triangular *Lagrangian points of a planet–satellite system. Three such satellites have been discovered by the *Voyager probes in the Saturnian system: two (*Telesto, **Calypso**) orbiting 60 degrees behind and ahead of *Tethys and one **(Helene)** 60 degrees ahead of *Dione. All these bodies have nearly circular orbits in the plane of Saturn's equator. They are around 10 km in diameter and are probably icy fragments.

tropical year *year.

tropopause, the boundary of the Earth's atmosphere between the *troposphere and the *stratosphere. It is the region where temperature ceases to fall with height through the troposphere. Depending on the latitude and the seasons, it lies at a height of approximately 10–18 km (6–11 miles). The temperature is about −50 °C and the tropopause acts as an efficient cold-trap for water and stops it escaping into the upper atmosphere, from where it would be lost to the planet. The region is associated with cirrus cloud and lies just above the peaks of the Earth's highest mountains.

troposphere, the lowest region of the Earth's atmosphere. The troposphere lies between sea-level and the *tropopause at a height of approximately 10–18 km (6–11 miles). The air temperature decreases with height at the rate of 10 °C per kilometre. The cold atmospheric gases that lie on top of the relatively warm surface of the planet are heated by convection, a process whereby hot regions of the lower atmosphere rise up bodily through a region of higher density and then transfer their thermal energy to the cooler overlying regions. The troposphere contains about 70 per cent of the total mass of the Earth's atmosphere and the vast majority of the complex weather systems. Much of it is dominated by stratus, cumulus, and altocumulus clouds.

Trumpler's classification, a method devised by the US astronomer Robert Trumpler of classifying open *clusters. Three factors are recorded: the number of stars in the cluster, the degree of concentration of the stars towards the centre, and the spread in their brightnesses. In so doing Trumpler found a relationship between his classification and the cluster size.

T Tauri star, a very young, irregularly *variable star associated with a diffuse *nebula in a region of star formation, and still contracting to become a *dwarf star. The *spectrum shows emission lines of neutral iron at 404.6 and 413.2 nm which are a defining characteristic of this type. Young irregular variables associated with nebulae are called **Orion variables**, irrespective of the spectrum. *Stellar associations containing many T Tauri stars are called **T associations**. T Tauri itself varies between *magnitudes 9.3 and 13.5; it was discovered by the British astronomer John Russell Hind in 1852 and illuminates a *reflection nebula called Hind's Variable Nebula. The star's light changes are due to violent activity in its atmosphere and obscuration by moving clouds of gas and dust. The nebula's changes are due to intrinsic variation in T Tauri and to the play of light and shade on distant material as clouds closer to the star are in motion. Many T Tauri stars are sources of infra-red radiation.

Tucana, the Toucan, a *constellation of the southern sky introduced by the Dutch navigators Pieter Dirkszoon Keyser and Frederick de Houtman in the late 16th century. Beta Tucanae is a binocular double of fourth and fifth magnitudes. The constellation contains the Small *Magellanic Cloud and 47 Tucanae (NGC 104), a globular *cluster visible to the naked eye as a fuzzy fourth-magnitude star, second in prominence only to Omega Centauri. 47 Tucanae lies 5,000 parsecs (15,000 light-years) away and is estimated to contain about 100,000 stars.

Tunguska Event, a gigantic explosion, the equivalent of 12.5 megatonnes of TNT, that took place at 7.17 a.m. local time on 30 June 1908 in the basin of the river Podkamennaya Tunguska in Central Siberia. The most likely explanation is that a small *comet weighing about 50 million kilograms entered the Earth's atmosphere at a velocity of about 45 km/s. The comet approached Earth in the dawn sky but would have been much too faint to be seen. The friable nature of the nucleus of ice and dust ensured that it did not reach the ground and most of its energy was released when it broke up in the atmosphere making a blindingly bright, pale blue bolide. Eyewitnesses up to 500 km (300 miles) away saw, in a cloudless sky, the flight and explosion of the bolide, 'which made even the Sun look dark'. Seismic and air pressure waves were registered around the world and analysis of these waves indicated that the nucleus exploded 8.5 km (5 miles) above the Earth's surface. The ancient trees of the Yenissi taiga were torn up by their roots and in places piled up by the shock wave, their trunks pointing radially away from the centre of the explosion. The devastation extended over an area of 30 to 40 km (18 to 25 miles) radius, and the centre of the area was ravaged by fire scorching trees 18 km (11 miles) away. The explosion left in its wake a thick dust trail, and its sound, like gun-fire, reverberated for thousands of kilometres around. It is conservatively expected that about one such comet will hit the Earth every two to four thousand years.

turn-off point, the point on the *Hertzsprung–Russell diagram where the stars of a *cluster leave the main sequence. All the stars in a globular or open cluster may be assumed to have been formed from the *interstellar medium at the same time. Their subsequent evolution brings them in a relatively short time to a stable state where they shine by the energy released by the conversion of hydrogen into helium. The stars, if plotted by luminosity against surface temperature, then form the main sequence in the Hertzsprung–Russell diagram. They remain on the main sequence until this hydrogen-to-helium phase is completed. The stars of greatest mass leave (evolve) first and it is found that there is a point where stars of a certain mass are just beginning to leave the cluster main sequence, more massive stars having already left. This point is termed the turn-off point. The theory of *stellar evolution predicts that the turn-off point becomes progressively lower and redder on the main sequence according to the elapsed time since the birth of the cluster. Thus the position of the turn-off point is indicative of the age of the cluster.

twenty-one centimetre line, an important radio *spectral line at twenty-one centimetres *wavelength, produced by atoms of neutral hydrogen. The existence of the line, technically a hyperfine transition, was predicted by the Dutch astronomer Hendrik van de Hulst in 1944, and the line was first observed astronomically in 1951 by George Ewan and Edward Purcell in the US. It was the first spectral line detected by *radio astronomy and it remains the most important. Neutral hydrogen atoms make up over half of the interstellar gas, and are widely distributed in the Galaxy. All regions of the sky radiate at twenty-one centimetres wavelength. The radio waves are not affected by interstellar absorption, and so the distribution of hydrogen throughout the Galaxy can readily be investigated. By studying the small shifts in wavelength produced by the *Doppler effect radio astronomers can also measure the rotation of the Galaxy. In a similar way the distribution of gas in other galaxies, and their rotation, can be studied using the twenty-one centimetre line. For distant galaxies the expansion of the universe leads to a substantial increase in wavelength, or *red shift. The red-shifted hydrogen line has been observed at wavelengths as long as seventy centimetres because of this effect.

twilight, in astronomy, the two periods, in the morning and the evening, when the Sun is set, but the centre of its disc is less than 18 degrees below the horizon. Near midsummer, astronomical twilight lasts continuously all night at latitudes greater than about 48.5 degrees, as then, even at midnight, the Sun is less than 18 degrees below the horizon.

The above definition is precise but arbitrary; other definitions are possible. For civil twilight the Sun's centre must be not more than 6 degrees below the horizon, whereas for nautical twilight, not more than 12 degrees.

twinkling *scattering.

two-body problem *N-body problem.

Tycho, a *crater 87 km in diameter in the southern highlands of the Moon. It is 4.3 km deep and was formed about 100 million years ago, making it the most recent major lunar crater. The radiating rays from this crater cross much of the face of the Moon visible from the Earth and dominate the images of the full moon. The crater has a prominent central peak. The lunar soil near the rim of Tycho was sampled by the Surveyor 7 spacecraft and it was found that this highland soil was not only lighter in colour but was also richer in calcium and aluminium (US aluminum), and poorer in iron and titanium than the *mare soil. Many secondary craters were formed by the *ejecta thrown up from the Tycho basin.

Tycho Brahe (1546–1601), Danish nobleman of Swedish extraction who devoted his life to astronomy. He was the most accurate astronomical observer since the time of *Hipparchus and obtained stellar, lunar, and planetary positions to an accuracy of approximately 10 arc seconds without optical aid, relying on the use of a *quadrant. He was aware of the need for good sturdy instruments, which could be corrected for errors, and also for careful and multiple cross-checked observations extended over long periods of time. The patronage of King Frederick II of Denmark enabled him to build his double observatory of *Uraniborg and Stjerneborg on the island of Hveen in the Danish Sound. Tycho discovered a supernova, now known as *Tycho's Star, in the constellation of Cassiopeia in 1572. The lack of parallax convinced him that this object was not within the Earth's atmosphere or the solar system but in the region of the fixed stars. His observations of the comet of November 1577 indicated that it also was beyond the Moon. In order to explain his inability to measure stellar parallax he introduced a modified Copernican system in which the Sun and Moon orbited the Earth but the other planets orbited the Sun. Tycho left Denmark in 1597, spent some time in Hamburg, and arrived in Prague in 1599. He had his instruments shipped from Hveen to Prague where he decided to settle. His major work was the *Tabulae Rudolphinae* published posthumously in 1627. This listed the accurate positions of 1,000 stars as well as a long series of lunar, solar, and planetary positions. These were used by his pupil and assistant, *Kepler.

Tychonic system, a theory of the universe introduced by the Danish astronomer *Tycho Brahe in 1588 in his book *De Mundi Aretherei Recentioribus Phaenomenis*. In the Tychonic system, the Moon and Sun were in orbit around the Earth but Mercury, Venus, Mars, Jupiter, and Saturn were in orbit around the Sun. It was based on Tycho's observations of the movements of the planets against the celestial background.

In an attempt to prove that the Earth was the centre of the universe, **Tycho Brahe** measured the positions of the planets to an unprecedented degree of accuracy, a major achievement since the telescope had yet to be invented. However, when Johannes Kepler analysed his results, far from confirming his expectation, they were found to support the Copernican system.

It was a middle option in 16th-century cosmology that enabled the more timid astronomers to steer a safe course between the ancient *Ptolemaic system of nested eccentric spheres, and the new heliocentric cosmos described by the *Copernican system. As in the Ptolemaic system the stars formed a fixed permanent backdrop. Their relative movements and minute annual changes in position due to *parallax were not observable with 16th-century instruments.

Tycho's Star, a *supernova seen in the constellation of Cassiopeia by *Tycho Brahe in 1572, and also known as B Cassiopeiae. The first recorded sighting was by Francesco Maurolyko, a Sicilian mathematician, on 6 November, and Tycho first saw it on 11 November. It reached *magnitude −4 and was at this stage visible in daylight, but it eventually faded to invisibility. Tycho's measurements of its position over eighteen months showed that it was fixed in position among the stars. It had no detectable daily or annual *parallax, and so must have been a very distant object. This observation helped to demolish the medieval doctrine of the immutability of the heavens. The supernova was probably of Type I and no stellar remnant has been found but there is a *supernova remnant forming an expanding spherical shell of gas. This is an X-ray and radio source, and part of it is detectable on photographs as a faint *nebula.

UFO (Unidentified Flying Object), a phenomenon observed above the Earth's surface for which there is no immediate explanation. Many thousands of sightings of these objects have been reported all over the world in the past fifty years. Careful investigations by various agencies have satisfactorily explained about 93 per cent of authentic sightings as aeroplanes, high-flying meteorological balloons, the planet Venus, phenomena due to temperature inversions, and so on. There is no reliable evidence that the remaining 7 per cent are spacecraft of extraterrestrial origin. Natural phenomena as yet outside the recognition of modern science may cause them. On the other hand the possibility of the existence of *extraterrestrial life makes it impossible to dismiss the speculation that intelligent beings may be observing our planet.

ultraviolet astronomy, the observation of extraterrestrial ultraviolet radiation. At the end of the 18th century it was discovered by experiment that the spectrum of sunlight extended beyond both sides of the visible colours. Having produced a solar spectrum by means of a prism, William *Herschel investigated its heat content with a thermometer and discovered in 1800 that the maximum reading occurred at a position where there was no apparent light just beyond the red end of the spectrum and the radiation there took the term **infra-red**. Three years later experiments were reported by the Silesian scientist Johann Ritter on the response of photosensitive crystals (silver chloride) to various parts of the spectrum and he found that the strongest blackening of the grains was obtained for the region beyond the blue end of the spectrum and so discovered ultraviolet radiation. Both infra-red and ultraviolet are unseen by the human eye. Significant interest in the ultraviolet radiation emitted by astronomical bodies did not develop for another 150 years, when new technologies opened the necessary avenues. Most of the ultraviolet radiation is absorbed by the *ozone layer in the Earth's atmosphere and measurements of it can only be made from rocket and satellite platforms. Important results coming from the ultraviolet domain are on the brightness of stars as seen at these wavelengths, on the shapes of spectral lines which have indicated the presence of stellar winds, and the discovery of an absorption hump round about 220 nm caused by the dusty grains of the interstellar medium. *Infra-red astronomy explores the infra-red region of the spectrum.

umbra *eclipse.

Umbriel, a satellite of Uranus discovered by the British amateur astronomer William Lassell in 1851. It has a nearly circular orbit above the equator of Uranus at 266,300 km from the planet's centre. It is about 1,200 km in diameter and its density of 1,400 kg/m^3 suggests that the interior contains a mixture of ice and rock. Photographed by *Voyager 2, its surface is darker and more uniform than those of the other large Uranian moons; it is also heavily cratered.

Universal Time (UT), an alternative name for *Greenwich Mean Time (GMT), the mean *time kept by the

Greenwich time zone. Although this method of measuring the passage of time is related to the mean *diurnal motion of the Sun, the process of accurately deriving it is complicated by a number of factors, including the irregular rate of the Earth's rotation.

universe, the collection of all astronomical objects and the space between them. In the past, the term was sometimes used less comprehensively to mean the Galaxy or an external galactic system. In *cosmology today, the properties of model universes are compared with observations of the real universe to decide what kind of universe we live in.

Uranian inner satellites, ten small satellites orbiting Uranus inside the orbit of *Miranda, discovered by *Voyager 2. In order of increasing distance from the planet, their names are: **Cordelia, Ophelia, Bianca, Cressida, Desdemona, Juliet, Portia, Rosalind, Belinda**, and **Puck**. The largest one, Puck, is about 170 km in diameter, has a nearly uniform albedo of 0.07, and is almost spherical in shape. Cordelia and Ophelia orbit near either side of Uranus's outermost ring, and they probably act as *shepherding satellites confining the ring and retaining its sharp edges.

Uraniborg, an observatory built by *Tycho Brahe in the 16th century on the island of Hveen in Denmark. Set in a landscaped park, the building provided observation platforms for highly accurate *quadrants and circles by which Tycho and his staff observed the positions of planets and stars for twenty years. From these data *Kepler deduced his three laws of planetary motion.

Uranus, the seventh *planet from the Sun and the fourth most massive planet of the solar system. Uranus is generally not visible to the naked eye. It was discovered in 1781 by William *Herschel who initially thought it was a comet. Because of its slow motion, it became apparent that its orbit was nearly circular, and it was then clear that it was actually a planet. Uranus had, in fact, appeared in contemporary star catalogues but because it moves so slowly, it was thought to be a star. This previous information quickly led to a good knowledge of its orbit. Moreover, its computed distance from the Sun of 19.2 astronomical units agreed well with that expected on the basis of *Bode's Law, and this renewed interest in the attempt to fill the empty place in the sequence at 2.8 astronomical units, which resulted in the eventual discovery of the asteroid *Ceres. In the first part of the 19th century, the observed motion of Uranus departed from that computed on the basis of the then-known planets and the analysis of the required perturbations led to the discovery of *Neptune. If the formation of Uranus, like that of Neptune, took much longer than Jupiter and Saturn, allowing the light elements such as hydrogen and helium in their vicinity to be removed by the already formed Sun, this might explain the smaller mass and higher density with respect to Jupiter and Saturn. Among the outer planets, Uranus is characterized by several unique properties. Its equator is inclined at 97 degrees to the plane of its orbit, its satellites are all small to medium in size and are in nearly circular equatorial orbits, and it is surrounded by nine very narrow dark *rings, some of which are elliptical. Uranus apparently lacks an internal energy source comparable to those of the other outer planets and the line joining its magnetic poles is inclined at about 60 degrees to the rotation axis, a value that is larger than for any other planet.

Ursa Major, the Great Bear, the third-largest *constellation, lying in the northern hemisphere of the sky. It was one of the forty-eight constellations known to the ancient Greeks and according to mythology it represents Callisto, a secret love of Zeus who was changed into a bear. Its most distinctive feature is a group of seven stars that form the shape commonly known as the *Plough. Two stars in the bowl of the dipper point to the north celestial pole. The **Pointers** are Alpha and Beta Ursae Majoris, named Dubhe and Merak, each of second magnitude. The second star in the handle of the plough is Zeta Ursae Majoris, known as *Mizar. Xi Ursae Majoris was the first double star to have its orbit computed, which was accomplished by the French astronomer Félix Savary in 1828; it consists of fourth and fifth magnitude stars orbiting every 60 years. *Lalande 21185 is a nearby red dwarf. Some notable nebulae and galaxies are to be found in the constellation. M97 is the *Owl Nebula. M81 (NGC 3031) is a beautiful seventh-magnitude spiral galaxy visible in small telescopes with an unusual companion, M82 (NGC 3034), once thought to be an exploding galaxy but now believed to be a spiral encountering a cloud of dust. M101 (NGC 5457) is a face-on spiral galaxy.

Ursa Major Cluster, an open *cluster principally in the constellation of Ursa Major. It has more than 100 stars spread over an area of the night sky wider than 16 degrees. Five of its stars belong to the Plough, while other members lie in the constellations of Leo, Auriga, Eridanus, and Corona Borealis. A notable and rather surprising member is *Sirius which at first sight seems to be quite unconnected with Ursa Major.

Ursa Minor, the Little Bear, a *constellation at the north celestial pole, one of the forty-eight figures known to the ancient Greeks. In mythology it represents Ida, one of the two nymphs who nursed the infant Zeus. The constellation

This woodcut of the constellation of **Ursa Minor** was published in Johann Bayer's *Uranometria* in 1603. Now known as the Little Bear, to the Greeks this constellation represented Ida, one of the nymphs who hid the infant Zeus from his father Cronus who threatened to devour him.

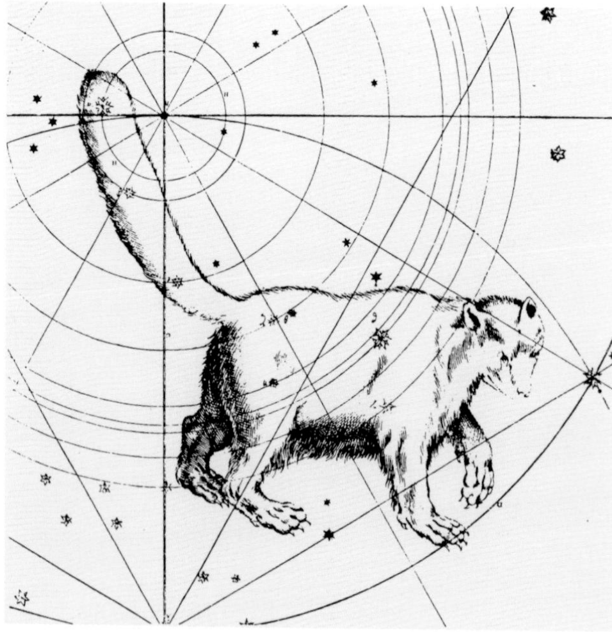

forms a ladle shape that is popularly known as the **Little Dipper**. At the end of the dipper's handle lies the north pole star, *Polaris. The two brightest stars in the bowl of the dipper, second-magnitude Beta and third-magnitude Gamma Ursae Minoris, named Kochab and Pherkad, are sometimes referred to as the guardians of the pole. Pherkad has a wide fifth-magnitude companion visible with the naked eye or binoculars. An *asterism near Polaris is known as the Engagement Ring.

Ursa Minor System, a faint elliptical galaxy which is a member of the *Local Group of galaxies. It is 67 kiloparsecs away and its approximate mass is only 10^5 times that of the Sun, in contrast to our own galaxy which is 2×10^{11} times the mass of the Sun, and the Andromeda Galaxy (M31) which is 3×10^{11} times the mass of the Sun.

Ursids, a meteor shower that reaches its peak around 23 December. It was first announced in 1945 by the Czech astronomer Antonin Bečvář. The maximum number of meteors per hour varies widely from year to year, and the shower is associated with the *comet Tuttle (1980 XIII) which has a period of 13.7 years.

Utopia Planitia, one of the northern plains of Mars that appeared from orbiting spacecraft to be covered in wind-driven dust. It was chosen to be the landing place of the *Viking 2 spacecraft and was found to be heavily littered with rocks. This spacecraft touched down on 3 September 1976 and worked through two Martian winters until 11 April 1980.

V

Valles Marineris, a vast system of canyons on Mars. The system was initially known as Coprates Canyon but was renamed in honour of the US *Mariner 9 space project. It stretches 4,000 km along the equatorial zone, eastwards from the *Tharsis Ridge. Individual canyons are 200 km wide and as much as 7 km deep. The canyon walls exhibit layering which was probably caused by the sequence of lava flows that initially made up the plains into which the canyons have cut. Most astronomers think that the canyons were initially produced by *tectonic processes in which the Tharsis Ridge volcanic activity extended the Martian crust and caused it to fracture. The resulting valleys were further enlarged by wind erosion. Ground water and subsurface ice also had a role to play in the canyon's evolution.

Van Allen radiation belts, doughnut-shaped regions surrounding the Earth containing energetic electrically charged particles, mainly protons and electrons. They were

Valles Marineris, probably the most dramatic feature on the surface of Mars, is a deep canyon stretching nearly a third of the way around the planet. Crossing the bottom half of the picture is the main canyon, the far wall of which shows several large landslides; a series of branch channels cut into the plateau from the near wall. This mosaic was constructed from photographs taken by the Viking 1 orbiter.

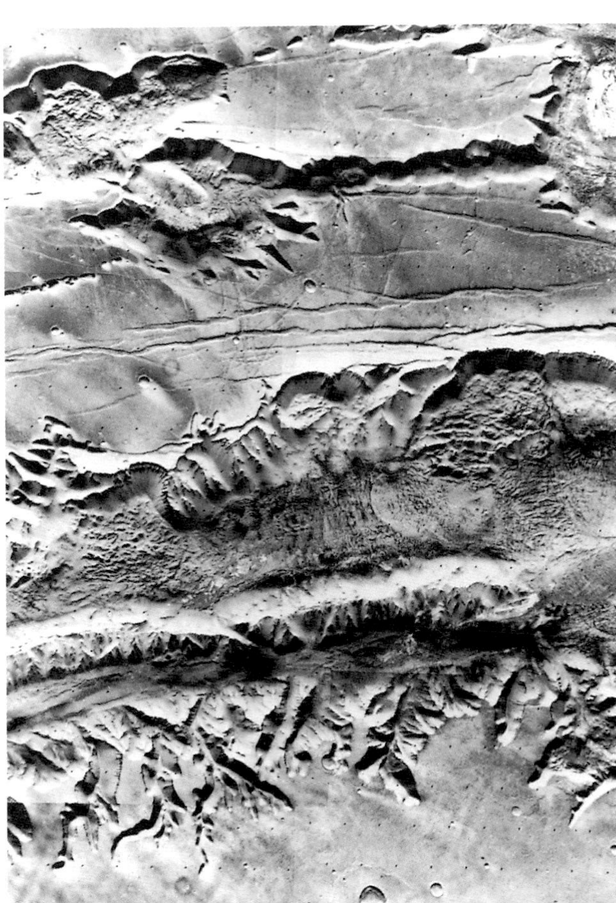

discovered by the US scientist James Van Allen in 1958 when cosmic ray detectors on board the first US satellite Explorer 1 registered signals too high to record when they entered the belts at a height of 2,000 km. The particles in the belts originate both from the *solar wind and from *cosmic rays and are trapped by the magnetic field of the Earth.

variable star, a star whose *magnitude changes over time. Over 30,000 such stars have been catalogued and about eighty types and subtypes are recognized. These may be grouped into **intrinsic variables**, whose changes are due to physical processes in or near the stars themselves, and **extrinsic variables**, whose changes are a geometrical effect due to rotation or orbital motion. The most important types of intrinsic variables are as follows.

Pulsating variables: *Cepheid variable stars, *Mira Ceti, *RR Lyrae stars, *RV Tauri stars, and *red giant semi-regular and irregular variables.

Eruptive variables: *shell stars; Orion variables, which are young irregular variables often associated with *nebulae, and include *T Tauri stars; *supergiant S Doradus stars or *Hubble–Sandage variables and the *R Coronae Borealis stars, both of which types are rare but of high luminosity; and red dwarf UV Ceti flare stars, such as *Proxima Centauri, which are of low luminosity but probably the commonest variables in the Galaxy.

*Cataclysmic variables: these include *novae, *symbiotic stars, and *supernovae.

*X-ray binaries: a few dozen have been identified as optically variable objects.

These categories account for 98 per cent of known intrinsic variables. Extrinsic variables include *eclipsing binaries, and rotating variables such as *pulsars and magnetic variables like *Cor Caroli. The Sun is a rotating variable, since its magnitude changes very slightly as sunspots move into and out of view at its limb.

Van Allen belts

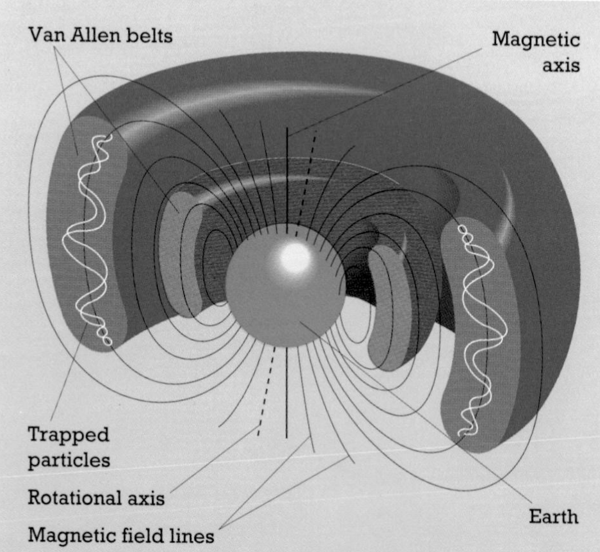

The inner van Allen belt, containing protons and electrons, is located around 2,000 km above the equator. The outer belt which is between 15,000 and 25,000 km above the equator curves in towards the poles. As a result of these charged particles the van Allen belts pose a potential hazard to the electrical systems of any Earth-orbiting spacecraft.

variation, a periodic change in the Moon's motion resulting from the gravitational attraction of the Sun and Earth. The period of variation is 14.77 days, or one-half the *synodic month. It is one of many such changes in the Moon's motion that make the creation of a satisfactory *lunar theory such a difficult task.

variation of latitude, the slight shift in latitude of a point on the Earth's surface because of displacements of the Earth's crust with respect to the Earth's axis of rotation. Careful measurements of the *zenith distances of *circumpolar stars as they transit across the observer's *meridian enable these slight changes of latitude to be measured.

Vega, a star in the constellation of Lyra, and also known as Alpha Lyrae, with an apparent magnitude of 0.03. It is a hot *dwarf star at a distance of 8 parsecs, with about fifty times the luminosity of the Sun. Infra-red radiation detected by the IRAS satellite in 1983 is due to a disc of cool material that may be forming planets. Vega was the first star to be photographed, at the Harvard Observatory in 1850.

Veil Nebula, a *supernova remnant in the constellation of Cygnus. It is the gaseous debris of a star that exploded some 30,000 years ago in a supernova outburst. The Veil Nebula forms a giant loop of material and is part of the *Cygnus Loop, a wispy shell of dust and gas more than three degrees across.

Vela, the Sails, a *constellation of the southern sky, one of the three parts into which the Greek constellation of Argo Navis was divided by the 18th-century French astronomer Nicolas Louis de Lacaille. Gamma Velorum is a double star of second and fourth magnitudes, divisible through binoculars and small telescopes; the primary is the brightest *Wolf–Rayet star. Vela lies in a rich area of the Milky Way, containing the *Gum Nebula and the *Vela pulsar.

Vela pulsar (Vela X-2), a *pulsar in the constellation of Vela that was discovered to be an *optical pulsar in 1977. Located in the remnant of a *supernova explosion some ten thousand years ago, it is one of the faintest known *variable stars. It has a magnitude in the blue part of the *spectrum ranging between 23.2 and 25.2 every 0.089 seconds, but it is the strongest gamma-ray source in the sky. Also in the constellation Vela is the pulsating *X-ray binary Vela X-1.

velocity, the rate of change of position with respect to time. It comprises both speed and direction. Speed is the distance moved in a specific time; velocity requires the direction of motion to be specified. All astronomical bodies are moving with respect to each other. In the solar system the planets are moving in elliptical orbits around the Sun and their speeds decrease as the inverse square root of their distance from the Sun. The Earth has an orbital speed of 29.8 km/s whereas Neptune's is only 5.43 km/s. The Sun is orbiting the galactic nucleus at a speed of about 250 km/s, the *escape velocity in the solar neighbourhood being around 360 km/s. Galaxies are moving away from each other and the universe is expanding. *Hubble found that the velocity of recession of a galaxy was directly proportional to its distance from the observer. The velocity of a celestial body along the line of sight can be measured using the *Doppler

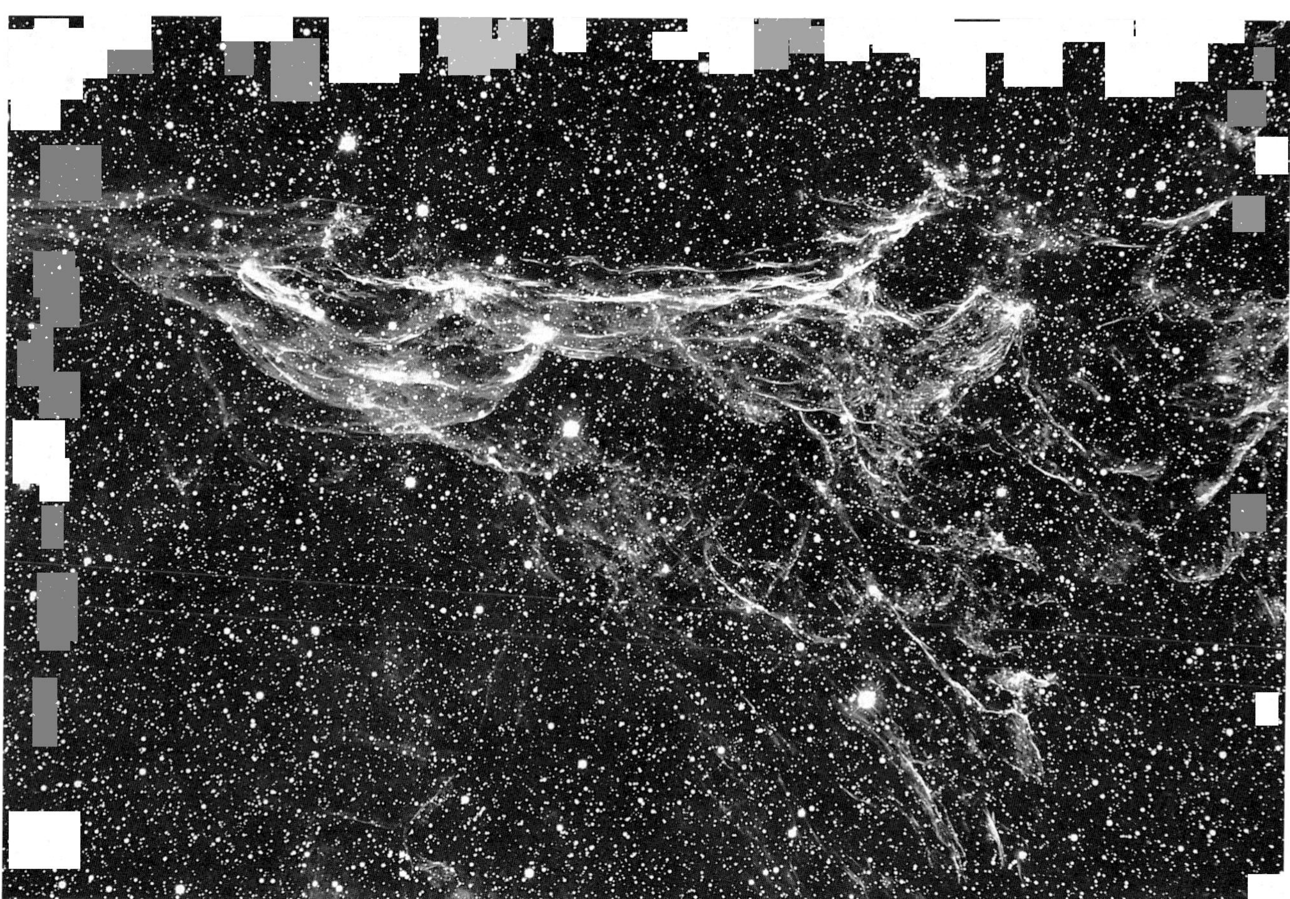

The **Veil Nebula**, discovered by William Herschel in 1784, is an old supernova remnant that is now dispersing into the interstellar medium. This huge filamentary arc, rich in oxygen and neon but with little hydrogen, is part of the gaseous shell of the exploding star. The arc is cooling and its rate of expansion is slowing down, being about 100 km/s at present.

effect. If a stationary source emits a *wavelength λ_u it will emit a wavelength λ_m when moving along the line of sight at a velocity of v. Doppler found that $(\lambda_m - \lambda_u)/\lambda_u = v/c$, where c is the velocity of light. Velocities perpendicular to the line of sight are measured by comparing photographic plates taken many years apart.

velocity–distance relation *Hubble constant.

velocity of light, the speed at which light and other *electromagnetic radiation travels in a vacuum. That the propagation of light was not instantaneous was first clearly shown in 1675 when the Danish astronomer *Roemer analysed eclipses of Jupiter's moons. The eclipses were sometimes early, sometimes late. Roemer deduced that the differences came from different light travel times depending on whether Jupiter was close to the Earth or distant. In 1728 Roemer's conclusion was confirmed when *Bradley discovered the *aberration of starlight. This effect is a slight annual variation in stellar directions which results from the Earth's orbital motion and the finite speed of light. From its size, the velocity of light, c, can be calculated. Neither Roemer's nor Bradley's value of c was accurate because the size of the Earth's orbit was poorly known. The first accurate measurements of c were laboratory ones made in the mid-19th century. Stroboscopic-like methods were employed, based on rotating toothed wheels or mirrors. In the 20th century ever more precise techniques involving electro-optical shutters, and, ultimately, stabilized lasers, led to measurements of c with a precision of a few parts in a billion. This accuracy surpassed that of the standard metre length, and led to imprecision in activities such as lunar laser ranging. In 1983 the metre was abandoned as a fundamental unit when, by international agreement, the speed of light in a vacuum was defined to be precisely 299,792,458 m/s (186,282.397 miles per second). Measurements of the velocity of light in standards laboratories, coupled with the atomic definition of time, now determine the length of the metre. Electromagnetic theory predicts that c in a vacuum is the same for all wavelengths of light. This is confirmed observationally by the simultaneous eclipses at all wavelengths of *eclipsing binary stars in external galaxies. It was the realization that c is the same for all experimenters that led Einstein to develop the theory of *relativity.

Venera, a highly successful series of sixteen Soviet spacecraft launched between 1961 and 1983 to explore the planet Venus. The early spacecraft simply flew past the planet, but later ones went into orbit around Venus. Some of these were in two parts; one part remained in orbit while the second part descended through the dense Venusian atmosphere to the surface. Measurements were made of temperature and atmospheric density and pressure, and samples of surface material were analysed. These measurements, as well as panoramic pictures of the surface, were relayed back to Earth via the orbiting module. The major surface features such as plateaux, mountain ranges, and volcanoes were mapped by radar from orbiting spacecraft.

Venus, the second major *planet in order of distance from the Sun. Also known as *Hesperus, the **evening star**, or *Phosphorus, the **morning star**, Venus can be the brightest object in the sky after the Sun and Moon and follows a near-circular path some 106 million kilometres from the Sun, taking 225 days to complete one orbit. It shows phases in the same way as Mercury and can also pass in front of the Sun during a *transit. The last transits were in 1874 and 1882 and the next are not until 7 June 2004 and 5 June 2012. In visible light the planet shows an almost featureless cloud-shrouded globe. However, in photographs taken with ultraviolet light, it shows dark patches and streaks, which are temporary breaks in the dense lower cloud levels. These race round the planet in only four days, carried by a high-altitude 350 km/h gale which never abates. The clouds closely resemble natural terrestrial fogs or industrial smogs and are composed of sulphuric, hydrochloric, and hydrofluoric acid droplets. The atmosphere is extremely dense and hot, made up of over 90 per cent carbon dioxide. The surface pressure is 91 times that of the Earth and the global surface temperature is kept permanently at about 475°C by the *greenhouse effect. The solid globe, with a mean density of 5,240 kg/m^3, is 12,104 km in diameter and rotates slowly from east to west, that is, in a retrograde manner, with a period of 243 days so that its day is longer than its year. The Soviet spacecraft *Venera 9 and 10 gave the first glimpses of the surface in 1975. Both spacecraft soft-landed and returned panoramic photographs showing hot stony desert landscapes scattered with rocks of all sizes. Recent radar studies suggest that this is typical of the whole planet and show that Venus has a cratered surface. Radar maps reveal two high continent-sized areas, Ishtar Terra and Aphrodite Terra, and a very high mountain feature, **Maxwell Montes**. There is also evidence for the presence of long canyons and volcanoes, though these seem to be inactive at present. Venus has no natural satellites.

This topographic Mercator map of **Venus** was constructed from radar data returned by the Pioneer Venus spacecraft. Highlands are shown in red, yellow, and green, and lowlands in dark blue. The huge area of light blue and blue-green is a plain covering 60 per cent of the planet's surface. Maxwell Montes, 11 km above the plain in the Ishtar Terra highlands, is the highest mountain on Venus.

vernal (spring) equinox *equinox.

vernier, a small movable graduated scale that slides along a fixed scale in order to obtain fractional parts of the divisions on the main fixed scale. It enables a quantity such as an angle to be measured with greater accuracy. On a sextant, for example, use of the vernier scale permits the altitude of the Sun or other celestial object to be measured to an accuracy of 10 arc seconds.

Very Large Array (VLA), the world's most powerful *radio telescope. Situated at Socorro in New Mexico, USA, it consists of twenty-seven radio dishes in a gigantic Y-shaped configuration 36 km (23 miles) across. The telescopes weigh 200 tonnes each, and can be moved to different positions along the arms of the Y using a specially designed transporter. The telescopes are linked to make use of *interferometry, with 351 different pairings of telescopes. The large number of telescopes, together with the Y-shaped design, enable the VLA to map a radio source in only a few minutes, whereas a conventional *aperture synthesis array

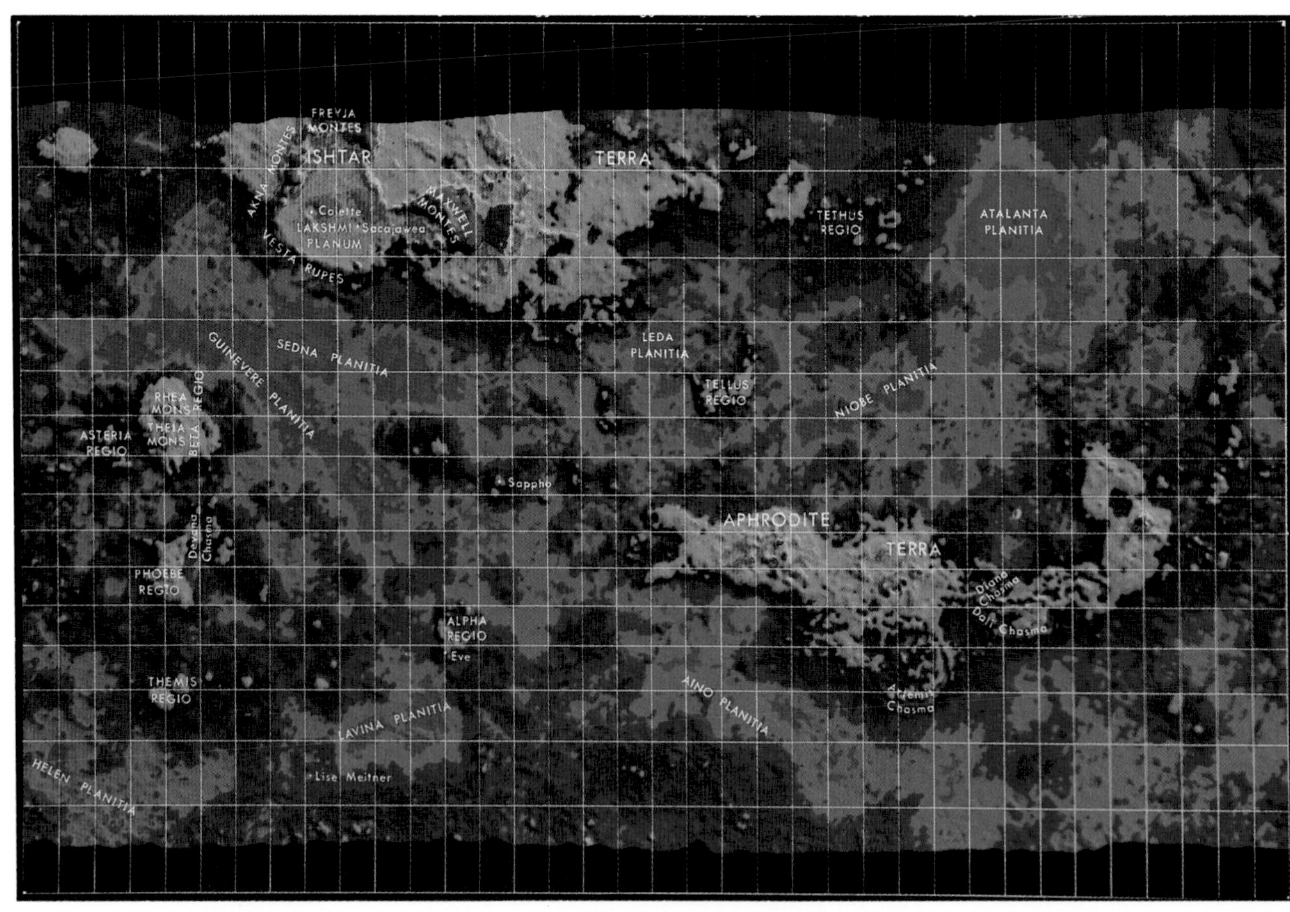

Each of the twenty-seven dishes of the **Very Large Array** radio telescope in New Mexico is 25 m in diameter. The data obtained by the dishes is combined, using a computer, to form a single radio image so that together the dishes effectively form one giant radio telescope.

would take many hours. The radio maps produced by the VLA are as rich and beautiful as optical photographs of the heavens. The VLA was completed in 1980, and is operated by the National Radio Astronomy Observatory.

Vesta, the brightest and third-largest *asteroid, first observed by the German philosopher and astronomer Heinrich Olbers in 1807. Its orbit has a semi-major axis of 2.362 astronomical units, an eccentricity of 0.0897, and an inclination of 7.14 degrees to the *ecliptic. Vesta probably has a flattened, nearly symmetrical shape, with a diameter of about 550 km, a rotation period of 5.34 hours, and a density close to 3,000 kg/m^3. Analysis of Vesta using a *spectrophotometer indicates that it has a basaltic crust of igneous origin, and therefore an interior made up of several different materials. Vesta's surface is not uniform, however, but has distinct patches of high and low albedo that cause brightness variations over every rotation period.

Viking, two successful US spacecraft sent to Mars. Launched in 1975, they went into orbit around Mars in 1976 a few months apart. Each consisted of an orbiter and a lander. The orbiters sent about 52,000 pictures of the Martian surface to Earth; they also relayed the information collected by the landers on the surface. Each of the two landers carried two television cameras, a seismometer, a meteorological station, and a small laboratory in which soil samples collected by a sampling arm and scoop could be analysed. A major concern in planning the Viking missions was to search for any signs of life on Mars. The results from the lander experiments were ambiguous and have been interpreted as being more probably due to very unusual chemical reactions rather than to the presence of life.

Virgo, the Virgin, a *constellation of the zodiac, and the second-largest constellation of all. In Greek mythology it represents Dike, goddess of justice. Its brightest star is *Spica. Gamma Virginis, named Porrima, consists of a pair of fourth-magnitude yellow stars that orbit each other every 171 years. About 5 degrees north of Gamma Virginis is the brightest *quasar, 3C 273, of thirteenth magnitude, about a billion parsecs (3 billion light-years) away. The *Virgo Cluster of galaxies lies in the northern part of the constellation,

In 1976 two identical **Viking** spacecraft transmitted back to Earth a wealth of scientific information about the planet Mars. The orbiters mapped the surface and the landers returned the first detailed data about the geology and meteorology of Mars. The simulated Martian terrain seen here behind this full-scale model of a lander is based on the data gathered by the original landers.

spilling over into *Coma Berenices. M104 (NGC 4594), the *Sombrero Galaxy, is not a member of the Virgo Cluster but is somewhat closer, about 11 million parsecs (35 million light-years) away. The Sun passes through the constellation between mid-September and early October.

Virgo A, a powerful radio source in the *Virgo Cluster of galaxies and associated with the centre of the giant elliptical galaxy M87. A jet of matter emerges from the centre at high speed and it is thought that the enormous energy required for this jet might come from a massive *black hole in the nucleus of M87, and into which stars may be falling.

Virgo Cluster, a group of several thousand galaxies in the constellation of Virgo at a distance of 11 megaparsecs. The Virgo Cluster is the centre of the *Local Supercluster of galaxies to which the *Local Group of galaxies belongs. The French astronomer Gérard de Vaucouleurs suggested that this supercluster may contain as many as 50,000 galaxies.

visual binary *binary star.

VLBI (very long baseline interferometry), a technique for combining signals from widely separated *radio telescopes in order to study the fine detail of radio sources. Even a large radio telescope has poor *resolving power compared to that of a small optical telescope so that long baseline *interferometry is used to increase the overall system's resolving power.

Volans, the Flying Fish, an inconspicuous *constellation of the southern sky introduced by the Dutch navigators Pieter Dirkszoon Keyser and Frederick de Houtman in the late 16th century. Gamma and Epsilon Volantis are both attractive double stars for small telescopes. Volans lies near the Large *Magellanic Cloud and contains no stars brighter than fourth magnitude.

This false colour image of the giant elliptical galaxy **Virgo A** was mapped at a wavelength of 2 cm by the Very Large Array radio telescope in New Mexico. The red spot on the left is the nucleus of the galaxy, which is thought to contain a black hole. Both optical and radio observations clearly show the narrow jet emanating from the nucleus.

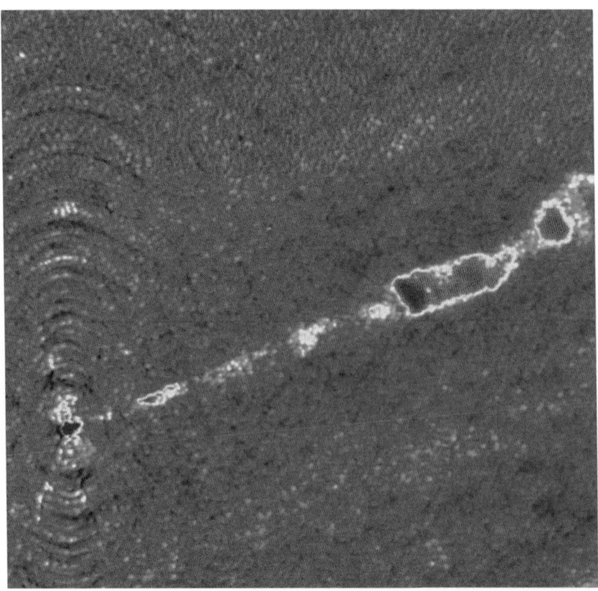

Voyager

Both Voyager 1 and 2 are equipped with eleven measuring instruments. They can each measure magnetic fields, low-energy charged particles, cosmic rays, and the characteristics of plasma between the planets. The large parabolic antenna on each probe is always pointing towards Earth, enabling permanent radio communication.

The electrical power required to run the measuring instruments and other equipment on board is supplied by thermoelectric generators.

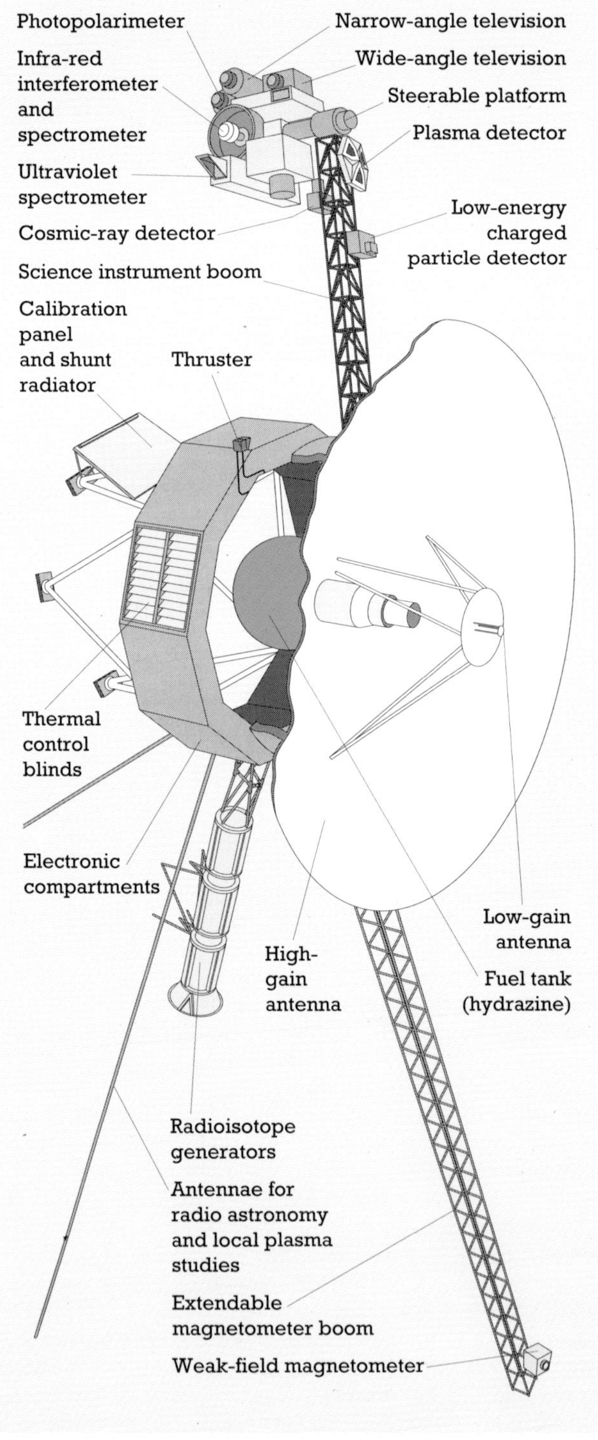

After **Voyager** 1 (orange track) was launched on 5 September 1977, it encountered Jupiter on 5 March 1979, Saturn on 12 November 1980, and then turned sharply and headed out of the solar system. Voyager 2 (red track) was launched on 20 August 1977, before Voyager 1. It visited Jupiter on 9 July 1979, Saturn on 25 August 1981, Uranus on 24 January 1986, and Neptune on 24 August 1989. It is now approaching the edge of the solar system.

Voyager, two unmanned US spacecraft designed to explore the *giant planets. Voyager 1 was launched in 1977 and reached Jupiter in 1979 before using Jupiter's gravitational field to put it on a new trajectory to Saturn, which it reached in 1980. Voyager 1 is now leaving the solar system. Voyager 2, launched in 1977, reached Jupiter in 1979 before following Voyager 1 to reach Saturn in 1981. It has since passed by Uranus in 1986 and Neptune in 1989. Like Voyager 1 it is destined to leave the solar system. The Voyager missions have been spectacularly successful in the wealth of new scientific information they have transmitted back to Earth, especially in the multitude of pictures taken of Jupiter, Saturn, Uranus, and Neptune and of their natural satellite systems. Many small satellites have been discovered, the surface features of the big satellites have been revealed, and the intriguing fine detail of Saturn's rings has been closely observed. In addition, active volcanoes on Jupiter's satellite Io have been discovered.

Vulcan, a hypothetical planet moving inside the orbit of *Mercury, long searched for by 19th-century astronomers. The existence of Vulcan was assumed to explain the anomalous precession by 43 arc seconds per century of the perihelion of Mercury. *Le Verrier in particular applied to the orbital perturbations of Mercury the same analytical techniques which had led him to the discovery of Neptune, suggesting that an unknown planet inside Mercury's orbit could be responsible. Claims that Vulcan had been detected during total solar eclipses and alleged *transits over the solar disc were raised several times in the second half of the 19th century, but never confirmed by subsequent observations. Photographic searches, carried out at the beginning of this century, concluded that any object with a Vulcan-type orbit cannot be larger than about 100 km in diameter. However, it is possible that smaller, asteroid-like bodies named **vulcanoids** actually orbit inside Mercury's orbit; they may be best detectable at infra-red wavelengths. The anomalous precession of Mercury's perihelion was finally accounted for by Einstein's theory of general *relativity.

Vulpecula, the Fox, a *constellation of the northern sky, introduced by the Polish astronomer Johannes Hevelius (1611–87) under the original title of Vulpecula cum Anser, the Fox and Goose. Brocchi's Cluster (also called Collinder 399) is an attractive group of sixth- and seventh-magnitude stars, popularly termed the Coathanger on account of its shape. Vulpecula lies in a rich area of the Milky Way and includes the *Dumb-bell Nebula. In 1967 the first *pulsar was found in Vulpecula.

Waterbearer *Aquarius.

Water Snake *Hydra.

wavelength, the distance between neighbouring minima or maxima in waves existing in a medium. For example, the distance between troughs or peaks in waves in a pond of water is a wavelength. In the case of *electromagnetic radiation, which travels with the velocity of light and was once thought to be oscillations in a hypothetical medium called the *ether, the wavelength λ multiplied by the frequency ν (the number of waves passing a given point each second) equals the velocity of light c:

$$\nu\lambda = c.$$

In electromagnetic radiation the wavelengths range from hundreds of metres (for radio waves) to as small as 10^{-16} of a metre (for gamma radiation). The observed wavelength depends upon the relative velocity of the source and observer and is described by the *Doppler effect.

wave mechanics *quantum theory.

weight, the *force of gravitational attraction exerted on a body. This concept is usually applied to objects that are on the surface of a planet or satellite. The force is directed towards the centre of mass of the planet and has a magnitude equal to the mass of the body multiplied by the surface gravity of the planet. The mass of a body is related to the quantity of material that it contains. The surface gravity changes considerably from planet to planet. The value on the surface of the Moon is only 16.5 per cent of that on the surface of the Earth and so objects weigh six times less there.

weightlessness, the effect of zero *gravity. It occurs in free fall. Consider a person inside a spacecraft that is falling freely. The rate of acceleration of the person and the spacecraft are the same, so the person seemingly has no weight and floats freely. During most phases of space travel the astronauts are weightless. The human body is not accustomed to this state and on long space flights special exercises must be carried out to ensure that there are no long-term effects. Some Soviet cosmonauts have spent around a year under conditions of weightlessness and it seems that no long-term damage results.

Weizsacker's theory, a theory of the origin of the planets. The *Laplace nebular hypothesis proposed that a rotating disc of dust and ice is formed in the pre-planetary nebula. In 1944 the German physicist Carl Friedrich von Weizsacker suggested that the disc broke up into a pattern consisting of rings of turbulence-induced eddies. These were spinning clockwise in a system that was as a whole spinning anticlockwise. Material at the boundaries between the eddies moving with high velocity and friction between the opposing motions of this boundary material caused it slowly to stagnate and then condense to form a collection of *planetesimals around the ring boundary. A judicious choice of the number of eddies per ring enabled the theory to match the radii of the rings of planetesimals with the present-day radii of the planetary orbits and the requirements of *Bode's Law.

Whale *Cetus.

Whipple's Comet, any of six *comets discovered by the US astronomer Fred L. Whipple. Whipple has been studying comets, meteors, and cosmic dust all his life and in 1951 he wrote two key scientific papers which concluded that the cometary nucleus must be a spinning dirty snowball. This model was the foundation for considerable theoretical work in the following decades and was proved to be essentially correct when *Giotto flew past *Halley's Comet in 1986. Whipple's Comet of 1933 is a faint comet that returns every 7.5 years and has been seen eight times. Brighter ones were Whipple–Bernasconi–Kulin, discovered in 1942, and Whipple–Fedtke–Tevzadze, discovered in 1943. The three names indicate that there were three independent discoverers. Whipple–Fedtke–Tevzadze had a tail that not only changed rapidly within a few hours but also broke up. The spectrum of this comet showed that it was almost purely gaseous with little dust.

Whirlpool Galaxy, an eighth-magnitude *spiral galaxy in the constellation of Canes Venatici. It has a small satellite galaxy at the end of one spiral arm. Because it lies face-on to us and is comparatively near (6 megaparsecs), its structure has been extensively studied with large telescopes.

white dwarf, a hot star (typically 30,000 K) with low luminosity and very high density (100,000 to 100 million times that of water). A white dwarf has exhausted its nuclear fuel and is composed of *degenerate matter, so that the only remaining source for the energy radiated is the star's residual heat. The diameter of a typical white dwarf is similar to that of the Earth, but the mass is about that of the Sun; the upper limit is the *Chandrasekhar limit of 1.44 times the

The striking spiral structure of the **Whirlpool Galaxy** is the result of the gravitational interaction between the large galaxy (NGC 5194) and the smaller galaxy (NGC 5195) which orbits around it. The nucleus of NGC 5194 contains hot, rapidly moving gas and dust, as well as many young hot stars. NGC 5195 appears to be developing into a barred spiral galaxy, perhaps because of the gravitational interaction.

Sun's mass. White dwarfs are the most common closing stage of *stellar evolution and several are found in the vicinity of the Sun, but they are faint objects and none are visible to the naked eye. The best known and first to be discovered is the companion to *Sirius. Many of them are pulsating *variable stars. Single white dwarfs will eventually cool down to become non-luminous bodies. The presence of white dwarfs in close binary systems gives rise to the *cataclysmic variable stars, and to the outbursts of Type I *supernovae.

white hole, a theoretical region of space in which matter is created and ejected into the universe. This is in contrast to a *black hole which drags matter in its vicinity out of existence from the universe. It has been speculated that, in a manner not at all understood, a black hole and a white hole, connected by a tunnel, may act as a means of transporting matter from one point in our universe to another.

Wien's Displacement Law, the relation between the *wavelength of peak emission from a *black body, λ_{peak}, and the temperature, T, of the black body. The mathematical expression is $\lambda_{peak}\ T$ = constant, and can be derived from *Planck's Law of black body radiation, which historically it predates. The value of the constant, known as Wien's constant, depends on details of exactly how the radiated *spectrum is measured, but a value of 0.003 metres Kelvin is representative. Although celestial bodies rarely radiate precisely like black bodies, the Displacement Law can nevertheless be used to indicate near which wavelengths most of their radiation is emitted. Examples include: (1) the *microwave background radiation, $T = 2.7$ K, $\lambda_{peak} = 1$ mm; (2) a cool, red star, $T = 3{,}000$ K, $\lambda_{peak} = 1\ \mu$m in the infra-red region of the spectrum; (3) the Sun, $T = 6{,}000$ K, $\lambda_{peak} = 500$ nm in the green part of the visible spectrum; (4) the hottest normal stars, $T = 30{,}000$ K, $\lambda_{peak} = 100$ nm in the satellite ultraviolet; and (5) *planetary nebula nuclei and proto-*white dwarf stars $T = 100{,}000$ K, $\lambda_{peak} = 30$ nm in the extreme ultraviolet spectral range. The Law won the 1911 Nobel Prize for physics for its discoverer, the German physicist Wilhelm Wien (1864–1928).

Wolf–Rayet star, a very hot star (up to 50,000 K) whose *spectrum contains broad emission lines, mainly of neutral and ionized helium, and also ionized carbon (WC stars) or nitrogen (WN stars). They may be the exposed helium-rich cores of evolved *giant stars that have lost their outer hydrogen envelope before continuing their *stellar evolution to the *white dwarf stage. The emission lines may originate from material ejected in stellar winds of very high velocity (about 2,000 km/s). The mass outflow is often unstable and slight variations in magnitude are frequently observed. Wolf–Rayet stars are rare as the stage is short-lived, but they have high luminosity and can be detected at great distances. The brightest example is Gamma^2 Velorum, of apparent magnitude 1.8 and absolute magnitude −6.7, at an estimated distance of 460 parsecs. Wolf–Rayet stars are *Population I stars and are strongly concentrated towards the plane of the Galaxy. The central stars of *planetary nebulae have spectra similar to those of Wolf–Rayet stars, but have lower luminosity and are old Population II disc stars. The Wolf–Rayet stars are named after the French astronomers Charles Wolf and Georges Rayet who discovered them in 1867.

world line, the path of a particle through four-dimensional *space–time. The path describes the motion of the particle through space as well as its evolution in time. In the four-dimensional universe all particles, including those making up larger bodies, have their own paths from the point of creation to the point of annihilation. Therefore, for example, the world line of a planet orbiting the Sun would be made up of the individual world lines of its constituent particles and would form a fixed spiral tube in space–time.

world model *cosmology.

wrinkle ridge, a ridge-like feature that is generally found on the surface of a lunar *mare. Concentric wrinkle ridges occur in the major ringed basins such as *Imbrium, Crisium, Humorum, Moscoviense, *Serenitatis, Nubium, Fecunditatis, and *Tranquillitatis. Wrinkle ridges are typically several tens of metres high, several kilometres wide, and can be hundreds of kilometres long. They could have been formed during the cooling, shrinking, and subsidence of the lava that filled the mare basins, and may be a result of the deformation, compression, and folding of the original mare surface. Some wrinkle ridges are formed along the rims of buried crater basins. Others might have been produced by the eruption or slow extrusion of lava along ancient cracks associated with the underlying crater topography. This process would lift up the overlying solidified lava surface.

X-ray binary, a close *binary star in which one of the components is a *white dwarf, *neutron star, or a *black hole. The other star may be a normal star from which material is drawn into an *accretion disc around its smaller companion. As a consequence the binary emits X-rays, the intensity of which depends upon the nature of the larger (accreting) component. The most luminous sources are binaries with a black hole or neutron star component. The temperature and luminosity are both proportional to the mass of the accreting component and inversely proportional to its radius. Examples are Centaurus X-3 and *Cygnus X-1. In some cases the X-ray emission pulsates, indicating the presence of a rotating magnetized neutron star.

X-ray burster *burster.

X-ray satellite, an artificial satellite containing one or more *X-ray telescopes. In 1970 the first of a series of X-ray satellites was launched carrying X-ray telescopes to survey the sky in the X-ray region of the electromagnetic spectrum. These satellites, which include Uhuru and Einstein (High Energy Astrophysics Observatory-B), have revealed that in addition to many individual X-ray sources in the universe, there seems to be a faint background of this radiation whose origin is at present uncertain.

Clusters of galaxies are often strong emitters of X-rays. The X-rays originate from hot gas, typically tens of millions of degrees, that is stripped from the member galaxies by their mutual gravitational interaction and which then accumulates in intergalactic space. This false colour image of the cluster Abell 2256 was taken by ROSAT, the Roentgen **X-ray satellite**, launched on 1 June 1990. The colour coding runs from red (the most intense emission), through yellow, to blue (the weakest emissions).

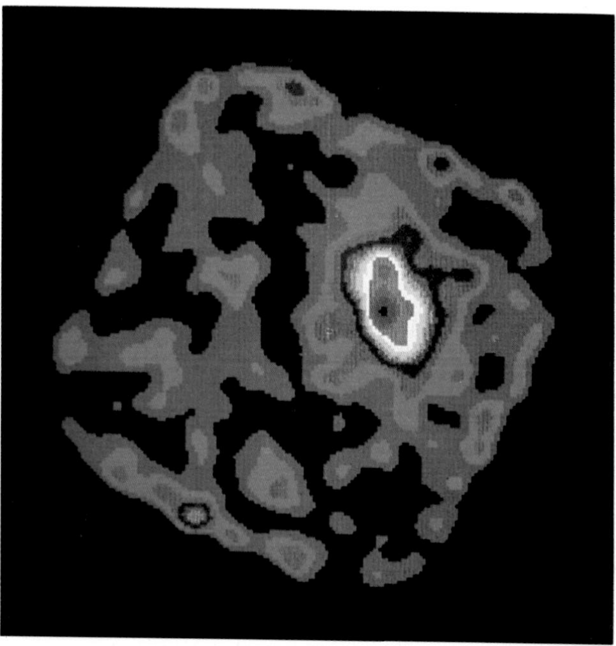

X-ray telescope, a device used in *XUV astronomy for detecting and measuring X-rays from distant sources and for determining the positions of those sources. The extreme penetrating power of X-rays makes it difficult to design telescope-like apparatus but techniques have been developed in recent years utilizing coincidence detectors, collimators, and occultation devices. X-rays are focused by reflecting them at very small angles. In some telescopes the X-rays are collected through a honeycomb of metal tubes and the *field of view depends on the length and diameter of the tubes. X-ray telescopes have been flown in an *X-ray satellite to map out the X-ray universe, that is, the distribution of sources in the sky that emit radiation in the X-ray region of the spectrum.

XUV astronomy, astronomy using instruments capable of detecting that region of the electromagnetic spectrum involving ultraviolet and X-rays. This recently developed field of astronomy is mapping the number of sources in the universe emitting such radiation. In particular, the artificial Earth satellite ROSAT (Roentgen satellite), which came into operation in January 1991, has pin-pointed a number of *Seyfert galaxies as strong X-ray emitters. During its first year of operation ROSAT detected over a thousand XUV sources.

year, the time required for the Earth to complete one orbital revolution about the Sun. Depending on how the measurement of this revolution is made, four astronomical definitions of the year are possible.

The **sidereal year** is the Earth's orbital period with respect to the stellar background.

The **tropical year** (also known as the **astronomical**, **equinoctial**, or **mean solar year**) is the interval of time between two successive vernal *equinoxes. However because of *precession the equinoxes are not fixed in the *celestial sphere and so the tropical year differs slightly from the sidereal year.

The **anomalistic year** is the interval from one perihelion to the next. This definition is necessary when the perturbations of the other planets on the Earth's orbit are taken into account.

Finally, the **eclipse year** is defined as the interval between successive passages of the Sun through the same *node of the Moon's orbit. This interval, which differs significantly from the others, indicates the times when *eclipses are likely to occur.

The lengths of the four years defined above are shown in the table for time on page 192.

The **calendar year** is that used most commonly for the reckoning of time in civil matters. Based on a calendar, it contains either 365 or 366 days. The occurrence of leap years is arranged so that the average length of the year is as close as possible to the tropical year. Minor corrections to the length of the calendar year are needed from time to time to take account of changes in tide patterns and bulk movements in the Earth's interior, each of which has an effect on the relative periods of axial rotation and of orbit about the Sun. A *lunar calendar is based on the **lunar year** which is usually taken to be 12 *synodic months (354 days). It is therefore related to the motion of the Moon about the Earth rather than the Earth about the Sun.

Yerkes Observatory, an *observatory on the northern shores of Lake Geneva, at Williams Bay, Wisconsin and now part of the University of Chicago. It was founded in 1892 by the financier Charles Tyson Yerkes to create a telescope that was bigger than that at the *Lick Observatory. With a 40-inch (1.02 m) achromatic doublet lens, the telescope is still the largest *refracting telescope in the world. This record is unlikely to be broken, as it is now much cheaper to provide the equivalent light-gathering power with a *reflecting telescope. In the field of positional astronomy the Yerkes telescope is unsurpassed. The 19 m (62 feet) focal length provides a superb star image and the lens has not been moved from its cell for ninety years, thus allowing the accurate comparison of star positions over that period.

Z

Zeeman effect, the splitting of a *spectral line into several polarized components under the action of a magnetic field. By measuring the degree of *polarization and the splitting of spectral lines from a cosmic source, astronomers can estimate the strength of the magnetic field within the source. The Galaxy has an average magnetic field about 100 thousand times weaker than the magnetic field of the Earth. However, the magnetic field is higher in dense clouds of gas where new stars are forming. In *sunspots the magnetic field is up to 10 thousand times greater than the magnetic field of the Earth, giving rise to the Zeeman effect.

Zel'dovich, Yakov Borisovich (1914–87), astrophysicist from the former Soviet Union noted for his theories on the evolution of the universe. After an early career as a nuclear physicist when he pioneered research into uranium fission and radioactive decay, Zel'dovich turned to cosmology in the 1950s. In 1967, together with other physicists, he proposed that the universe was originally isotropic, that is, uniform at all points and in all directions, although as the universe has evolved this isotropy has decreased.

Zelenchukskaya telescope, a very large *reflecting telescope situated at the Zelenchukskaya Astrophysical Observatory in the Caucasus Mountains. Commissioned in 1977, it has a primary mirror of 6 m (236 inches) diameter and is supported on an alt-azimuth *mounting. The observatory is 2,100 m (6,900 feet) above sea-level and also includes the *RATAN 600 radio telescope.

zenith, that point on the *celestial sphere directly above the observer. This is the geometrical zenith obtained by inter-

The **Zelenchukskaya** Astrophysical Observatory in the Caucasus mountains contains two very large instruments: the RATAN 600 radio telescope, and the 6 m optical reflecting telescope, seen here in its protective dome. Since the completion of the 10 m Keck telescope at Mauna Kea, this is now the second largest optical reflecting telescope in the world.

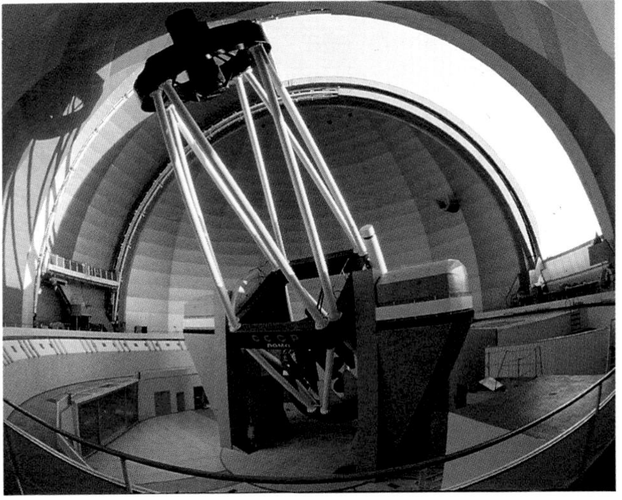

secting the celestial sphere by a line drawn from the Earth's centre through the observer's position. It is 90 degrees away from the *horizon and is in the direction opposite to the *nadir. Strictly speaking, because the Earth is not spherical, the plumb-bob zenith, that is, the direction opposite to which a suspended plumb-bob would hang, rarely coincides with the geometrical zenith.

zenith distance *celestial sphere.

zenith tube, a telescope that points vertically to the zenith. It is used for accurately measuring the positions of stars at or near the zenith to provide a precise determination of time. By keeping the telescope vertical the telescope tube does not deflect under its own weight as it would if it were not vertical. Furthermore, light from the zenith is not refracted as it passes through the Earth's atmosphere. Stars' positions can be recorded using the *photographic zenith tube.

zodiac, a band of the heavens close to the Sun's apparent annual path through the *celestial sphere, as viewed from Earth, and including about 8 degrees on each side of the *ecliptic. This band is divided into twelve equal parts, the **signs of the zodiac**, each named after a prominent constellation situated in it. The Sun appears to move through these signs at the approximate rate of one per month. For the signs of the zodiac, in order, with their symbols, see the table on page 199.

The zodiac, comprising stars immensely distant, appeared from Earth to be immutable. All the seven moving stars or planets known before modern times—the Sun, Moon, Mercury, Venus, Mars, Jupiter, and Saturn—moved within the zodiacal band, constantly changing their positions in relation to the Earth and each other, and these movements formed the basis of *astrology. Each sign was supposed to govern a different part of the body, and a planet's influence changed as it moved from one sign to another. Before the 19th century, many physicians thought it vital to choose the appropriate astrological moment in letting blood, giving medicines, or attempting surgery. The constellations are no longer located in the zodiacal signs named after them, a consequence of the fact that the Earth is not a perfect sphere and the polar axis changes slightly each year by *precession.

The twelve constellations of the **zodiac**, through which the Sun, Moon, and planets pass on their periodic journeys around the heavens, are well known to followers of astrology. These illustrations depict the mythological origins of the names of the zodiacal constellations.

zodiacal light, a faint cone of light extending from the horizon soon after evening twilight and prior to morning twilight. The axis of the cone is on or close to the *ecliptic. Although the effect appears to fade into the general brightness of the sky at an angular distance of about 90 degrees from the Sun, it continues along the ecliptic, growing again in brightness at a point opposite to the Sun, giving rise to the *gegenschein. In fact sensitive *photometers can detect the zodiacal light presence all over the sky. Giovanni Domenico *Cassini correctly interpreted the phenomenon as being sunlight reflected by interplanetary material in the form of a disc-shaped cloud surrounding the Sun mainly in the plane of the ecliptic. The high *polarization of the light suggested that the light was scattered by free electrons, that is, electrons not bound to atoms, but the chief constituent is now known to be *dust grains of between 1 and 10 microns in diameter.

zone of avoidance, a part of the *celestial sphere coinciding roughly with the Milky Way where no galaxies are seen. The light from any galaxies in that region is absorbed by clouds of dust in or near the galactic equator. The zone of avoidance also refers to a region of the celestial sphere, roughly circular, of diameter 74 degrees and centred on the position the south celestial pole occupied in 2300 BC. Because it was never visible from the northern hemisphere it contains no ancient constellations.

zone of totality *path of totality.

Zurich relative sunspot number *sunspots.

Tables

This section gives eight general tables which are relevant to the study of astronomy, and also seventeen tables which each relate to a particular entry in the text. The entries in question are indicated by asterisked cross-references.

The Greek alphabet

Letters		Name	Letters		Name
Α	α	alpha	Ν	ν	nu
Β	β	beta	Ξ	ξ	xi
Γ	γ	gamma	Ο	ο	omicron
Δ	δ	delta	Π	π	pi
Ε	ϵ	epsilon	Ρ	ρ	rho
Ζ	ζ	zeta	Σ	σς	sigma
Η	η	eta	Τ	τ	tau
Θ	θ	theta	Υ	υ	upsilon
Ι	ι	iota	Φ	ϕ	phi
Κ	κ	kappa	Χ	χ	chi
Λ	λ	lambda	Ψ	ψ	psi
Μ	μ	mu	Ω	ω	omega

Constants

Mathematical

π	the ratio of the circumference to the diameter of a circle	3.141 592 654
e	the base of natural logarithms	2.718 281 828

Physical (SI units)

Velocity of light	c	$2.997\,924\,56 \times 10^{8}$ ms^{-1}
Planck's constant	h	$6.626\,196 \times 10^{-34}$ Js
Boltzmann's constant	k	$1.380\,622 \times 10^{-23}$ JK^{-1}
Gravitational constant	G	6.670×10^{-11} Nm^{2} kg^{-2}
Stefan–Boltzmann constant	σ	5.669×10^{-8} Wm^{-2}K^{-4}
Wien's Law constant	$\lambda_{max}T$	2.90×10^{-3} mK
Temperature of the microwave background radiation		2.7K
Obliquity of the ecliptic	ϵ	23 degrees 26 minutes
Hubble's constant (estimated)	H	55 km/s per megaparsec

Time

Seconds

1 ephemeris second	= 1/31 556 925.975 of the length of the tropical year at 1900.0
1 mean solar second	= 1.002 737 909 mean sidereal seconds

Days

1 mean solar day	= 1.002 737 909 sidereal days
	= 24 hours 3 minutes 56.55 seconds mean sidereal time
	= 86 636.55 mean sidereal seconds
1 sidereal day	= 0.997 269 566 mean solar days
	= 23 hours 56 minutes 4.09 seconds mean solar time
	= 86 164.09 mean solar seconds

Months	days	hours	minutes	seconds
synodic	29	12	44	03
sidereal	27	07	43	12
anomalistic	27	13	18	33
tropical	27	07	43	05
nodical	27	05	05	36

Years	days	hours	minutes	seconds
tropical	365	5	48	46
sidereal	365	6	9	10
anomalistic	365	6	13	53
Julian	365	6	0	0
eclipse	346	14	52	57

Symbols of the Sun and planets of the solar system

Sun	☉	Jupiter	♃
Mercury	☿	Saturn	♄
Venus	♀	Uranus	⛢
Earth	⊕	Neptune	♆
Mars	♂	Pluto	♇

Angular measure

Normally, the circle is divided into 360 degrees of angle
60 arc minutes = 1 degree of angle
60 arc seconds = 1 arc minute

For right ascension the circle is divided into 24 hours
60 minutes = 1 hour
60 seconds = 1 minute

1 radian = 360/2π degrees = 57.295 827 degrees
= 3 437.747 minutes of arc = 206 264.8 seconds of arc

Solar parallex = 8.794 arc seconds

Basic masses

Mass of the electron	9.108×10^{-31} kg
Mass of the Earth	5.98×10^{24} kg
Mass of the Sun (M☉)	1.99×10^{30} kg
Chandrasekhar's limit	1.44 × M☉
Mass of the Galaxy	1.4×10^{11} M☉
Mean density of the Sun	1,400 kg m^{-3}
Mean density of the Earth	5,520 kg m^{-3}

Basic distances

1 Angstrom	10^{-10} m
1 Astronomical unit (a.u.)	149,596,000 km
1 Light-year	9.46×10^{12} km
1 Parsec	3.086×10^{13} km = 206,265 a.u.
Radius of the Earth	6,378.164 km
Radius of the Sun	696,000 km
Radius of the Galaxy	15,000 parsecs
Mean thickness of the Galaxy	2,000 parsecs
The distance of the Sun from the Galactic centre	8,500 parsecs

Conversion between SI and Imperial units

Length

1 km	= 0.621 371 miles	1 mile = 1.609 34 km
1 m	= 1.093 61 yards	1 yard = 0.914 4 m
1 cm	= 0.393 701 inches	1 inch = 2.54 cm

Mass

1 tonne	= 0.984 205 tons	1 ton = 1.016 05 tonnes
1 kg	= 2.204 62 pounds (lb)	1 lb = 0.453 592 kg
1 g	= 0.035 274 ounces (oz)	1 oz = 28.349 5 g

The biggest asteroids

see entry on *asteroid

The absolute magnitude is the magnitude an asteroid would have if it were observed at full phase from a position that was one astronomical unit from both the Sun and the Earth.

Asteroid	Number	Approximate radius (km)	Type	Absolute magnitude	Spin period (hours)	Orbital data semi-major axis (a.u.)	eccentricity	inclination (degrees)	period (years)
Ceres	1	512	C	4.35	9.075	2.767	0.079	10.6	4.60
Pallas	2	304	CU	5.08	7.811	2.772	0.234	34.8	4.62
Vesta	4	290	U	4.24	5.342	2.362	0.090	7.1	3.63
Hygiea	10	215	C	6.32	17.495	3.134	0.120	3.8	5.54
Interamnia	704	169	U	7.01	8.727	3.061	0.148	17.3	5.36
Davida	511	162	C	7.19	5.167	3.176	0.177	15.9	5.66
Cybele	65	154	C	7.75	6.070	3.434	0.105	3.6	6.36
Europa	52	146	C	7.29	5.631	3.108	0.103	7.5	5.48
Sylvia	87	141	P	7.88	5.183	3.484	0.083	10.9	6.50
Patientia	451	140	C	7.56	9.727	3.063	0.070	15.2	5.36
Euphrosyne	31	135	C	7.73	5.531	3.146	0.228	26.3	5.58
Eunomia	15	130	S	6.40	6.081	2.644	0.185	11.7	4.30
Bamberga	324	126	C	7.76	29.430	2.680	0.341	11.1	4.39
Juno	3	124	S	6.47	7.210	2.668	0.258	13.0	4.35
Psyche	16	124	M	6.91	4.196	2.923	0.134	3.1	5.00
Doris	48	123	C	7.79	11.900	3.113	0.068	6.5	5.49
Egeria	13	122	C	7.56	7.045	2.577	0.086	16.5	4.14
Eugenia	45	122	C	8.11	5.699	2.721	0.084	6.6	4.49
Hektor	624	116	U	8.53	6.924	5.172	0.023	18.2	11.76
Arethusa	95	114	C	8.80	8.688	3.072	0.144	12.9	5.38
Themis	24	114	C	7.90	8.830	3.130	0.134	0.8	5.54
Loreley	165	114	C	8.14	7.600	3.131	0.070	11.2	5.54
Hilda	153	111	U	8.40	–	3.973	0.142	7.8	7.92
Herculina	532	110	S	6.76	9.408	2.773	0.176	16.4	4.62

The Astronomers Royal

see entry on *Astronomer Royal

Name	Birth–death	Period as Astronomer Royal	Comments
John Flamsteed	1646–1719	1675–1719	Undertook accurate measurements of stellar positions using his 7-foot (2.1 m) equatorial sextant, and published the stellar catalogue *Historia Coelestis.*
Edmond Halley	1656–1742	1720–42	Introduced a 5-foot (1.5 m) transit instrument and associated clock, and an 8-foot (2.4 m) quadrant. Concentrated on lunar positional measurements.
James Bradley	1693–1762	1742–62	Expanded the Royal Greenwich Observatory and introduced new instruments, a second 8-foot (2.4 m) quadrant and an 8-foot (2.4 m) transit instrument.
Nathaniel Bliss	1700–64	1762–64	Introduced a small observatory with a revolving roof.
Nevil Maskelyne	1732–1811	1765–1811	Responsible for the first publication of the *Nautical Almanac* and the *Greenwich Observations.*
James Pond	1767–1836	1811–35	Troughton's 6-foot (1.8 m) mural circle was introduced and the time ball was installed.
George Biddell Airy	1801–92	1835–81	Founded new departments, including that for magnetism and meteorology, and introduced the transit circle that defines the present Greenwich Meridian.
William Henry Mahoney Christie	1845–1922	1884–1910	Enlarged the observatory and expanded its scope. Introduced the 28-inch (71 cm) refractor and constructed the New Physical Observatory.
Frank Watson Dyson	1868–1939	1910–33	Organized eclipse expedition to test Einstein's theory of general relativity and set up the BBC's 'six pip' time signal.
Harold Spencer Jones	1890–1960	1933–55	Installed the Yapp reflector and quartz clocks, and moved the observatory to Herstmonceux.
Richard van der Riet Woolley	1906–86	1956–71	Completed the move of the observatory and built the Isaac Newton 98-inch (2.5 m) telescope.
Martin Ryle	1918–1984	1972–82	One of the creators of radio astronomy and first Astronomer Royal not to be director of the Royal Greenwich Observatory.
Francis Graham-Smith	1923–	1982–90	Radio astronomer and director of the Royal Greenwich Observatory and then of Jodrell Bank.
Arnold W. Wolfendale	1927	1990–	Professor of astrophysics, Durham University, and specialist in cosmic ray studies.

Notable comets

see entry on *comet

The famous comets are those that have appeared the brightest, have given birth to dense meteoroid streams, or have played a key role in the history of cometary astronomy.

Bright comets, in reverse order of appearance

Name	Year of last appearance	Date of perihelion	Distance from the Sun (a.u.)	Eccentricity at perihelion	Absolute magnitude	Apparent magnitude (brightest)
West	1976	25 Feb	0.1966	0.999971	5.0	23
Bennett	1970	20 Mar	0.5376	0.996193	4.5	0
Ikeya–Seki	1965	21 Oct	0.0078	0.999915	6.2	24
Mrkos	1957	1 Aug	0.3549	0.999365	4.5	1
Arend–Roland	1957	8 Apr	0.3160	1.000168	5.5	2
Southern Comet	1947	2 Dec	0.1100	0.999548	6.0	25
Skjellerup–Maristany	1927	18 Dec	0.1762	0.999840	5.2	25
Great January Comet	1910	17 Jan	0.1290	0.999995	5.0	25
Great Comet	1901	24 Apr	0.2448	1.00	5.9	22
Cruls	1882	17 Sep	0.0077	0.999907	0.8	26
Coggia	1874	9 Jul	0.6758	0.998820	5.7	1
Great Southern	1865	14 Jan	0.0258	1.00	3.8	0
Donati	1858	30 Sep	0.5785	0.996295	3.3	1
	1843	27 Feb	0.0055	0.999914	4.9	27
Great Comet	1819	28 Jun	0.3415	1.00	4.0	1
Great Comet	1807	9 Sep	0.6461	0.995488	1.6	1
	1758	11 Jun	0.2154	1.0	4.0	22
	1744	1 Mar	0.2222	1.0	0.5	25
	1686	16 Sep	0.3360	1.0	5	1
	1677	6 May	0.2806	1.0	4.4	0
	1665	24 Apr	0.1065	1.0	4.9	22
	1618	8 Nov	0.3895	1.0	4.6	1
	1577	27 Oct	0.1775	1.0	0	27
	1556	22 Apr	0.4908	1.0	1	22
	1402	21 Mar	0.38	1.0	1	25
	1264	20 Jul	0.8249	1.0	3.5	0

Other famous comets, in alphabetical order

Name	Year of last appearance	Distance from the Sun at perihelion (a.u.)	Eccentricity	Period (years)	Number of appearances	Meteoroid stream
D'Arrest	1982	1.2911	0.6248	6.38	14	
Biela	1852	0.8606	0.7558	6.62		Andromedids
Encke	1984	0.3410	0.8463	3.31	53	Taurids
Giacobini–Zinner	1946	0.9957	0.7167	6.59	11	Giacobinids
Grigg–Skellerup	1982	0.9892	0.6657	5.09	14	
Faye	1984	1.5935	0.5782	7.34	18	
Halley	1986	0.5872	0.9673	76.1	30	Eta Aquarids and Orionids
Iras–Araki–Allcock	1983	0.9913	0.9901			
Kohoutek	1973	0.1424	1.0000			
Kreutz	1970	0.0089	1.00			
Lexell	1770	0.6744	0.7861	5.60		
Morehouse	1908	0.9453	1.0007			
Newton	1680	0.0062	0.9999			
Pons–Winnecke	1951	1.1605	0.6546	6.16	19	
Schwassmann–Wachmann 1	1974	5.4479	0.1054	15.0	5	
Schuster	1975	6.8808	1.0022			
Swift–Tuttle	1862	0.9626	0.9604	120		Perseids
Tebbutt	1861	0.8224	0.9851	409		
Tempel–Tuttle	1965	0.9817	0.9044	32.9	4	Leonids
Tuttle	1980	1.0149	0.8226	13.7		Ursids

Constellations

see entry on *constellation

Name	Abbreviation	Area (square degrees)	Order of size
Andromeda	And	722	19
Antlia	Ant	239	62
Apus	Aps	206	67
Aquarius	Aqr	980	10
Aquila	Aql	652	22
Ara	Ara	237	63
Aries	Ari	441	39
Auriga	Aur	657	21
Boötes	Boo	907	13
Caelum	Cae	125	81
Camelopardalis	Cam	757	18
Cancer	Cnc	506	31
Canes Venatici	CVn	465	38
Canis Major	CMa	380	43
Canis Minor	CMi	183	71
Capricornus	Cap	414	40
Carina	Car	494	34
Cassiopeia	Cas	598	25
Centaurus	Cen	1,060	9
Cepheus	Cep	588	27
Cetus	Cet	1,231	4
Chamaeleon	Cha	132	79
Circinus	Cir	93	85
Columba	Col	270	54
Coma Berenices	Com	386	42
Corona Australis	CrA	128	80
Corona Borealis	CrB	179	73
Corvus	Crv	184	70
Crater	Crt	282	53
Crux	Cru	68	88
Cygnus	Cyg	804	16
Delphinus	Del	189	69
Dorado	Dor	179	72
Draco	Dra	1,083	8
Equuleus	Equ	72	87
Eridanus	Eri	1,138	6
Fornax	For	398	41
Gemini	Gem	514	30
Grus	Gru	366	45
Hercules	Her	1,225	5
Horologium	Hor	249	58
Hydra	Hya	1,303	1
Hydrus	Hyi	243	61
Indus	Ind	294	49
Lacerta	Lac	201	68
Leo	Leo	947	12
Leo Minor	LMi	232	64
Lepus	Lep	290	51
Libra	Lib	538	29
Lupus	Lup	334	46
Lynx	Lyn	545	28
Lyra	Lyr	286	52
Mensa	Men	153	75
Microscopium	Mic	210	66
Monoceros	Mon	482	35
Musca	Mus	138	77
Norma	Nor	165	74
Octans	Oct	291	50
Ophiuchus	Oph	948	11
Orion	Ori	594	26
Pavo	Pav	378	44
Pegasus	Peg	1,121	7
Perseus	Per	615	24
Phoenix	Phe	469	37
Pictor	Pic	247	59
Pisces	Psc	889	14
Piscis Austrinus	PsA	245	60
Puppis	Pup	673	20
Pyxis	Pyx	221	65
Reticulum	Ret	114	82
Sagitta	Sge	80	86
Sagittarius	Sgr	867	15
Scorpius	Sco	497	33
Sculptor	Scl	475	36
Scutum	Sct	109	84
Serpens	Ser	637	23
Sextans	Sex	314	47
Taurus	Tau	797	17
Telescopium	Tel	252	57
Triangulum	Tri	132	78
Triangulum Australe	TrA	110	83
Tucana	Tuc	295	48
Ursa Major	UMa	1,280	3
Ursa Minor	UMi	256	56
Vela	Vel	500	32
Virgo	Vir	1,294	2
Volans	Vol	141	76
Vulpecula	Vul	268	55

Principal meteor showers

see entry on *meteor shower

The zenithal hourly rate is the maximum stream activity if the radiant were at the zenith.
Solar longitude is the longitude of the Sun at the meteor shower date.

Shower	Solar longitude (degrees)	Date of maximum activity	Dates of detectable meteors	Zenithal hourly rate	Geocentric velocity (km/s)	Radiant right ascension (degrees)	Radiant declination (degrees)
Quadrantids	282.6	3 Jan	1–6 Jan	87	41.5	230.1	+50
Delta Leonids	338	26 Feb	5 Feb–19 Mar	8	23	159	+19
Gamma Normids	352.0	14 Mar	9–18 Mar	7			
Delta Pavonids	8.0	30 Mar	17 Mar–1 Apr	7			
April Lyrids	31.5	22 Apr	19–25 Apr	12	47.6	271.4	+34
Alpha Scorpiids	38	3 May	11 Apr–12 May	10	35	240	–22
Eta Aquarids	42.4	5 May	24 Apr–20 May	45	65.5	335.6	–2
Lambda Sagittariids	81	11 Jun	8–16 Jun	12	52	304	–35
June Lyrids	84.5	16 Jun	11–21 Jun	9	31	278	+35
Delta Aquarids	125	29 Jul	15 Jul–20 Aug	19	41.4	333.1	–17
Iota Aquarids	134	7 Aug	15 Jul–25 Aug	8	33.8	333.3	–15
Perseids	139.4	12 Aug	23 Jul–20 Aug	78	59.4	46.2	+58
Aurigids	157.9	1 Sept		20	66.3	84.6	+42
Piscids	188	20 Sept	31 Aug–2 Nov	7	26.3	6	+00
October Draconids	197	9 Oct		260	20.4	262	+54
Orionids	208	22 Oct	16–27 Oct	25	66.4	94.5	+16
Taurids	222	4 Nov	30 Oct–30 Nov	8	30	56	+14
Leonids	234.6	17 Nov	15–20 Nov	200	70.7	152	+22
December Phoenicids	251.2	5 Dec		9	21.7	15	–55
Puppids Velids	257	9 Dec	27 Nov–9 Jan	9		135	–48
Geminids	261.3	14 Dec	7–16 Dec	60	34.4	112.3	+32
Ursids	270	23 Dec	17–25 Dec	9	33.4	217.1	+78

Characteristics of the Moon's orbit

see entry on the *Moon

Mean distance from the Earth (semi-major axis)	384,400 km
Mean eccentricity of the orbit	0.0549
Mean orbital inclination to the ecliptic	5 degrees 9 minutes
Sidereal period of revolution (sidereal month)	27 days 7 hours 43 minutes 12 seconds
Synodic period (lunation or synodic month)	29 days 12 hours 44 minutes 3 seconds

Planetary orbital data

see entry on *planet

Planet	Mean distance from Sun (a.u.)	Mean distance from Sun (10^6 km)	Minimum distance from Sun (a.u.)	Maximum distance from Sun (a.u.)	Eccentricity	Inclination to ecliptic (degrees)
Mercury	0.387	57.9	0.308	0.467	0.206	7.0
Venus	0.723	108.2	0.718	0.728	0.007	3.4
Earth	1.000	149.6	0.983	1.017	0.017	0.0
Mars	1.524	227.9	1.381	1.666	0.093	1.9
Jupiter	5.203	778.3	4.951	5.455	0.048	1.3
Saturn	9.539	1,427.0	9.009	10.069	0.056	2.5
Uranus	19.182	2,869.6	18.275	20.089	0.047	0.8
Neptune	30.058	4,496.6	29.800	30.317	0.009	1.8
Pluto	39.44	5,900.1	29.58	49.3	0.25	17.1

Planetary periods and motions

see entry on *planet

The rotation period given for Jupiter is at its equator. R = retrograde.
The rotation period for Uranus is that of its magnetic field.

Planet	Sidereal period	Mean synodic period (days)	Mean orbital velocity (km/s)	Sidereal mean daily motion (degrees)	Sidereal period of axial rotation	Inclination of equator to orbit (degrees)
Mercury	87.969 days	115.88	47.87	4.092	58.65 days	0.0
Venus	224.701 days	583.92	35.02	1.602	243.01 days (R)	177.34
Earth	365.256 days	–	29.79	0.986	23.934 hours	23.45
Mars	686.980 days	779.94	24.13	0.524	24.623 hours	25.19
Jupiter	11.862 years	398.88	13.06	0.083	9.842 hours	3.13
Saturn	29.457 years	378.09	9.65	0.033	10.233 hours	26.72
Uranus	84.010 years	369.66	6.80	0.012	17.24 hours(R)	97.86
Neptune	164.793 years	367.49	5.43	0.006	*c.* 18.4 hours	29.56
Pluto	248.5 years	366.73	4.74	0.004	6.39 days(R)	117.56

Physical data for the planets

see entry on *planet

Planet	Equatorial diameter[a] (km)	Mass (Earth = 1)	Volume (Earth = 1)	Mean density (10^3 kg/m^3)	Oblateness	Surface gravity (Earth = 1)	Escape velocity (km/s)
Mercury	4,878	0.06	0.06	5.43	0	0.377	4.25
Venus	12,104	0.82	0.86	5.24	0	0.902	10.36
Earth	12,756	1.00	1.00	5.52	0.0034	1.000	11.18
Mars	6,787	0.11	0.15	3.94	0.0052	0.379	5.02
Jupiter	142,800	317.83	1,323	1.33	0.065	2.69	59.6
Saturn	120,000	95.16	752	0.70	0.108	1.19	35.6
Uranus	50,800	14.50	64	1.30	0.030	0.93	21.1
Neptune	48,600	17.20	54	1.76	0.026	1.22	24.6
Pluto	2,250	0.002	0.01	*c.* 2	0	0.03	*c.* 1

[a] Polar radii: Earth 12,714 km, Mars 6,752 km, Jupiter 133,500 km, Saturn 107,100 km, Uranus 49,300 km, Neptune 47,300 km.
Source: *Norton's Star Atlas*, 18th edition, edited by I. Ridpath; Longman Scientific and Technical, 1989

Components of the Apollo spacecraft

see entry on *Project Apollo

	Height (m)	Diameter (m)	Mass (tonnes)
Command module (CM)	3.6	4	5.5
Service module (SM)	6.7	4	25
Lunar excursion module (LEM)	6	9	16

Landmarks in the Apollo programme

see entry on *Project Apollo

Dates	Apollo number	Crew	Achievements
21–27 December 1968	Apollo 8	Frank Borman, James A. Lovell, William A. Anders	Ten orbits of the Moon without landing; photography of the Earth and Moon, including lunar landing sites.
3–13 March 1969	Apollo 9	James A. McDivitt, David R. Scott, Russell Schweikert	Full check-out of the performance of the LEM in Earth orbit and of all necessary manœuvres such as docking and rendezvous.
18–26 May 1969	Apollo 10	Thomas P. Stafford, John W. Young, Eugene A. Cernan	Check-out of crew/space vehicle/mission support facility performance during a lunar mission; LEM flew to within 15 km of the lunar surface.
16–24 July 1969	Apollo 11	Neil A. Armstrong, Michael Collins, Edwin E. ('Buzz') Aldrin	Armstrong and Aldrin landed on the Moon in the Sea of Tranquillity; ALSEP was set out. 22 kg of samples were obtained.
14–24 November 1969	Apollo 12	Charles Conrad, Richard F. Gordon, Alan L. Bean	ALSEP was set out; samples and parts of the unmanned Surveyor 3 spacecraft were collected.
11–17 April 1970	Apollo 13	James A. Lovell, Fred W. Haise, John L. Swigert	No landing on Moon; the spacecraft was crippled by an explosion in the oxygen tanks of the SM. LEM was used as a 'lifeboat' and power source to orbit the Moon and return the crew safely to Earth.
31 January–9 February 1971	Apollo 14	Alan B. Shephard, Stuart A. Roosa, Edgar D. Mitchell	Visit to Fra Mauro crater in a hilly upland region; ALSEP was set out and samples were collected.
26 July–7 August 1971	Apollo 15	David R. Scott, Alfred M. Worden, James B. Irwin	Landed near Hadley Rille at the foot of the Apennine Mountains; the lunar roving vehicle (LRV) was used. ALSEP was set out and samples were collected.
16–27 April 1972	Apollo 16	John W. Young, Thomas K. Mattingly, Charles M. Duke	Landed north of Descartes crater; used the LRV. Samples were obtained and ALSEP was set out.
7–19 December 1972	Apollo 17	Eugene A. Cernan, Ronald E. Evans, Harrison H. Schmitt	Collected over 100 kg of lunar soil samples and set out ALSEP.

Planetary ring systems

see entry on *ring, planetary

Planet	Ring	Mean distance from centre of planet (10^3 km)	(planetary radii)
Jupiter	Halo	100–122	1.4–1.71
	Main ring	122–129	1.71–1.81
	Gossamer ring	129–215	1.81–3
Saturn	D ring	67.0–74.4	1.12–1.24
	C ring	74.4–91.9	1.24–1.53
	B ring	91.9–117.4	1.53–1.96
	A ring	121.9–136.6	2.03–2.28
	F ring	140.3	2.34
	G ring	170.0	2.83
	E ring	180–480	3–8
Uranus	Rings	41.9–51.1	1.65–2.0
Neptune	Galle	41.9	1.72
	Leverrier	53.2	2.19
	Adams	62.9	2.59

Partial source: *Norton's Star Atlas*, 18th edition edited by I. Ridpath; Longman Scientific and Technical, 1989

Planetary satellites

see entry on *satellite, natural

Orbital inclinations are relative to the planet's equator, except for the Moon and Phoebe, which are relative to the ecliptic. The Moon's inclination relative to the Earth's equator ranges from 18.28 degrees to 28.58 degrees. R = retrograde.

Planet and satellite	Mean distance from centre of primary (10^3 km)	(planetary radii)	Orbital period (days)	Inclination (degrees)	Eccentricity	Diameter (km)	Reciprocal mass (planet = 1)	Density	Mean magnitude (10^3 kg/m^3) at opposition
Earth									
Moon	384.4	60.27	27.3217	5.15	0.055	3,476	81.30	3.34	−12.7
Mars									
I Phobos	9.38	2.76	0.319	1.0	0.015	27 × 21.6 × 18.8		2.2	11.3
II Deimos	23.46	6.91	1.262	0.9–2.7	0.001	15 × 12.2 × 11.0		1.7	12.4
Jupiter									
XVI Metis	128	1.79	0.295	0	0	40			17.5
XV Adrastea	129	1.81	0.298	0	0	25 × 20 × 15			19.1
V Amalthea	181.3	2.54	0.498	0.4	0.003	270 × 166 × 150			14.1
XIV Thebe	221.9	3.11	0.675	0.8	0.015	110 × 90			15.6
I Io	421.6	5.9	1.769	0.04	0.004	3,630	21,400	3.57	5.0
II Europa	670.9	9.4	3.551	0.47	0.009	3,138	39,700	2.97	5.3
III Ganymede	1,070	15.0	7.155	0.21	0.002	5,262	12,800	1.94	4.6
IV Callisto	1,883	26.4	16.689	0.51	0.007	4,800	17,700	1.86	5.7
XIII Leda	11,094	155.4	238.72	26.1	0.148	16			20.2
VI Himalia	11,480	160.8	250.56	27.6	0.158	186			14.8
X Lysithea	11,720	164.2	259.22	29.0	0.107	36			18.4
VII Elara	11,737	164.4	259.65	24.8	0.207	76			16.8
XII Ananke	21,200	296.9	631 (R)	147	0.17	30			18.9
XI Carme	22,600	316.5	692 (R)	164	0.21	40			18.0
VIII Pasiphae	23,500	329.1	735 (R)	145	0.38	50			17.0
IX Sinope	23,700	331.9	758 (R)	153	0.28	36			18.3
Saturn									
XXVIII Pan	133.57	2.23	0.575			20			
XV Atlas	137.67	2.29	0.602	0.3	0.000	40 × 20			18
XVI Prometheus	139.35	2.32	0.613	0.0	0.003	140 × 100 × 80			16
XVII Pandora	141.70	2.36	0.629	0.0	0.004	110 × 90 × 70			16
XI Epimetheus	151.42	2.52	0.694	0.34	0.009	140 × 120 × 100			15
X Janus	151.47	2.52	0.695	0.14	0.007	220 × 200 × 160			14
I Mimas	185.52	3.09	0.942	1.53	0.020	392	12,500,000	1.17	12.9
II Enceladus	238.02	3.97	1.370	0.00	0.005	500	7,700,000	1.24	11.7
III Tethys	294.66	4.91	1.888	1.86	0.000	1,060	770,000	1.26	10.2
XIII Telesto	294.66	4.91	1.888	0	0	34 × 28 × 26			18.5
XIV Calypso	294.66	4.91	1.888	0	0	34 × 22 × 22			18.7
IV Dione	377.40	6.29	2.737	0.02	0.002	1,120	540,000	1.44	10.4
XII Helene	377.40	6.29	2.737	0.0	0.005	36 × 32 × 30			18
V Rhea	527.04	8.78	4.518	0.35	0.001	1,530	230,000	1.33	9.7
VI Titan	1,221.8	20.36	15.945	0.33	0.029	5,150	4,200	1.88	8.3
VII Hyperion	1,481.1	24.69	21.277	0.43	0.104	410 × 260 × 220	33,000,000		14.2
VIII Iapetus	3,561.3	59.36	79.330	14.72	0.028	1,460	300,000	1.21	10.2–11.9
IX Phoebe	12,952	215.9	550.48 (R)	177	0.163	220			16.5
Uranus									
VI Cordelia	49.77	1.96	0.335		0.0	50			
VII Ophelia	53.79	2.12	0.376		0.01	50			
VIII Bianca	59.17	2.33	0.435		0.0	50			
IX Cressida	61.78	2.43	0.464		0.0	60			
X Desdemona	62.68	2.47	0.474		0.0	60			
XI Juliet	64.35	2.53	0.493		0.0	80			
XII Portia	66.09	2.60	0.513		0.0	80			
XIII Rosalind	69.94	2.75	0.558		0.0	60			
XIV Belinda	75.26	2.96	0.624		0.0	50			
XV Puck	86.01	3.39	0.762		0.0	170			
V Miranda	129.39	5.09	1.413	4.2	0.003	480	500,000	1.26	16.3
I Ariel	191.02	7.52	2.520	0.3	0.003	1,158	55,500	1.65	14.2
II Umbriel	266.30	10.48	4.144	0.36	0.005	1,172	83,300	1.44	14.8
III Titania	435.91	17.16	8.706	0.14	0.002	1,580	14,700	1.59	13.7
IV Oberon	583.52	22.97	13.463	0.10	0.001	1,524	14,500	1.50	13.9
Neptune									
1989 N6 Naiad	48.23	1.99	0.30			60			
1989 N5 Thalassa	50.07	2.06	0.31			80			
1989 N3 Despina	52.53	2.16	0.34			180			
1989 N4 Galatea	61.95	2.55	0.43			150			
1989 N2 Larissa	73.55	3.03	0.56			190			
1989 N1 Proteus	117.64	4.84	1.12			415			
I Triton	354.29	14.58	5.877 (R)	159.0	0.0	3,800	770		13.5
II Nereid	5,511	226.8	360.2	27.6	0.75	300	5,000,000		18.7
Pluto									
Charon	19.13	12.75	6.387	94	0	1,200	4.5	c. 2	16.8

The nearest stars

see entry on *star

The nearest stars, in order of distance from the Sun, with the Sun for comparison.
var. denotes that the star is a variable star in which case the mean magnitude is shown.

Star	Right ascension (hours minutes)	Declination (degrees minutes)	Apparent magnitude	Approximate distance (light-years)	(parsecs)	Absolute magnitude
Sun	–	–	–26.72	–	–	4.8
Proxima (V645 Cen)	14 29.7	–62 41	11.05 (var.)	4.2	1.3	15.5
Alpha Cen A	14 39.6	–60 50	–0.01	4.3	1.3	4.4
B			1.33			5.7
Barnard's Star	17 57.8	+04 34	9.54	6.0	1.8	13.2
Wolf 359 (CN Leo)	10 56.5	+07 01	13.53 (var.)	7.7	2.4	16.7
BD +36° 2147	11 03.3	+35 58	7.50	8.2	2.5	10.5
UV Cet A	01 38.8	–17 57	12.52 (var.)	8.4	2.6	15.5
B			13.02 (var.)			16.0
Sirius A	06 45.1	–16 43	–1.46	8.6	2.7	1.4
B			8.3			11.2
Ross 154	18 49.8	–23 50	10.45	9.4	2.9	13.1
Ross 248	23 41.9	+44 10	12.29	10.4	3.2	14.8
Epsilon Eri	03 32.9	–09 28	3.73	10.8	3.3	6.1
Ross 128	11 47.8	+00 48	11.10	10.9	3.4	13.5
61 Cyg A (V1803 Cyg)	21 06.9	+38 45	5.22 (var.)	11.1	3.4	7.6
B			6.03			8.4
Epsilon Ind	22 03.4	–56 47	4.68	11.2	3.5	7.0
BD +43° 44 A	00 18.5	+44 01	8.08	11.2	3.5	10.4
B			11.06			13.4
L789-6	22 38.5	–15 19	12.18	11.2	3.5	14.5
Procyon A	07 39.3	+05 13	0.38	11.4	3.5	2.6
B			10.7			13.0
BD +59° 1915 A	18 43.1	+59 38	8.90	11.6	3.6	11.2
B			9.69			11.9
CoD –36° 15693	23 05.9	–35 51	7.35	11.7	3.6	9.6

Source: Alan Batten, *The Royal Astronomical Society of Canada Observer's Handbook* (1989)

The brightest stars

see entry on *star

The brightest stars, in order of apparent magnitude, with the Sun for comparison.
var. denotes that the star is a variable star in which case the mean magnitude is shown.
[a] denotes the combined magnitude if the star is a double star;
[b] denotes the distance to the Orion Cluster;
[c] denotes the distance to the Scorpio Cluster.

Star		Right ascension (hours minutes)	Declination (degrees minutes)	Apparent magnitude	Approximate distance (light-years)	(parsecs)	Absolute magnitude
Sun		–	–	–26.72	–	–	4.8
Sirius	Alpha CMa	06 45.1	–16 43	–1.46	8.6	2.6	1.4
Canopus	Alpha Car	06 24.0	–52 42	–0.72	74	23	–2.5
Rigil Kentaurus	Alpha Cen	14 39.6	–60 50	–0.27	4.3	1.3	4.1
Arcturus	Alpha Boo	14 15.7	+19 11	–0.04	34	10	0.2
Vega	Alpha Lyr	18 36.9	+38 47	0.03	25	8	0.6
Capella	Alpha Aur	05 16.7	+46 00	0.08	41	13	0.4
Rigel	Beta Ori	05 14.5	–08 12	0.12	1,400[b]	400	–8.1
Procyon	Alpha CMi	07 39.3	+05 13	0.38	11.4	3.5	2.6
Achernar	Alpha Eri	01 37.7	–57 14	0.46	69	21	–1.3
Betelgeuse	Alpha Ori	05 55.2	+07 24	0.50 (var.)	1,400[b]	400	–7.2
Hadar	Beta Cen	14 03.8	–60 22	0.61 (var.)	320	100	–4.4
Acrux	Alpha Cru	12 26.6	–63 06	0.76[a]	510	160	–4.6[a]
Altair	Alpha Aql	19 50.8	+08 52	0.77	16	5	2.3
Aldebaran	Alpha Tau	04 35.9	+16 31	0.85 (var.)	60	18	–0.3
Antares	Alpha Sco	16 29.4	–26 26	0.96 (var.)	520[c]	160	–5.2
Spica	Alpha Vir	13 25.2	–11 10	0.98 (var.)	220	70	–3.2
Pollux	Beta Gem	07 45.3	+28 01	1.14	40	12	0.7
Fomalhaut	Alpha PsA	22 57.6	–29 37	1.16	22	7	2.0
Becrux	Beta Cru	12 47.7	–59 41	1.25 (var.)	460	140	–4.7
Deneb	Alpha Cyg	20 41.4	+45 17	1.25	1,500	500	–7.2
Regulus	Alpha Leo	10 08.4	+11 58	1.35	69	21	–0.3
Adhara	Epsilon CMa	06 58.6	–28 58	1.50	570	177	–4.8
Castor	Alpha Gem	07 34.6	+31 53	1.57[a]	49	15	0.5[a]
Gacrux	Gamma Cru	12 31.2	–57 07	1.63 (var.)	120	37	–1.2
Shaula	Lambda Sco	17 33.6	–37 06	1.63 (var.)	330	102	–3.5

Source: Robert F. Garrison, *The Royal Astronomical Society of Canada Observer's Handbook* (1989)

The signs of the Zodiac

see entry on the *zodiac

Aries (Ram)	♈
Taurus (Bull)	♉
Gemini (Twins)	♊
Cancer (Crab)	♋
Leo (Lion)	♌
Virgo (Virgin)	♍
Libra (Scales)	♎
Scorpio (Scorpion)	♏
Sagittarius (Archer)	♐
Capricorn (Goat)	♑
Aquarius (Waterbearer)	♒
Pisces (Fishes)	♓

Acknowledgements

Photographs

Abbreviations: *t* = top; *b* = bottom; *r* = right; *l* = left

Andromeda Oxford Ltd. 7, 55*br* (Hale Observatories), 76, 96*b* and 145 (NASA)

© Anglo-Australian Telescope Board, photography by David Malin 24*r*, 52, 61, 68, 82, 89, 125, 139, 167, 169, 173

Archiv fur Kunst und Gershichte, Berlin 176

The Bridgeman Art Library (The British Library Board) 32, 44

European Space Agency 60

Hale Observatories 50, 63, 186

Michael Holford 9 (Science Museum, London)

Lick Observatory 8, 14, 34, 42, 64*t*, 101 all four), 109, 162

Mary Evans Picture Library 190

National Aeronautics and Space Administration/NASA 20, 37*t* and *b*, 46, 53, 77, 81, 91*t*, 92*t*, 106, 108*b*, 119, 129, 138, 179, 183*b*

National Maritime Museum, London 57, 66

National Optical Astronomy Observatory/NOAO 39, 55*bl*, 132*l*

National Portrait Gallery, London 108*t*

Novosti Press Agency 99

Oxford University Press 4, 178

Rex Features 64*b* (Bryn Colton)

Ann Ronan Picture Library 3, 84

Royal Greenwich Observatory 121, 164

Royal Observatory, Edinburgh 26, 54, 112

Science Photo Library 5 (NASA), 13 (Jack Finch), 24*l* (Starlink Facility, Rutherford Appleton Laboratory), 36 (NRAO/AUI), 41 (Earth Satellite Corporation), 55*tl* and *tr* (Lick Observatories), 56 (Jerry Schad), 70 (NASA), 74 (NASA and US Geological Survey), 78 (David Parker), 79 (Gary Ladd), 83*t* (Roger Ressmeyer), 83*b* (David Parker), 91*b*, 92*b* (NASA), 93 (Roger Ressmeyer), 96*t* (NOAO), 105 (Roger Ressmeyer/Starlight), 110, 123 (Tony Ward), 124*t* and *b* (Kim Gordon), 126 and 128 (NASA), 132*r* (Dr Stephen Unwin), 134 and 135 (Roger Ressmeyer/Starlight), 141 (David Parker), 149*t* and *b* (NASA), 155 (NASA), 157 (Freeman D. Miller), 163 (NASA), 166 *t* and *b* (NOAO), 168 (US Geological Survey), 174 (NASA), 181, 182 (NASA), 183*t* (Peter Menzel), 184 (NRAO/AUI), 185 (Julian Baum), 188 (NASA), 189 (Roger Ressmeyer)

Yerkes Observatory, Chicago 80 (all three).

Picture researcher: Suzanne Williams

Illustrations

Russell Birkett, Information Design Unit: all line artwork

National Aeronautics and Space Administration/NASA 69, 81, 184

Le Grand Atlas Universalis de l'Astronomie, Encyclopaedia Universalis, 1990 28, 130, 138, 152, 3, 180

NSTS 69

© 1965 Falk Verlag, Germany 102

The Sun Iain Nicolson, 1982, Mitchell Beazley International Ltd. 164

Endpapers

Front: Information Design Unit.

Back: © Bartholomew, Edinburgh. Reproduced by kind permission.

Tables

Royal Astronomical Society of Canada Observer's Handbook, 1989, Alan Batten: 199

Royal Astronomical Society of Canada Observer's Handbook, 1989, Robert F. Garrison: 199

Norton's Star Atlas, 18th edition (1989), edited by Ian Ridpath, Longman: 196, 197, 198

The publishers have made every attempt to contact the owners of material appearing in this book. In the few instances where they have been unsuccessful they invite the copyright holders to contact them direct.

0 h R.A.

PISCES
CETUS
Menkar
Hamal
ARIES
TRIANGULUM
Mirach
M31
ANDROMEDA
Alpheratz
Markab
PEGASUS
Scheat
Enif
EQUULEUS
DELPHINUS
LACERTA
Almach
Algol
Pleiades
TAURUS
Hyades
Aldebaran
PERSEUS
Mirfak
Ruchbah
H VI 33 34
Schedar
Caph
CASSIOPEIA
CYGNUS
Deneb
Altair
AQUILA
Galactic Equator
CEPHEUS
LYRA
Vega
Capella
AURIGA
Elnath
Bellatrix
ORION
Betelgeuse
CAMELOPARDALIS
Eltanin
DRACO
Polaris
URSA MINOR
Kochab
HERCULES
M92
M13
Rasalhague
OPHIUCHUS
M35
Galactic Equator
MONOCEROS
LYNX
GEMINI
Castor
Pollux
CANIS MINOR
Procyon
CORONA BOREALIS
Alphecca
SERPENS CAPUT
Mizar
Alioth
Alkaid
Dubhe
Merak
Phecda
URSA MAJOR
CANES VENATICI
M3
BOOTES
M5
Arcturus
N. Galactic Pole
COMA BERENICES
LEO MINOR
CANCER
M44
HYDRA
LEO
Regulus
Denebola
SEXTANS
VIRGO
Ecliptic
Equator
12

NORTHERN SKY

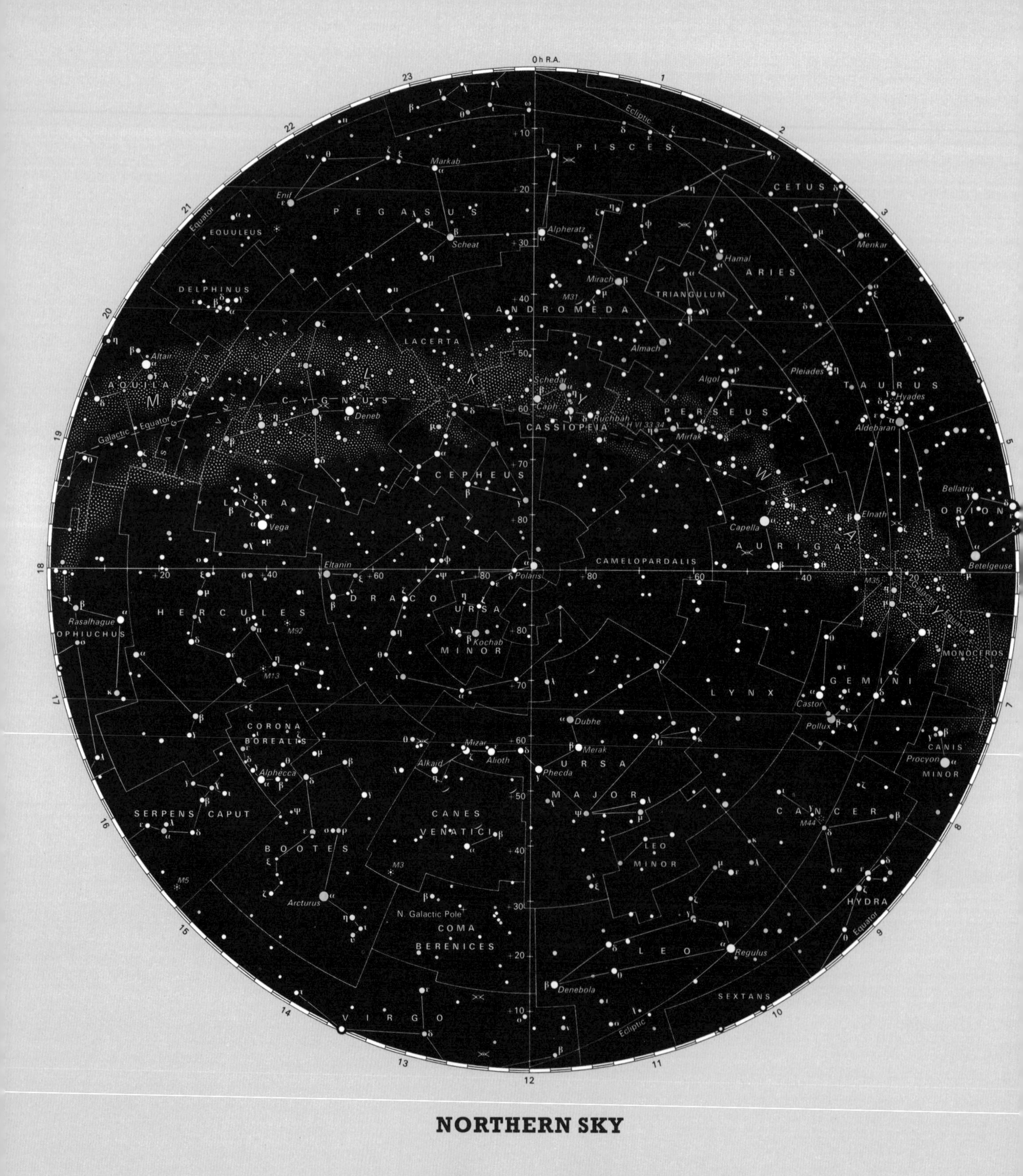

Star Magnitude

Each unit of magnitude indicates a difference of brightness of 2.512 times. The brightest star is Sirius (mag. −1.45)

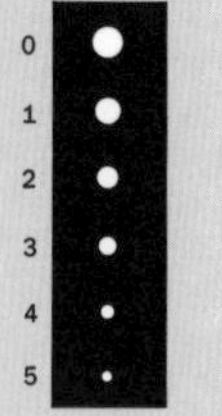

Star Colours and Spectral Types

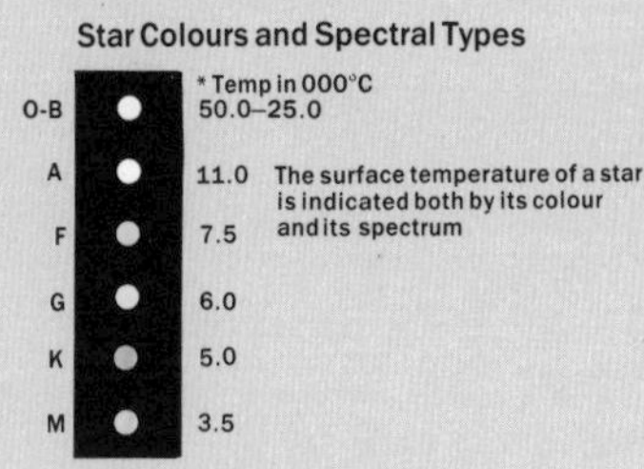

Type	* Temp in 000°C
O-B	50.0–25.0
A	11.0
F	7.5
G	6.0
K	5.0
M	3.5

The surface temperature of a star is indicated both by its colour and its spectrum